Sir Arthur Davies
K.B.E., M.Sc., Dr. (h.c.), Dr. ès Sc. (h.c.), D.Sc. (hon.)
Secretary-General Emeritus, WMO

FORTY YEARS OF PROGRESS AND ACHIEVEMENT

A HISTORICAL REVIEW OF WMO

Edited by

Sir Arthur Davies
Secretary-General Emeritus, WMO

Contributors (see p. VIII):

O.M. Ashford
S. Gupta
P.J. Meade
H. Taba
G. Weiss

WMO – No. 721

Secretariat of the World Meteorological Organization - Geneva - Switzerland
1990

ISBN 92-63-10721-1

NOTE

The designations employed and the presentation of material in this publication do not imply the expression of any opinion whatsoever on the part of the Secretariat of the World Meteorological Organization concerning the legal status of any country, territory, city or area, or of its authorities, or concerning the delimitation of its frontiers or boundaries.

CONTENTS

		Page
Editor's Foreword		VII
Brief notes on the contributors		VIII
Chapter I:	In the beginning	1
Chapter II:	First Congress sets WMO on its course	11
Chapter III:	World Weather Watch Programme	17
Chapter IV:	Instruments and Methods of Observation Programme	27
Chapter V:	Tropical Cyclone Programme	33
Chapter VI:	Marine Meteorology and Associated Oceanographic Activities Programme	39
Chapter VII:	Aeronautical Meteorology Programme	45
Chapter VIII:	Agricultural Meteorology Programme	51
Chapter IX:	Hydrology and Water Resources Programme	57
Chapter X:	WMO Research Programme	73
Chapter XI:	Protection of the environment	87
Chapter XII:	World Climate Programme	97
Chapter XIII:	Education and Training Programme	111
Chapter XIV:	Technical Co-operation Programme	121
Chapter XV:	Regional Programme	133
Chapter XVI:	Collaboration with the United Nations System	139
Chapter XVII:	Organizational machinery	151
Chapter XVIII:	Further outlook	165

APPENDICES:

A:	Convention of the World Meteorological Organization	173
B:	Agreement between the United Nations and the World Meteorological Organization	185
C:	Members of the World Meteorological Organization	189
D:	Presidents and Secretaries-General of the World Meteorological Organization	191
E:	Presidents and sessions of Regional Associations	193
F:	Presidents and sessions of Technical Commissions	197
G:	Winners of the IMO Prize	201

List of abbreviations 203

EDITOR'S FOREWORD

On 11 October 1947, the representatives of 31 countries signed a document which provided for the creation of a new governmental organization to be known as the World Meteorological Organization (WMO). The occasion was the 12th Conference of Directors of the non-governmental International Meteorological Organization (IMO) then in session in Washington, DC. The Conference had been convened by IMO in order to prepare the way for the transfer of its own responsibilities, as the focal point of international meteorology, to a new body more suited to meet the changing needs of the countries of the world in the post-World War II era. The document was the text of the WMO Convention.

Now, some 40 years later, the record of WMO's activities shows how well the founding fathers succeeded in their task. WMO was duly launched on its course and has successfully met the many and complex challenges which it has inevitably had to face. Major progress has been made in advancing scientific knowledge of the atmospheric processes and unprecedented developments have been achieved in applying such knowledge for practical purposes in many fields of human endeavour—and all this made possible by the Member countries (now numbering 161), working together under the aegis of WMO. Indeed, the most significant factor in the success story of WMO has been that of maintaining a spirit of full and friendly international co-operation between all countries—co-operation first developed by IMO and which WMO was privileged to inherit from its predecessor.

WMO has been required to provide greatly improved services in fields of activity of long-standing meteorological concern—e.g. agriculture, shipping, aviation, water resources, etc. In addition, however, many new world problems of direct concern to WMO have arisen and these new responsibilities have been met with no less success. A particularly relevant development in recent years has been the widespread realization of the dangers to the human race of excessive changes in the natural environment. This in turn has created an urgent need to understand the nature and extent of such changes as well as the means of controlling them. The basic activities of WMO since its inception (and in some respects, those of IMO before it) provide essential information in this context, while the special programmes adopted by WMO for studying relevant aspects of the environment—programmes established long before the dangers became so widely accepted—all constitute a major contribution to identifying the further action necessary.

In these circumstances, it is hardly surprising that a general wish has arisen that the events and activities of the past 40 years should be placed on permanent record and that a historical review of WMO should be prepared and published. The Secretary-General therefore suggested that I invite a small number of meteorologists, all with long experience in WMO affairs, to prepare such a volume. All readily accepted and the WMO "History group"* was established. It was recognized from the outset that the WMO Secretariat would need to make essential contributions as regards both the text itself and the arrangements for its publication.

It may be of interest to explain the procedure adopted by the group. A list of chapters was drawn up, each dealing with a specific aspect of WMO's activities. The initial drafting of each chapter was then assigned to a member of the group with, in some cases, other members designated to help as required. The first draft of each chapter was then circulated for comment to all other members and also to appropriate members and ex-members of the WMO Secretariat. Highly valuable and constructive comments and, in some cases, revised sections of the actual texts were received. As the work progressed it became clear that in addition to the changes necessary in the original first drafts, some amendments to the original list of chapters were needed.

Special mention must be made of one such additional chapter. As the work on the other chapters proceeded it became evident that the volume, when completed, would comprise a story of great success and achievement in meeting past and present responsibilities. It became equally clear however that in the years to come the Organization would be involved in further and different responsibilities. It was therefore felt that a chapter should be added, explaining the future challenges as far as they could be foreseen as well as the preparations being made to meet them. It was recognized that the Secretary-General in office was the most competent person to prepare such a chapter and he kindly agreed to do so. Chapter XVIII has therefore been added as the concluding chapter with the appropriate title of "Further outlook".

As regards the form of presentation of the information in each chapter, the aim has been to produce a text that is not only accurate but couched in language that will make the significance of the activities described readily comprehensible to all readers, whether meteorologists or not. A further aim has been to make each chapter largely self-contained and hence easily readable without frequent cross-references to other chapters.

In conclusion, I would like to say that I am very happy, indeed proud, to have served as the leader of the "History group". The members have carried out their respective responsibilities in an excellent manner and in the spirit of friendly co-operation so typical of a WMO project. I am grateful to them all. I feel sure, however, that they will all understand if I single out one member of the group for special mention—Mr P. J. Meade. During a period when I was unable to meet fully my responsibilities as editor, Mr Meade devoted much time and effort to assist me and did so in an extremely efficient manner—such work being in addition to the other duties allocated to him as a member of the group, including the drafting of several of the chapters added at a very late stage of the preparatory work. For all these reasons I am particularly grateful to him.

I should also like to acknowledge with much appreciation the essential role played by the Secretary-General and his staff in ensuring that this very worthwhile project has been brought to a successful conclusion by the publication of this book. As already mentioned, the Secretary-General, Professor G. O. P. Obasi, was kind enough to prepare the final report. Mention must also be made of Professor Rudolf Czelnai, Assistant Secretary-General, whose great attention to this project has contributed largely to its success.

D. Arthur Davies

D. Arthur Davies
Secretary-General Emeritus, WMO

* See note overleaf.

BRIEF NOTES ON THE CONTRIBUTORS

Sir Arthur Davies	President, Regional Association for Africa, 1951-1955 Secretary-General of WMO, 1956-1979 Secretary-General Emeritus Awarded IMO Prize 1985
Oliver M. Ashford	WMO Secretariat, 1952-1977. Posts included: Director, Meteorological Applications Department, 1972-1973 Programme Planning and UN Affairs, 1973-1977 Editor, WMO *Bulletin*, 1952-1975
Sunil Gupta	WMO Secretariat, 1959-1981. Posts included: Executive Assistant to the Secretary-General, 1963-1981
Patrick J. Meade	Formerly Deputy Director-General, UK Meteorological Office Attended numerous WMO meetings and chaired various panels and committees
Hessam Taba	WMO Secretariat, 1960-1984. Posts included: Director, Education and Training Department, 1976-1977 Programme Planning and UN Affairs, 1977-1979 Technical Co-operation Department, 1979-1980 Languages, Publications and Conferences Department, 1980-1984 Editor, WMO *Bulletin*, 1979-1984
Gottfried Weiss	WMO Secretariat, 1961-1986. Posts included: Chief, Networks and Telecommunications Division, Scientific and Technical Department, 1968-1974 Director, World Weather Watch Department, 1974-1986

CHAPTER I

IN THE BEGINNING

The events which immediately preceded the creation of WMO and the record of its subsequent activities are presented in the chapters which follow. There had been, however, many important and, in the present context, highly relevant developments in the science and practice of meteorology over a long period prior to the decision to create WMO itself and a knowledge of these early developments greatly facilitates a full understanding of the challenges, the difficulties and the achievements which were to follow. The purpose of the present chapter is therefore to present a brief review of these early developments and thereby to provide an appropriate background for the more detailed picture which follows.

METEOROLO.

Tomus ſextus operum

ARISTOTELIS

STAGIRITAE,

PERIPATETICORVM PRINCIPIS.

Naturalis eam Phyloſophiæ adferens ſectionem,

QVAE AD METEOROLOGICORVM,

ſeu ſublimium corporum, & vegetabilium quidditates ſpecies, & paſsiones, Animalium vero Hiſtoriam, partes, & Inceſſum ſpectare dignoſcitur.

CVNCTA VSQVE AD EO CLARA, NITIDA, Illuſtrata, diſtinctaq; prodeunt vt Ariſtotelem, Auerroemq; ipſum, viuentes hæc differere videatur.

QVAE HOC CLAVDVNTVR TOMO ſequens indicat pagina.

Cum ſummi Pontificis, Gallorum Regis, Senatusq; Veneti decretis.

VENETIIS MDLX.

Title page of a Latin edition of Aristotle's work on meteorology, printed in Venice in 1560

Meteorology emerges as an international science

From the earliest times, Man has been dependent upon his natural environment and particularly upon the environmental processes we call weather and climate. He has constantly sought to acquire knowledge of these processes and to apply such knowledge to safeguard and improve his way of life. Records of the earliest civilizations contain innumerable references to weather and climate. The ancient civilizations of Mesopotamia, Chaldea, China, India and Egypt abound with such references—as do the records of other regions of the world. The seasonal flooding of the River Nile was, for example, a phenomenon well known to the ancient Egyptians and formed an essential feature of their way of life.

The writing by Aristotle of his work *Meteorologica* in the fourth century BC was a major step forward in the study of the atmosphere and may be said to have removed the subject from the realms of mythology. That work and the small treatises on wind and the signs of the weather by Theophrastus, his pupil and successor, were based on observation and reason rather than on fantasy. These works of Aristotle and Theophrastus thus comprise a major development in meteorology and indeed explain the origin of the word "meteorology" which has long since been generally accepted. The still-standing Tower of the Winds constructed in Athens in the first or second century BC is a further indication of the systematic study of this subject by the ancient Greeks.

The Tower of the Winds in Athens was built in the second or first century BC

But progress was still slow. The earliest known systematic recording of local weather in Europe is believed to date from the early fourteenth century when William Merle, Rector of Driby in England, made daily records during a period of seven years. But two more centuries had still to elapse before major advances were to be made. The Renaissance in Italy and northern Europe, the voyages of Columbus and the opening of the sea routes between Europe and the Far East all gave an immense stimulus to European trade and industry and created a need for objective information about meteorological phenomena affecting them.

The need for meteorological information for practical purposes was thus becoming clear. Fortunately meteorology as a science also began to make progress. The first network of observation stations is thought to be that created by Ferdinand II of Tuscany in 1653. His *Accademia del Cimento* (Academy of Experimentation) established seven meteorological stations in northern Italy and four outside Italy. The invention of certain instruments for measuring physical elements assisted in the work of the Academy and made further progress possible. The air thermometer was invented about 1600, probably by Santorio, although Galileo is said to have claimed to have been the inventor. Castelli invented a new raingauge in 1639 and a little later, in 1644, Evangelista Torricelli, a pupil of Galileo, invented the barometer. At about this time also various forms of hygrometer and anemometer appeared, one version of the latter being invented by the English physicist Robert Hooke in 1664.

The development of instrumentation in turn enabled another wave of scientists in the seventeenth and eighteenth centuries to determine certain basic physical laws without which there could be little further progress in understanding meteorological phenomena. Robert Boyle, in 1659, enunciated his famous law on the relation of volume to pressure of a gas, the first step in understanding the dynamics of the atmosphere. Hadley, in 1735, gave an explanation of the relationship between trade winds and the rotation of the Earth. Franklin, in 1752, worked on atmospheric electricity. Lavoisier, in 1783, and Dalton, in 1800, laid the physical basis of meteorology as a science with their findings concerning the nature, condition and composition of the mixture of gases we call the atmosphere.

In 1771 Lambert proposed a fully standardized world network of meteorological stations and in 1780 the *Societas Meteorologica Palatina*, Mannheim, started its activities to that end. A network of 40 weather observing stations was in due course established in Germany and other European countries and a small number even in the United States—in other words, the society set up an international network of meteorological stations. Each station was equipped with comparable instruments including a barometer, a thermometer, a hygrometer and in some cases a wind vane and a raingauge. Standard instructions for their use were also provided. The siege and capture of Mannheim in 1795 halted this worthy effort.

All these developments soon led to the possibility of preparing maps showing the weather over a large area. The first systematic attempt at preparing a weather map was made by Brandes in Leipzig in 1820. His first weather map was based on data assembled by the *Societas Meteorologica Palatina* many years previously in 1783. He later prepared maps to show the storms in Europe in 1820 and 1821. About the same time Redfield, of New York, prepared the first series of charts of hurricanes, showing their rotary and progressive motion. Adding to this work, Espy, of Philadelphia, and the British scientists Piddington and Reid were able to establish the existence of characteristic patterns of pressure, wind and weather as well as empirical rules for their development and the accompanying sequence of weather changes.

All the investigations referred to above were of course based on observations collected long after the event, and hence far too late for any useful forecasts to be made, but they were highly important in that they established a firm foundation on which further progress could be made. The need now was for frequent and accurate measurements of meteorological elements over large areas of the Earth's surface and for the rapid transmission of such data to meteorological centres where they could be assimilated with a view to preparing weather forecasts.

As regards the first of these requirements, the developments referred to above resulted in good progress being made in the expansion of networks of meteorological observing stations. As regards the second, it was the invention by Samuel Morse of the electric telegraph in 1843 (when he transmitted his famous message "What hath God wrought" over a line between Washington, DC, and Baltimore) which opened the door to unprecedented progress in this field and indeed revolutionized the possibilities of weather forecasting.

The first weather maps based on telegraphic data were publicly displayed in Washington, DC, in 1850 and in France in 1855. The emphasis which public opinion was now giving to forecasting was in fact a reflection of the increasing demands of a society in which many changes were taking place. Industrial development was increasing and with it demands placed on science and technology as a whole—and not least on meteorology. Of particular significance to meteorology at that time was the vast expansion of international trade and hence of merchant shipping. The security and efficiency of maritime transportation called for reliable and regular information about the weather over all sea areas of the world. This not only underlined the importance of meteorology but also made it clear that international co-operation in this science was a *sine qua non* to progress. It was therefore no accident that an international meteorological conference was convened and that it concerned itself largely with maritime meteorological problems. This was the First International Meteorological Conference, which took place in Brussels in August 1853.

In the context of this historic event, the name of Lieutenant Matthew Fontaine Maury of the US Navy will long be remembered. It was thanks to his efforts that the Conference was convened and was so well organized. The Conference was attended by 12 delegates (mainly naval officers) from nine countries (Belgium, Denmark, France, Great Britain, the Netherlands, Norway, Portugal, Sweden and the USA)—a small gathering by modern standards of international conferences but a highly significant occasion at the time, not only because of its successful outcome from the point of view of marine meteorology, but also because it demonstrated very clearly the important benefits to be derived for meteorology from international co-operation.

The kernel of Maury's proposal to the Brussels Conference was that:

> ... the navies of all maritime nations should co-operate and make these meteorological observations in such a manner and with such means and implements, that the system might be uniform and the observations made on board the public ship be readily referred to and compared with the observations made on board all other public ships, in whatever part of the world. And, moreover, as it is desirable to enlist the voluntary co-operation of the commercial marine, as well as that of the military of all nations in this system of research, it becomes not only proper, but politic, that the forms of the abstract log to be used, with the description of the instruments to be employed, the things to be observed, with the manipulation of the instruments and the methods and modes of operation should be the joint work of the principal parties concerned.

The proposal was adopted by the Conference and has indeed formed the basis of maritime observations ever since. Stimulated by the success of the Brussels Conference in the field of international marine meteorological co-operation, efforts were soon made to develop international co-operation in the many other fields of meteorology. In particular, many ideas were put forward for an international system for terrestrial meteorological observations by, *inter alia*: Professor Dove, of Prussia (1860); Professor Quetelet, of Belgium (1867); Professor Jelinek, of Austria (1867); Dr Renou and Dr Marie-Davy, of France (1868).

During this same period, the importance of providing meteorological information to assist many other activities on the national level was becoming increasingly recognized and many countries were establishing national Meteorological Services, including networks of observing stations, in order to provide the advice required. The developments made even clearer the need for international co-operation and with highly commendable foresight the leading meteorologists of Austria (Professor Jelinek), Russia (Professor Wild), and Germany (Professor Bruhns) arranged a meeting in Leipzig in 1872 to discuss what further steps should be taken. The letter of invitation stated clearly the purpose of the meeting in the following terms:

> At the present time, the increasing interest in meteorological research shown by all civilized countries has led to a demand for far-reaching co-ordination and standardization of the methods and procedures in use in different countries. Such suggestions have been put forward and discussed so frequently (for example, by C. H. D. Buys Ballot in his paper 'Suggestions on a uniform system of meteorological observations', Utrecht, 1872) that the undersigned consider it both feasible and timely to propose the convening of a meteorological conference. . .

The achievements of the Leipzig Conference were twofold. It brought together most of the world's foremost meteorologists who were able, in large measure, to reach agreement on standardized methods of observation and analysis, including the use of a single set of symbols. It also prepared the way for holding, in Vienna in the following year, the First International Meteorological Congress. The documentation which it prepared for that Congress contemplated the establishment of a permanent body to deal with meteorological problems common to the international community.

International Meteorological Organization

The Vienna Congress was convened by the Government of Austria in September 1873. Invitations were extended through diplomatic channels to the governments of those countries that had established national Meteorological Services. Virtually all accepted and 32 delegates from 20 governments took part. The agenda included all existing and foreseeable problems of the science and practice of meteorology. A study of the record of the Congress in the light of subsequent events reveals the astonishing skill and foresight shown by the delegates in reaching appropriate decisions on the international procedures to be followed—as regards both technical and organizational questions.

Most important of all was the acceptance on all sides of the need to establish a permanent international body in order to ensure continued progress in the science of meteorology and also to ensure that all nations could reap the practical benefits that such progress would make possible. In other words, the concept of an International Meteorological Organization was born. It may be mentioned parenthetically that exactly 100 years later, and in the same conference room in Vienna, the organization which was to succeed IMO held a centenary celebration attended by 210 participants, including representatives of 73 countries and 17 international organizations and in the presence of the President of Austria.

To return to the Vienna Congress, the agenda was long, with most items referring to such practical matters as calibration and checking of instruments, hours of observations, scales and units, and the mutual exchange of meteorological observations by telegraph Ever present at all the discussions was the awareness of the need for a permanent machinery to ensure the continuity of effort in the international organization of meteorology and the shrewd far-sightedness shown in achieving this goal. The Congress gave preliminary consideration to the statutes and the appropriate structure of the organization. To ensure continuity of effort, the Congress created a Permanent Committee of seven members with the presiding officer of the Congress itself, Professor Buys Ballot, as President. It is of interest to note that the Committee established the principle that work in international meteorology would be carried out on a voluntary basis—a principle which has continued to be of importance in the development of international meteorology since that time.

C. H. D. Buys Ballot, First President of IMO (1873–1879)

The first task of the Permanent Committee, whose members were Bruhns, Buys Ballot, Cantoni, Jelinek, Mohn, Scott and Wild (all, it is important to note, directors of Meteorological Services), was to communicate the decisions of the Vienna Congress to the governments which had participated in it. Their next task was to prepare a work programme which would ensure that these decisions were carried out. The Permanent Committee in fact wasted no time in getting down to its work: it began its first meeting in Vienna just two hours after the Congress had ended.

The Committee set about its allotted tasks in a highly efficient manner at this first meeting and also at the three subsequent meetings which followed it—Utrecht, 1874; London, 1876; Utrecht, 1878—all under the presidency of Buys Ballot. By this time, the stage had been reached when the Second International Meteorological Congress could appropriately be held. This was convened in Rome in April 1879. Some 40 scientists from 18 countries attended. In the absence of Buys Ballot, due to grave illness in his family, the presidency was assumed by Professor G. Cantoni, Director of the Italian Central Meteorological Office.

The Congress gave attention to all aspects of the tasks to be faced and established an International Meteorological Committee (to replace the previous Permanent Committee) with responsibility for ensuring that its decisions were implemented; Professor H. Wild was elected President. The IMC met at intervals to review progress and to take any necessary further action. Meetings were thus convened in Berne (1880), in Copenhagen (1882), in Paris (1885) and in Zurich (1888).

Heinrich Wild
1879–1896

E. Mascart
1896–1907

W. Napier-Shaw
1907–1923

With the needs of weather forecasting in mind as well as other applications of meteorological information, great attention was given to promoting and advancing work on the standardization of instruments, on methods of observation and on the publication of observations. The IMC published or assisted in the publication of a number of studies and reports of wide general interest to meteorologists. These included International Meteorological Tables (published eventually in Paris in 1889 through the efforts of Professor Mascart); a report and instructions on the observation of the motion of clouds; catalogues of published and unpublished observations in a number of countries as well as general bibliographies; a report on the Intercolonial Meteorological Conference held in Sydney in November 1879 to apply the Rome resolutions in the Meteorological Services in Australia and New Zealand; and a report on the establishment of mountain meteorological stations in Europe and the United States.

The IMC also stimulated the holding in Austria in September 1880 of a Conference on Agricultural and Forestry Meteorology. While successful in encouraging the extension of regular observations in countries and regions where these were lacking, particularly in China, Japan, Korea, Argentina, Brazil, New Guinea, Congo and South-West Africa, its efforts to develop telegraphic services for the collection and distribution of the observational data experienced some initial difficulties, because of the high costs involved.

Mention must also be made of the arrangements by IMO for the International Polar Year (1882–1883). Prompted by its originator, the Austrian naval officer Weyprecht, a number of countries sent 13 expeditions to the Arctic and two to the Antarctic, to undertake observations into meteorological and magnetic phenomena over a period of one whole year. The report of this successful undertaking was edited by Professor Wild and published by the St. Petersburg Academy of Sciences.

The IMC had been asked by the Rome Congress to arrange an intergovernmental Third Meteorological Congress and consideration was given to convening such a Congress in 1885. By that time, however, it was becoming clear that meetings of this kind could with advantage be arranged on a less formal basis. Thus, after due consideration, the IMC decided to convene instead a Conference of Directors of Meteorological Services on a non-governmental basis. In a speech winding up the work of the IMC, and after noting the form of its successor, its President, Professor Wild, summarized the situation of international meteorological co-operation in 1888. He observed that the two Congresses and their subsidiary bodies:

> ... have doubtless contributed to the advancement of meteorology, by uniting the efforts of the different countries in certain directions, by facilitating the researches of meteorologists by means of the unification of the methods of observation and of the publication of observations in different countries, and by notifying the most important questions to be settled by fresh investigations. Much still remains to be done in these different directions. Let us hope that the future Conference of the heads of all the Meteorological Services which are in future to replace the Congresses and their Committees will be able still better to realize the agreement between different countries, so desirable for meteorology, which is a science truly international.

The first such Conference took place in Munich in 1891. Invitations were extended to some 80 directors of Meteorological Services and observatories in all regions of the world, of whom 31 found it possible to attend, despite the now non-governmental nature of the gathering. The President of the Conference was Dr C. Lang, Director of the Central Meteorological Station of Bavaria.

In addition to discussing the many scientific questions which presented themselves, the Conference paid great attention to the structure of the Organization, being well aware that the future of international meteorology would depend upon the decisions taken. The outcome was a decision that the system of Conferences of Directors should continue and that an International Meteorological Committee should replace the previous Permanent Committee but with essentially similar responsibilities, i.e., conducting the affairs of the Organization between the Conferences of Directors. Subsequent developments were to show that the system put into operation in Munich in 1891 continued to be successfully applied up to 1914 when World War I arrived and inevitably interfered with such activities.

Due attention was also given at the Munich Conference to scientific and technical matters and, as a result, much progress was made in many fields during the period 1891–1914. A particularly significant aspect of such progress was the development of the Technical Commission machinery. It was realized that if groups of specialists in the respective fields were set up, they could, through their joint efforts and with an exchange of skills and experience, advance the frontiers of meteorological knowledge in a way which would not otherwise be possible. A start had in fact been made at the Munich Conference itself when the Commission for Terrestrial

E. van Everdingen
1923–1935

Th. Hesselberg
1935–1946

Sir Nelson K. Johnson
1946–1951

Magnetism and Atmospheric Electricity had been set up. With the subsequent improvements in instruments and technology, as well as basic knowledge of the atmospheric processes, the needs and possibilities of specialization were created and by 1914, no fewer than 25 Technical Commissions had been established, of which two had been dissolved before 1914 as their tasks had been completed.

During this same period (1891–1914), the Conference of Directors met in Paris (1896) and Innsbruck (1905). The Innsbruck Conference was particularly noteworthy because it concerned itself with the codification of the large number of internationally agreed rules which had been developed over the decades. At the meeting of the International Meteorological Committee which followed (Paris, 1907), the IMC accepted the draft prepared by Professor Mascart. World War I, however, prevented its approval by the Conference of Directors until 1919. The thoroughness and the vision shown in its preparation are evident from the fact that the first edition of the Technical Regulations published in 1955 by the successor to IMO (the World Meteorological Organization) followed closely the pattern set 50 years previously.

During the same period, the use of telegraphic communications for meteorological reports became possible. Every effort was made of course to rationalize and economize on such communications. Discussions with the International Telegraphic Bureau in Berne did much to overcome the inevitable difficulties and the Commission for Weather Telegraphy succeeded in introducing, and then accelerating, the international exchange of weather telegrams. At the Conference of Directors in Innsbruck (1905), the question came up for the first time of coastal stations receiving weather reports from ships at sea by wireless telegraphy, and so began an essential element in synoptic meteorology.

In all such activities, the International Meteorological Committee played an important role. It met in Uppsala (1894), in St. Petersburg (1899), in Southport, UK (1903), in Paris (1907), in Berlin (1910) and in Rome (1913). In addition to the many scientific and technical questions which the Committee had to discuss, there were also problems of an administrative nature to be dealt with. The Committee accordingly gave much time and thought to the establishment of a paid Secretariat; but the mood of governments was such that no progress could be made to that end. The burden of the work had therefore largely to be carried by the Secretary. On occasion, colleagues from other Services helped him with the translation and publication of proceedings and reports. Among the selfless scientists who gave of themselves without stint to furthering the cause of international meteorology, the name of R. H. Scott of the UK shines brightly. During the 28 years he served as Secretary of the International Committee and to the similar bodies which had preceded it, he made invaluable contributions to international co-operation in meteorology. He retired in 1900 when his place was taken by Dr H. Hildebrandsson, of Uppsala. He occupied the post until 1907, when he was succeeded by Professor D. G. Hellmann, of Germany.

A study of the reports of the International Meteorological Committee and its subsidiary bodies during the quarter-century preceding World War I reveals that the advancement of international co-operation in meteorology gradually became largely the task of the Technical Commissions of the Organization. It was a period when the science and practice of meteorology was on the threshold of great and revolutionary events. The next period of progress had, however, to await the end of the war.

The Organization was dormant during World War I and international meteorological co-operation virtually ceased from August 1914 until the early part of 1919—but it revived rapidly and with its reawakening, a new era of rapid progress commenced. The President of the International Meteorological Committee at that time was Sir Napier Shaw (UK), who at once called an informal meeting of that body to consider the action needed to revive the Organization. Arrangements were made at that meeting for the Conference of Directors to hold its first post-war meeting in Paris in September 1919. At this meeting, the previous structure of the Conference of Directors and the International Meteorological Committee was confirmed as were also the previous Technical Commissions, but now with a new one added—the Application of Meteorology to Aerial Navigation.

The Organization then pressed ahead on all scientific and technical fronts. The Technical Commission system which had become an integral part of international co-operation in meteorology was expanded and improved to exploit the rapid development of radio and aviation. This manifested itself most noticeably at the Conference of Directors in Copenhagen in 1929. A system of collective radio weather broadcasts for Europe and North America was developed, using revised meteorological codes. When the Copenhagen meeting closed, the expansion of the Organization's role in international co-operation in meteorology was graphically illustrated by the number and subject-matter of its extremely active Technical Commissions. The Commissions covered the following broad subjects and activities:

- Terrestrial Magnetism and Atmospheric Electricity
- Solar Radiation
- Exploration of the Upper Atmosphere
- Synoptic Weather Information (formerly Commission for Meteorological Telegraphy)
- Maritime Meteorology
- Agricultural Meteorology
- Application of Meteorology to Aerial Navigation
- Investigation of the Waves of Explosion
- Study of Clouds
- The Polar Year
- Climatology
- *The Réseau mondial* and Polar Meteorology

As regards research activities, while meteorology in the involved countries during the First World War was channelled into purposes directly related to each side's war effort, the situation was different in some of the neutral countries. In Norway, in particular, cut off from the major centres of meteorological work, intensive theoretical work was undertaken. Under the leadership of Professor Wilhelm Bjerknes, assisted by his son J. Bjerknes, T. Bergeron and H. Solberg, the Bergen school worked out the then revolutionary methods of weather study involving air mass and frontal analysis. This work and its development by later meteorologists introduced the third dimension into meteorological practice, enabling thermodynamic and hydrodynamic principles to be applied to forecasting.

During the inter-war period, meteorology made extensive and rapid use of technical advances to help the international observing and reporting systems. The first use of the aeroplane for upper-air sounding took place in the early 1900s, but aero-meteorographs came into their own in the 'twenties and 'thirties. They were gradually replaced by the radiosonde, the first being developed by Bureau and Idrac in France in 1927; later a practical radio transmitter was developed by P. A. Moltchanoff in the USSR in 1930. During that period, IMO maintained its identity and succeeded in fostering that remarkable spirit of friendly international co-operation which has characterized relations between meteorologists from the mid-nineteenth century.

The status of IMO is reviewed

All these activities were encouraged and developed by regular meetings of the Conference of Directors (Utrecht, 1923; Copenhagen, 1929; Warsaw, 1935) and by the more frequent meetings of the International Meteorological Committee—starting in London (1921) and culminating in Berlin (1939). At all these meetings, the status of the Organization in world affairs became increasingly relevant to its continued progress and success.

The status question was not at all abstract and posed a choice between two systems. Should international co-operation in meteorology take place as an intergovernmental activity controlled by an International Meteorological Bureau assisted by a small, paid Secretariat, whose decisions would be binding on governments? Or should these matters continue to be dealt with by heads of Meteorological Services in an unofficial capacity? Governments as well as meteorologists had different views on this important matter. Some meteorologists took an unfavourable view of government officials being involved in discussions and decisions which they (the meteorologists) considered outside the understanding of non-scientists. Other meteorologists saw things in a quite different light. They saw little possibility for a full exploitation of meteorological discoveries through international co-operation—and this would involve considerable expenditure of the taxpayers' money— without the direct intervention and support of governments. At the same time, a number of governments seemed unaware that the world was changing rapidly and that they were living at the beginning of a scientific and technological revolution. They were distrustful of scientists and favoured letting things go on quietly as they had since Vienna 1873 and Rome 1879 (even though these Congresses had been at the governmental level). This is probably the reason why the Organization turned down an offer by the intergovernmental League of Nations—its Committee for Intellectual Co-operation had offered space for an International Meteorological Bureau at its offices in Paris.

A step forward was made, however, when it was agreed at a meeting of the International Meteorological Committee (Vienna, 1926) that a small Secretariat should be established. It would have no policy-making functions and its principal task would be to relieve the President and Secretary of the Organization of their administrative and conference-servicing burdens. It would publish the reports and proceedings of the International Meteorological Committee and its Commissions, a task which had previously been carried out by the Members on a voluntary basis. It would also serve as a documentation centre.

The modest budget for the permanent Secretariat was apportioned among the countries whose directors of Meteorological Services participated in the Conference of Directors. The first chief of the Secretariat was the Netherlands meteorologist Dr H. G. Cannegieter—he was appointed in 1928. He and his Secretariat were under the direct control of the President of the IMC. Pending a final decision on the location of the Secretariat—it was agreed that it should be in one of the smaller countries of Europe—it was decided to install it initially at the headquarters of the Royal Meteorological Institute at De Bilt in the Netherlands.

While the Vienna decision temporarily settled the question of the Secretariat, the question of the status of the Organization was by no means in abeyance. The sixth Conference of Directors, held in Copenhagen in 1929, confirmed the structure of the Organization: a Conference of Directors, an International Meteorological Committee with an Executive Council, Technical Commissions and a Secretariat. But it also adopted a resolution, which was sent to governments through proper channels, on the desirability of the Organization becoming intergovernmental.

The seventh Conference of Directors, which met in Warsaw in 1935, studied further the structure of the Organization and, in addition to the bodies previously approved, took the important and far-sighted step of establishing regional bodies to be termed Regional Commissions. The Warsaw meeting also came back with vigour to the question of whether the Organization should become intergovernmental and decided that invitations to future meetings of the Conference of Directors should be sent to governments. The invitations would request governments to designate directors of their national Meteorological Services to represent them at the meetings and to vote on their behalf. It was hoped that this cautious move towards official recognition would enhance the status of the Organization.

At the Warsaw meeting, Dr Th. Hesselberg, Director of the Norwegian Meteorological Service, was elected President of the IMC. He was a strong supporter of an improved international status for meteorology. Indeed, Dr Hesselberg played an active role in drafting, in co-operation with Mr P. Wehrle, Director of the French Meteorological Service, a World Meteorological Convention which, if accepted by governments, would secure official status for IMO. This draft was carefully examined and discussed at the twenty-second session of the IMC, which met in Berlin on the eve of the Second World War. In presenting his preliminary draft to the Committee, Dr Hesselberg gave some of the reasons which he believed called for a change in the Organization's status:

> In view of the steadily increasing practical importance of meteorology, it is desirable that the governments of the various countries should have a greater influence on the work of the Organization. The resolutions of the Organization should be binding on the countries to a greater extent. The Organization must be able to rely on adequate resources so that efficient co-operation should not be hampered by financial difficulties. It is abnormal for one of the Organization's commissions (the International Commission for Aeronautical Meteorology, which had intergovernmental status—Ed.) to have a more official status than the Organization itself. Similar organizations (the International Commission for Air Navigation, the International Union of Geodesy and Geophysics and others) have a more official status than IMO, a circumstance which has its drawbacks. Governments have not sufficient control over the choice of representatives from their countries.

Dr Hesselberg's proposals were wide-ranging. Annexed to the draft of his Convention were detailed regulations covering the Technical Commissions. Some changes in other bodies were also proposed. After working through the draft Convention, article by article, the Committee produced a document which became known as the Berlin Draft. This draft was referred to a commission which was to refine and polish it. In definitive form, it was to be considered by the Conference of Directors planned to meet in Washington in 1941.

Subsequent events were to confirm the vision and wisdom of Dr Hesselberg and his colleagues in preparing the Berlin Draft—but their hopes for its speedy acceptance were in vain. A few weeks after the meeting in Berlin, World War II commenced and the activities and plans of IMO had inevitably to be suspended. When peace returned some six years later, however, consideration of the Berlin Draft was promptly revived and its salient features were incorporated into the Convention of a new body which was to replace IMO: the World Meteorological Organization.

Creation of the World Meteorological Organization

It will be clear from the preceding paragraphs that IMO, as a non-governmental organization, served the cause of international meteorology well for about three-quarters of a century and that governments and meteorologists alike were, in general, well satisfied with such an arrangement. Towards the end of that period, however, it was becoming increasingly evident, at least to most meteorologists, that such a status for the focal point of international meteorology was incompatible with the importance which meteorology was then assuming in the context of the vast economic and technological developments being made.

By 1939, a draft of a new World Meteorological Convention (the Berlin Draft) had been prepared but further action on its approval was inevitably delayed by World War II. Some further progress was, however, possible—even during this difficult period—as the Secretariat was transferred to a neutral country, Switzerland, in 1939, shortly after the Berlin meeting. Dr G. Swoboda, who had been appointed in 1938 to succeed Dr Cannegieter as Chief of the Secretariat, was thus able to maintain contact with Dr Th. Hesselberg, President of IMO, who was moreover the prime mover in the preparation of the Berlin Draft.

It is interesting to note that the Secretariat was also able to continue some other activities of the Organization in this difficult period. Some Commissions whose activities were of a purely scientific nature, such as those dealing with terrestrial magnetism and radiation of the Sun, continued to function on a limited scale, while statistical tables in climatology, marine meteorology, the *Réseau mondial* and oceanography continued to be prepared, using the information available. Reports of the meetings of the Commissions which had met in the years 1938 and 1939, as well as the report of the meeting of the Committee in Berlin in 1939, were all duly edited and were then circulated in the years 1941 to 1944. In addition, a number of previous publications were revised and reissued. As already explained, all such activities became possible because the Secretariat was now based in a neutral country—and also, it must be added, because the annual financial contributions continued to be received from the national Meteorological Services throughout the war years.

But perhaps the most significant feature of the activities of the Secretariat during the wartime disruption was that they had ensured that the overall structure of IMO remained intact and that, when hostilities ceased, the basis on which international meteorological activities could be resumed without delay was thereby available—as was also the basis on which plans for its successor could readily be formulated. Thus, when the war was over, steps were promptly taken to convene an Extraordinary Conference of Directors in London in February 1946.

The immediate task, as expressed in a Secretariat document, was "to bring IMO back into operation, to ensure its co-operation with other international organizations and to resume the study of constitutional and other questions, the settlement of which has been prevented by the war". This task had of course to be undertaken in the light of the enormous scientific and technological developments which had taken place during the war and which had created many new opportunities for meteorology—and many new responsibilities as well.

The President (Dr Hesselberg), in his report to the Conference, gave particular attention to the Draft Convention. He reported fully on the additional preparatory work which had been made on the original Berlin Draft. The text was then discussed further and when the Conference closed, the newly established International Meteorological Committee (IMC) was specifically charged with the preparation of a definitive draft. The Conference had further decided that this draft should be presented to the next Conference of Directors to be held in Washington, DC, the following year, 1947. The Committee wasted no time in setting about its appointed task and held its first meeting in London immediately after the conclusion of the Conference; Sir Nelson Johnson (UK) was elected President, and Dr Hesselberg, Dr F. W. Reichelderfer (USA) and Mr A. Viaut (France) were the Vice-Presidents. (All were persons who later played prominent roles in the new organization when in due course it came into existence.) A second meeting of the International Meteorological Committee was held a few months later (July 1946) in Paris, at which a revised draft of the Convention was produced which became known as the Paris Draft. It was this draft which was submitted the following year to the Conference of Directors in Washington, DC.

Before referring to the outcome of the Washington Conference, some comments on the other items discussed at the London and Paris meetings are offered lest the impression be gained that the technical problems at that time were being overlooked in the spirit of enthusiasm for the creation of a new organization.

The London Conference considered at length all aspects of the technical problems arising in the post-war situation. It recommended that certain new meteorological codes should be used temporarily until the codes as a whole had been examined by the Commission for Synoptic Weather Information. It believed, too, that there were urgent matters arising out of the recent war which called for an early meeting of this Commission. The Conference also instructed the Commission to prepare, in co-operation with the newly established Regional Commission for Europe, a plan for the issue of continental collective weather messages for all European countries. It took steps to ensure the recovery of meteorological material lost during the war and the publication of meteorological observations collected during the same period. Expressing awareness of the enormously important role that meteorology was then playing in everybody's life, the Conference urged that the profession and practice of meteorology be given wide publicity. It also urged that assistance be given to countries lacking qualified staff.

Similarly, the Paris meeting which followed gave great attention to immediate technical problems. No fewer than 60 resolutions were adopted, many of which dealt with such important matters as the reorganization of European collective transmissions, the extension of teleprinter connections, regular exchange of weather information between Europe and North America by high-powered transmitters, and the expansion and improvement of the network of reporting stations, including stationary weather ships at sea. A number of the resolutions recognized the potential of technical inventions and developments during the war and their applicability to meteorology. In particular, one resolution recommended further investigation of the radar detection of rain while another recommended the use of aircraft weather reconnaissance flights, particularly for the direct observation of the development, progress and decay of tropical revolving storms.

Another subject calling for immediate attention after the war was the relationship between IMO and other organizations, both governmental and non-governmental, with an interest in meteorology. In the past decades, such relations had been sporadic and somewhat *ad hoc*. The immense expansion of technology during the war increased the role and activity of many of these organizations and called for urgent attention by IMO to avoid duplication of work and to ensure fruitful collaboration for mutual benefit. The Conference of Directors that met in London in 1946 gave particular thought to collaboration with the Provisional International Civil Aviation Organization (PICAO), the International Telecommunication Union (ITU) and the International Ice Patrol Service. It suggested that the character of the collaboration should be defined by an agreement in each case.

The Paris meeting of the IMC in July 1946 gave special attention to the problem of aeronautical meteorology. A resolution on co-ordination of activities with PICAO set out guidelines for developing collaboration. It provided for reciprocal representation of both organizations at meetings and the preparation of joint regulations. In questions of general meteorology, PICAO (later ICAO) would accept the recommendations of IMO. The IMC also established terms of reference of the new Commission for Aeronautical Meteorology, which would be set up later in the year and which would replace the old intergovernmental *Comité international de météorologie aéronautique* (CIMAé). In its final resolution, the IMC also considered the relation of the Organization to the newly created United Nations.

In addition to the London and Paris meetings, there was another event of great relevance to the deliberations at the Washington Conference. It was the convening of sessions of the ten Technical Commissions and the six Regional Commissions in Toronto, Canada, in August 1947. The meetings of these bodies were extraordinarily fruitful. No fewer than 400 resolutions were produced and these greatly facilitated the decision-making process of the Washington Conference, which opened shortly afterwards (23 September 1947).

Turning now to the outcome of that Conference, the most significant event was that agreement was reached on a text for the new World Meteorological Convention. The Paris Draft had of course to be considered by the Conference. It had two annexes: the draft General Regulations of the new Organization and the draft Technical Regulations. These three documents had previously been submitted for comment to governments and Meteorological Services. The reactions of certain government authorities to the three documents showed that there were important areas of disagreement. Thus at the Washington Conference, four new drafts were submitted—by Canada, the UK, France and the USA. What many governments were concerned about was the safeguarding of the Organization's independence and world-wide character; others attached particular importance to retaining professional representation—i.e., they wished to be represented at meetings of the new body by directors of Meteorological Services. After lengthy and at times animated discussion, agreed texts of the Convention and of its annexes were eventually produced and were approved unanimously.

The new body would be known as the World Meteorological Organization. Its purposes, as defined in the Convention, would be to facilitate world-wide co-operation in the establishment of networks of meteorological observing stations and the establishment of appropriate centres charged with the provision of meteorological services; it would in addition promote the establishment of systems for the rapid exchange of meteorological information and the standardization of observational and other procedures; it would further the application of meteorology to aviation, shipping, agriculture and other human activities; and it would encourage research and training.

The approved Convention also specified the constituent bodies of the Organization. The supreme body is the World Meteorological Congress, which normally meets once every four years to adopt the policy, programme and budget for the ensuing four-year period. The Executive Committee (now the Executive Council*) meets annually; its tasks include reviewing the implementation of the decisions taken by Congress, approving the annual expenditure and taking action on rec-ommendations of the other constituent bodies, namely the Regional Associations and Technical Commissions. The Regional Associations correspond broadly to the Regional Commissions of IMO, their main functions being to promote the execution of WMO resolutions in their respective Regions and to co-ordinate meteorological and related activities in these Regions. The Technical Commissions, similar to those of IMO, consist of technical experts who are called upon to study and make recommendations to Congress and the Executive Council on subjects within their terms of reference as laid down by Congress.

To ensure as wide a Membership as possible of WMO, the Convention provides for two broad classes of Members, States and Territories. In brief, States are countries which are fully responsible for the conduct of their international relations while Territories are dependent countries, each of which however

* The Executive Committee was renamed Executive Council by Ninth Congress (Resolution 42 (Cg-IX)). For reasons of consistency, the new name is used throughout this publication; reference is made to this note where appropriate.

maintains its own Meteorological Service. The difference between the two classes of Members is of greatest significance in voting at sessions of Congress: the Convention lays down a list of subjects on which only States are entitled to vote, such as amendments to the Convention and requests for Membership of WMO. In all other respects, however, and particularly in technical activities, the two classes of Members work together on an equal footing.

And so, on 11 October 1947, the Convention of the new World Meteorological Organization was signed in Washington by the representatives of 31 countries—a major event in the history of international meteorology. But further action was still necessary to bring this agreement into effect. The Convention itself stipulated that it would come into force only on the thirtieth day after the date of the deposit of the thirtieth instrument of ratification or accession. On 21 February 1950 the Government of Iraq deposited its instrument of ratification with the State Department in Washington. Thus, on 23 March 1950, the World Meteorological Organization was formally established. To mark this important event, the World Meteorological Organization a few years later declared 23 March as World Meteorological Day—a day which has since been appropriately celebrated each year.

Although WMO came into existence on 23 March 1950, further action was still necessary before it would become effective as a new world organization. Steps had first to be taken formally to declare that IMO no longer existed and to make detailed arrangements for WMO to assume its appointed functions.

This final action was taken in Paris in March 1951. The final Conference of Directors of IMO opened on 15 March 1951; it closed two days later, on 17 March. At the concluding meeting, the President, Sir Nelson Johnson, formally declared that IMO had ceased to exist and that WMO had taken its place. His words on that occasion were:

> Thus came to an end one of the pioneer organizations for international co-operation. The torch it has kindled is not extinguished, but is handed on to a new organization to maintain and foster. . . . We say goodbye to the IMO with affection but no regret, and we go forward with confidence in the WMO to apply meteorology more fully to the service of mankind. Gentlemen: the old IMO is dying, long live the WMO.

CHAPTER II

FIRST CONGRESS SETS WMO ON ITS COURSE

Just two days after the end of the final Conference of Directors of IMO, on 19 March 1951, the First Congress of WMO opened in the Palais d'Orsay, Paris. Sir Nelson Johnson, the retiring President of the Conference of Directors of IMO, was elected President of the First Congress.

The Membership of WMO was then 66 (46 States and 20 Territories). No fewer than 59 Members (49 States and 10 Territories) had sent delegates. Representatives of the United Nations and many other international organizations were also present. A group photograph of participants at this historic gathering is shown below. It is interesting to note that by the tenth Congress, 36 years later, the total membership had risen to 159 (154 States and five Territories) and the number of those represented at the Congress was 138 (134 States and four Territories).

A long and complicated agenda was before the participants. Some of the items dealt with matters with which they were familiar, such as the functions of the Technical Commissions and Regional Commissions set up by IMO; others were of a new character—such as the future relations with the United Nations and especially the proposal that WMO should become a Specialized Agency of the United Nations and thereby participate in its technical assistance and other relevant programmes. The United Nations Convention on Privileges and Immunities of the Specialized Agencies was another such subject. The future budget of the Organization and the structure and location of its Secretariat were also items which, if not entirely new, had now assumed new proportions. Several weeks of long and complicated discussion and decision-making therefore ensued and by 28 April 1950, the prescribed closing date, First Congress had successfully completed its task and had taken decisions on all the issues before it. Subsequent events were to prove that First Congress had in fact laid a solid foundation on which WMO, as a Specialized Agency of the United Nations, could be firmly established. These decisions are summarized below.

Membership questions

The number of States and Territories represented at First Congress has already been given. As recommended in the Convention (see Chapter XVII), most principal delegates were directors of national Meteorological Services, but several delegations included diplomats or Foreign Office (or equivalent) officials to advise on political issues. Many of the participants had of course been present at the preceding Conference of Directors of IMO, but in a different capacity; there they had been individuals from national Meteorological Services but now they were serving as official representatives of their governments. Some were doubtless wondering how this change of status would affect the manner of conducting business, the hitherto friendly spirit of understanding and of course the nature of decisions taken.

One important consequence of the change of status of the Organization now became apparent. On the second day of the session, the chairman of the Credentials Committee, presenting his first report, stated that, with one exception, the credentials of all delegations were recommended for approval.

The exception referred to China, the question being whether the delegation of Nationalist China or the People's Republic of China should be recognized. After an animated debate Congress, by a majority decision, gave recognition to Nationalist China, thus following the policy at that time of the United Nations itself. By another majority decision, however, it

Participants at the First World Meteorological Congress (Paris, 1951)

was decided that "the representatives of the Chinese People's Republic be invited to take part in the capacity of observers in this Congress".*

This was the first indication that the new Organization, WMO, would need to take into account factors of a political nature. It was equally an indication that this new responsibility in no way detracted attention from the basic scientific and technical objectives for which the Organization had been created. The chapters which follow show clearly that these same objectives have been given appropriate attention by all subsequent Congresses.

Relationship with the United Nations and other Specialized Agencies

An urgent matter arising from the change of status of the Organization was the negotiation of an agreement with the United Nations by virtue of the provisions of Article 25 of the WMO Convention. This was in fact the first agenda item to be taken up at First Congress after the report of the Credentials Committee. A draft agreement had been prepared in advance by a special committee of the IMO Conference of Directors and had been sent to Meteorological Services for comment. Amended in the light of the suggestions received, the draft was further revised during Congress by a small group of representatives of WMO and UN. The revised draft agreement was approved by Congress without opposition and was finally adopted by the UN General Assembly in December 1951.

In this way, WMO became officially recognized as a Specialized Agency of the United Nations. There were many advantages to both organizations: for example, WMO was committed "to co-operate with and to render all possible assistance to the UN" and in its turn was eligible to partake fully in what was then known as the UN Expanded Programme of Technical Assistance for Economic Development of Under-developed Countries, later to become the United Nations Development Programme.

The discussions at First Congress on the subject of technical assistance were largely based on a report by the President of the IMO Executive Council outlining the various forms which it could take. These included professional training of meteorologists, provision of technical meteorological information, and missions of technical experts to advise on the creation or development of national Meteorological Services. The representative of the United Nations pointed out that other organizations participating in the UN expanded programme also had their own normal technical assistance programmes, financed under their own budgets. The implication was that it would facilitate WMO participation in the UN programme if it were to have its own regular programme of technical assistance. A token sum was in fact earmarked by Congress for this purpose. Even during the session it was possible to decide on the first country to receive an offer of technical assistance from WMO, namely Libya, which was on the point of becoming independent. The UN General Assembly, in Resolution 387(V), had urged Specialized Agencies to extend to Libya "such technical financial assistance as it may request in order to establish a sound basis for economic and social progress". It was in the light of this request from the United Nations that Congress directed the Executive Council** to "arrange for the preparation of a plan for a Meteorological Service in Libya capable of meeting local and regional requirements".

This was the very modest beginning of the WMO Technical Co-operation Programme, surely one of the Organization's most valuable contributions to the development of meteorology throughout the world. Its rapid and impressive growth is the subject of Chapter XIV while Chapter XIII deals with the closely associated work within the WMO Education and Training Programme.

Being a UN Specialized Agency also facilitated WMO collaboration with other members of the UN family, such as Unesco, ICAO and FAO, all of which were represented at First Congress. There it was decided that the Executive Council** should enter into working arrangements with such bodies. Special importance was attached by Congress to collaboration with ICAO, which had its own Meteorology Division in view of the vital nature of the influence of weather on the safety of air navigation. We shall see in Chapters VII and XVI how the relationship which had already existed between IMO and ICAO (and its predecessor, Provisional International Civil Aviation Organization, PICAO) developed over the years. In Chapter XVI information is also given on the working arrangements which have been entered into between WMO and other Specialized Agencies.

These formal decisions on inter-agency collaboration have, in many cases, led to the active participation of WMO in projects of other agencies—and vice versa; examples of this will be found in the following chapters. During First Congress there was already one specific proposal for this kind of activity, in relation to the Arid Zone Programme of Unesco. This project was aimed at a study of the scientific and technical problems of the arid and semi-arid areas of the world with a view to the development of their resources in such a way as to improve the living conditions of the people. Congress adopted a resolution in which the Executive Council** was directed "to provide advice on meteorological problems" connected with the Unesco programme.

Before leaving the subject of WMO's relations with other Specialized Agencies, mention must be made of the International Telecommunication Union, whose history, like that of WMO, goes back well into the nineteenth century. Links with this Organization were clearly of great importance to WMO because of the vital role played by international telecommunications in the regular and rapid exchange of meteorological data for weather-forecasting services. Even before First Congress, ITU itself had recognized the need for close relations with IMO, as one of the three other Specialized Agencies with "the broadest mutual interest in telecommunications matters"; the others were ICAO and the proposed Inter-governmental Maritime Consultative Organization (which was later to become the International Maritime Organization). Congress approved the procedures for consultation and co-operation recommended by ITU and instructed the Secretary-General to take appropriate action for its implementation by WMO. The collaboration between WMO and ITU has continued on a friendly basis and contributed significantly to the success of what is now known as the WMO Global Telecommunication System, a major component of the World Weather Watch, the planning and implementation of which is discussed in Chapter III.

Delegates look to the future

The consideration by First Congress of WMO's relationship with the United Nations and Specialized Agencies led to a discussion which is of particular significance in the context of the future scientific activities of the Organization. It was a discussion which assumed a much wider scope than had initially been envisaged—namely, the evolution of WMO over the

*It may be noted that when some years later the United Nations changed its policy and gave recognition to the People's Republic of China, WMO promptly followed the same course. It is interesting to note that the Secretary-General of WMO was at that time the first representative of all the UN bodies to be invited to China for discussions on the changed situation.

**See note on p. 8.

long term, not only as a means of supporting the national Meteorological Services of its Members but also as an important element in the UN system as a whole. At the time Unesco was considering a proposal for the establishment of an international research institute of the arid zone which, in view of the interdisciplinary nature of the project, would inevitably have involved WMO in its climatological aspects. This proposal therefore linked up with an item on the Congress agenda regarding the setting up under WMO auspices of an International Meteorological Institute, an updated version of a proposal originally considered by IMO as long ago as 1873.

The report of the debate on this subject at First Congress is of particular interest because it highlights the different attitudes of delegates to policies which should be followed to fulfil the purposes of WMO as laid down in its Convention. While the resolutions approved by Congress pointed the way for the development of WMO during the next few years, this debate looked much further ahead and provided a range of opinions that could indicate the policies likely to achieve an acceptable measure of agreement in the course of time. First Congress revealed a maturity of outlook that was quite remarkable in the early stages of a highly important change from IMO to an intergovernmental organization entrusted with official responsibilities and objectives. As the years have passed it has been clear that the development of WMO has been determined not only by ensuring the development and efficiency of national Meteorological Services but also by the need to react increasingly to overall progress and interrelationships in such matters as population growth, economic and social advances, and world energy requirements, all of which have brought fresh and increasing demands. The present range of meteorological applications and the importance and urgency of the associated requirements were not entirely foreseen by delegates at First Congress but they in no way precluded any further changes that would be recognized as essential in the years to come.

In opening the debate the President invited Congress to consider the desirability of WMO co-operating with other international organizations and whether, as a general rule, WMO would wish to take part in all external programmes whenever they included meteorological aspects. Dr F. W. Reichelderfer (USA), who towards the end of First Congress was elected President for the ensuing four-year period 1952–1955, said there should be general agreement that WMO would have an interest in the meteorological requirements or proposals of other international organizations. WMO could at least give an assessment of a problem and say what it would be possible to do, taking into account scientific limitations as well as the priorities of the various schemes under consideration. He understood the reluctance of many delegates to state their views on general policy matters and likened the situation of the Congress to "that of a child in front of a table loaded with good things, wanting to taste all and not knowing where to begin". He stressed the importance of meteorological research for the improvement of meteorological services but thought it unlikely, and perhaps inappropriate, that WMO should have the resources, in terms of scientific staff and finance, to undertake fundamental research in a better manner than was already being carried out at existing institutes. He thought that WMO should restrict itself to the encouragement of fundamental research but play an active role in applied research, particularly by ensuring that all Members were made aware, by means of news-letters or regular bulletins, of the results of applied research carried out in various countries. To this end, he advocated an increase in the staff of the Secretariat and an extension of its powers and authority.

The delegate of Spain, Mr L. de Azcárraga, who was to serve as First Vice-President from 1959 to 1967, drew attention to budgetary realities and suggested that WMO should begin with limited objectives which would permit the Organization to develop gradually as and when required. With regard to the proposed International Meteorological Institute, he supported a written suggestion by Venezuela that the WMO Secretariat should simply be concerned with the encouragement of research by providing bibliographic material, by acting as an information centre and by helping to co-ordinate research activities undertaken by Meteorological Services in joint programmes.

Mr A. Viaut (France), who was to succeed Dr Reichelderfer as President of WMO for two four-year periods from 1955 to 1963, said his country was prepared to provide accommodation near Paris for an International Institute for Meteorological Studies and Research. Referring to the obstacles of a financial nature mentioned by other delegates, he exhorted Congress to show "cautious daring" in deciding on the future programme of WMO and the size of the Secretariat.

Some very helpful guidance was given to Congress in a speech by the representative of the United Nations. He explained the implications for WMO of being granted by UN the status of a Specialized Agency. By definition, he said, a Specialized Agency was an intergovernmental body with extensive international responsibilities. Thus the United Nations—in other words, the governments of the world—had recognized the importance of meteorology and had accepted WMO as providing the best basic machinery for the promotion of meteorology as an international science with numerous practical applications contributing to the development and prosperity of all countries. It seemed clear to him that in replacing the informal but highly regarded IMO, the new Organization should be given wider responsibilities and should undertake more extensive activities. The rate at which these developments would proceed would be determined by the governments which had ratified the WMO Convention and would try to furnish WMO with resources commensurate with the responsibilities entrusted to it.

Many other delegates took part in the debate, for the most part commenting on suggestions put forward by other speakers. The delegate of the USSR, Mr K. Pogosjan, spoke of the research already in progress at the national level and saw no need for an international institute. In his view the technical work of WMO should be assigned to the Technical Commissions, a procedure that had operated successfully under IMO.

At the conclusion of the discussion, the President noted the absence of any general agreement on the need for an international research institute. He said the proposal could remain in abeyance but available for further discussion if desired at a future Congress. The President then pointed out that Congress had registered virtually unanimous agreement in various matters of major importance. For example, WMO should be prepared to participate in any UN programmes which had a meteorological content and, of these, programmes of technical assistance should be accorded high priority. It was also clear that a technical section was required in the WMO Secretariat.

A review of this important discussion, after the lapse of nearly four decades, might give rise to such questions as to whether some, perhaps too many, of the delegates tended to be inward-looking and somewhat fearful of exploring widely the potentially enormous field of meteorological applications. An inference of this nature would be unjustified because it would ignore the circumstances prevailing at the time. When First Congress took place in 1951 most countries were still recovering from the disruption and other effects of World War II and

F. W. Reichelderfer
1951–1955

André Viaut
1955–1963

A. Nyberg
1963–1971

lacked sufficient resources to deal with national requirements even of the greatest urgency. The delegates, as officially accredited representatives of their countries, were therefore reluctant to approve schemes which would involve large expenditure without achieving early results of world-wide benefit. In any event the delegations included (usually as Principal Delegate) the directors of national Meteorological Services, who were able to assess priorities since they had acquired vast experience in the provision of weather forecasts and climatological information to assist in the planning and operations of outside interests such as shipping, aviation and public utilities.

However, in spite of the financial problems besetting the delegations, there were some persuasive attempts, notably by Dr Reichelderfer and Mr Viaut, to lead delegates to an expansive vision. In addition, the United Nations representative made it abundantly clear that WMO as a Specialized Agency would not be able to function within a narrow ambit but would find itself involved in a great variety of multidisciplinary projects as the UN family, including other Specialized Agencies, produced programmes and called for collaboration. His remarks were extremely perceptive and it may be regarded as fortunate that the expanding commitments which he envisaged were initially of slow growth and did not greatly embarrass WMO in its formative years.

In the international field scientific and technical projects are often of slow initial growth before proceeding almost explosively to maturity and fulfilment. It is no criticism of First Congress, which met in 1951, to state that most of the developments which have taken place in WMO could not then have been foreseen. For example, World Weather Watch could not have become a realistic concept until satellites were sent into orbit in space more than a decade later. A comparison between the position of WMO at First Congress and its position at Tenth Congress (in 1987), a span of 36 years, shows progress on a scale that can be qualified as remarkable. Whereas in 1951 First Congress approved a small technical division in the Secretariat, by 1987 Tenth Congress found itself reviewing the reports and future plans of separate departments of the Secretariat concerned with World Weather Watch, World Climate, Research and Development, Applications of Meteorology, Hydrology and Water Resources, Education and Training and with Technical Co-operation. The existence of these departments, nearly all of long standing by 1987, and their major programmes of world-wide scope provide a tribute to the delegates at First and subsequent Congresses who showed flexibility in outlook and a determination that WMO should play a full part as a Specialized Agency in the work of the United Nations.

Technical questions

In reading the *Proceedings* of First Congress, it is at first sight surprising that there is little discussion of the technical aspects of telecommunications and of meteorological observations. This does not in any way reflect a lack of interest in improving the national networks of meteorological observing stations, which of course provide the weather data needed by every country in the world for their day-to-day weather forecasting services. The reason for the absence of much discussion of this topic was simply that the procedures to be followed by national Meteorological Services in carrying out their weather observations had already been developed to a high degree of sophistication by IMO. In the absence of any specific proposals for urgent changes in the system, all that was required was for Congress formally to decide that the relevant technical decisions of IMO should remain in force "until such time as they may be repealed or amended by the appropriate body of the Organization". In the case of the networks of stations for synoptic purposes, the appropriate body to initiate any technical changes was the Commission for Synoptic Meteorology, which was duly established by First Congress with terms of reference similar to those of the former IMO Commission for Synoptic Weather Information.

As already mentioned, one of the advantages of the change of status from non-governmental to governmental organization—indeed, one of the main reasons for the change—was that the relevant decisions of WMO could be made *mandatory* whereas those of IMO were at most *advisory*. The WMO Convention contains provisions for the adoption by Congress of Technical Regulations and the terms of reference for all of the Technical Commissions include making recommendations to Congress on draft regulations. At First Congress it was agreed that one of the most urgent tasks of the Organization was the preparation of provisional Technical Regulations, based in part on those technical resolutions of the former IMO which were considered to be of a permanent nature. A very useful report on the incorporation of IMO resolutions in the WMO Technical Regulations had been prepared for First Congress by an IMO working group. The proposals of this group were adopted by

M. F. Taha
1971–1979

R. L. Kintanar
1979–1987

Zou Jingmeng
1987–

Congress with the minimum of discussion and thereafter rapid progress was made in the drafting of the Technical Regulations by all the Technical Commissions.

Closely related to the Technical Regulations are the WMO *Guides*. In 1947 IMO initiated the preparation of a *Guide to International Meteorological Instrument and Observing Practices*, aimed at providing information on the methods to be applied in order to obtain correct weather observations and to keep observing stations up to international standards. First Congress decided to accept responsibility for this publication and the task of completing it and keeping it up to date fell naturally to the Commission for Instruments and Methods of Observation.

Technical Commissions and Regional Associations

Mention has already been made of one of the eight Technical Commissions established at First Congress, namely the Commission for Instruments and Methods of Observation (CIMO). The technical activities of this and the other Commissions are discussed in more detail in subsequent chapters while the organizational aspects are described in Chapter XVII.

As can readily be seen from a study of their terms of reference, the eight Technical Commissions set up in 1951 cover virtually the whole science of meteorology and its applications—with one major exception. Within IMO there had for a short time been a Commission for Hydrometeorology, responsible for those aspects of hydrology most closely related to meteorology. This Commission had not however been very active and at the Conference of Directors in 1947 it was discontinued. Nevertheless, the list of Technical Commissions submitted to First Congress by the IMO Executive Council once again included a Commission for Hydrological Meteorology and this was a subject of considerable debate during the session. Delegates from countries with a combined Meteorological and Hydrological Service generally favoured the establishment of such a commission while others felt that the proposed terms of reference could be allocated as appropriate to the Commission for Climatology (CCl) or the Commission for Agricultural Meteorology (CAgM). The view which prevailed was that, at least in its early stages of development, WMO should not embark on extensive hydrological activities. The only specific reference to hydrology in the terms of reference of the eight Technical Commissions was for CCl to be responsible for the "arrangement of climatological data in order to meet the needs of hydrology". We shall see in Chapter IX that this was by no means the end of the story of WMO's involvement in hydrology.

First Congress also had to decide on the establishment of Regional Associations to replace the former IMO Regional Commissions. The responsibilities of the Regional Associations, as laid down in the WMO Convention, include ensuring the compliance by their Member countries with WMO procedures and the co-ordination of meteorological and associated activities in their respective Regions. The IMO Executive Council had proposed to Congress that there should be six of these Associations with geographic limits corresponding closely to those of the six IMO Regional Commissions. These proposals were accepted by Congress with very little debate and with only a few small adjustments to the geographic limits of the Regions. The subsequent activities of the Regional Associations are reviewed in Chapter XV.

The Secretariat

At that time of First Congress, the IMO Secretariat was located in Lausanne (Switzerland), having moved in 1940 from De Bilt (Netherlands). During First Congress, invitations were received from France to have the WMO Secretariat in Paris and from Switzerland to locate it either in Lausanne or Geneva. As pointed out by the representative of the United Nations at the session, WMO was extremely fortunate to be able to make a choice from three such excellent offers. The choice proved in fact to be somewhat contentious. An *ad hoc* group of three delegates was set up to report on the financial, administrative and technical features of each of the offers and the Secretariat was asked to provide climatological information on the three sites. Even with this additional information Congress found it extremely difficult to make a decision and the subject had to be debated at great length in plenary meetings—most questions being discussed in detail in committee and then settled fairly quickly in plenary.

Many delegates referred to the advantages of having the Secretariat in a neutral country like Switzerland, but others spoke of the compensating advantages of being in a large city, such as Paris, with excellent library and other facilities. To some, Geneva had the advantage of being the headquarters of the European office of the United Nations, while others spoke of the value of being located near Unesco in Paris. On the third day of the debate, the Swiss delegate, Professor J. Lugeon, withdrew the candidature of Lausanne and this led to a straightforward vote between Geneva and Paris. There were 28 votes for Geneva and 21 for Paris, which meant that neither received the necessary two-thirds majority. Finally, to facilitate matters, the French delegation proposed that there be a second vote in which his country would abstain. This magnanimity was greatly appreciated and the Geneva offer was then accepted by 38 votes to one. Further information about the Secretariat building will be found in Chapter XVII.

There was also considerable discussion at First Congress about the number of staff to be employed in the WMO Secretariat—at that time the IMO staff amounted to 24. Some delegates were in favour of considerable expansion, believing that this would be essential to enable the new Organization successfully to discharge its responsibilities; in particular, a new Technical Division would be required to provide support for the Technical Commissions, and especially their presidents, all of whom gave their services to WMO on an entirely voluntary basis. A joint report by two of the Congress committees proposed a total strength of 37, including the Secretary-General and his Deputy. This figure was considered to be much too large by the USSR delegate; he felt that the IMO Secretariat had "not been overworked" and that a staff of 25 would be adequate provided that 12 of them were in the Technical Division. This proposal was rejected by 17 votes to 11. After further debate, a compromise was eventually approved by 27 votes to five, namely that the number of staff appointed should not exceed 25, but that the Executive Council* could approve an increase to a maximum of 35. Here it must be remembered that Congress was taking decisions to cover a period of four years. As explained in Chapter XVII, further increases in staff were found to be necessary to enable the Secretariat to discharge the additional tasks assigned by successive sessions of Congress. In 1979 Congress approved the establishment of 246 posts in the Secretariat and this number has not been changed by any subsequent Congress.

The size of the WMO Secretariat had of course a major impact on the budget, another item of extensive debate during First Congress. At that time, the annual budget of IMO was running slightly above US $100 000, which would imply an expenditure approaching US $500 000 over the 4 3/4-year period under consideration. As the estimated staff costs in themselves exceeded this figure, it was clear that a substantial increase would be necessary. Detailed discussions in committee led to a proposed maximum expenditure of US $1 359 000 for the financial period. Major items were staff (US $856 000), meetings (US $160 000) and the technical programme (US $200 000). The debate was reopened when the committee report came before the plenary meeting. Some argued that as the functions of WMO were hardly more extensive than those of IMO, it would be sufficient to double the IMO budget. Others felt that the provisions for the technical programme were inadequate; it was specifically proposed that an additional sum of US $10 000 should be provided for comparisons of instruments, which had hitherto been carried out entirely at the expense of the participating Meteorological Services. It was finally agreed by 32 votes to nine that the maximum expenditure should be US $1 340 000; this included an increase of US $10 000 for the technical programme and a reduction of US $31 000 in the cost of the Secretariat. The ability of the WMO to function successfully over the next four years with such modest resources reflects of course the very substantial support provided by Members outside the budget. For example, there was no provision for WMO to help to finance meetings of working groups, an essential part of the activities of the Technical Commissions and Regional Associations; the expenses of experts attending these meetings had to be met by their parent governments. Again, many individuals provided technical and scientific advice to the Organization, including the preparation of *Guides* and *Technical Notes*, without cost to the WMO.

*See note on p. 8.

Disposal of IMO financial assets

The question of the disposal of the financial assets which IMO had available and which it had transferred to its successor, WMO, was another item which had to be considered by First Congress. These assets amounted to about US $77 000. Various views were put forward and discussed, but Congress finally decided to refer the matter to the Executive Council* with a request that specific proposals be submitted to Second Congress. As is explained in Chapter XVII, the funds were in fact used to improve the pensions of the staff of the IMO Secretariat and to establish the now well-known and highly esteemed IMO Prize.

Conclusion

In this account of the proceedings of First Congress attention has been concentrated on those discussions which illustrate the attitudes of delegates to the new Organization and on the decisions which were to have a major impact on its future development. In his closing speech, Sir Nelson Johnson listed the main achievements of the session: the establishment of WMO; the agreement with UN; the formulation of general policy; the setting up of various constituent bodies; the establishment of General, Staff and Financial Regulations. He also made reference to the election of his successor, Dr F. W. Reichelderfer, and to the first Secretary-General of WMO, Dr G. Swoboda. Mentioning the discussions at Congress, Sir Nelson remarked:

> We have had our arguments during this session of Congress, and I feel that one of the most refreshing things is that many of us are getting to understand each other better than when we met and I hope that this state of affairs will continue. In spite of the arguments and battles I feel confident that Congress can break up, certain that no ill feeling rests. That is an enormous achievement in itself and I offer you my congratulations.

Looking back on First Congress in the light of what has since transpired, it can be seen that the decisions then taken took full advantage of the experience gained by IMO over the preceding 77 years. The basic structure of IMO was maintained and the main scientific and technical strength of the Organization rested in the hands of Members. The Organization could thus operate at the highest possible scientific standard with the minimum budget. Full advantage was also taken of the new status of WMO as a Specialized Agency of UN, particularly the possibility it presented of giving assistance to developing countries in the implementation of WMO programmes and in deriving the full benefits from these programmes. The technical resolutions of IMO which were considered to be permanent in nature were kept in force. Among other things this meant that there was no break in the world-wide system of making and exchanging the weather observations that are needed for day-to-day weather forecasting. The next chapter shows how this system thrived in the hands of WMO and developed into what we now call the World Weather Watch.

CHAPTER III

WORLD WEATHER WATCH

Introduction

The steps taken by WMO's predecessor, IMO, to establish a system for the standardization of meteorological observations and for the international exchange of such information have been explained in the preceding chapters. When WMO took over IMO's responsibilities, the importance of such activities was fully recognized. IMO had established the Commission for Synoptic Weather Information (CSWI) in 1923 to attend to such matters and First Congress re-established it under the title Commission for Synoptic Meteorology (CSM) with extended responsibilities. CSM set about its task promptly and effectively and in the early years of WMO great progress was made in expanding and improving observational techniques and in making the fullest possible use of new telecommunications techniques as they became available. New means of processing the data so acquired were also fully applied including in particular the use of electronic computers for this purpose with the consequent introduction of numerical weather prediction.

Important practical benefits resulted from the progress achieved (particularly in weather forecasting), with corresponding benefits in such practical fields as aviation, shipping, agriculture, operational hydrology, information for the general public, etc. Another important development in those early days was the emergence of many newly independent Member States which not only contributed significantly to the observational and telecommunications requirements as a whole but also benefited greatly from the assistance which the improved meteorological services provided to their economic development.

Much of the progress achieved by CSM was a result of the improved observational system it developed. The land network of stations was increased in many countries while over the ocean areas merchant shipping continued to play a highly important role—a role which had started in the days when IMO itself had been created. In some areas special ocean weather ships were used. Floating meteorological ocean buoys also began to be used. In addition, civil aviation also provided valuable upper-air information on the routes flown. But despite such progress there were still large areas of the Earth's surface which were inadequately covered from the meteorological point of view. The meteorologist's prayer for a truly global coverage was however soon to be answered.

On 4 October 1957 the USSR launched the first artificial Earth-orbiting satellite, SPUTNIK-1. The ninth session of the WMO Executive Council* was in progress at the time. The news created great interest and, even at that early stage, informal discussions took place between Members as to future possibilities of this new technology being applied in due course to meet the meteorologist's long-felt need for an observational system covering the whole globe. The launching of SPUTNIK–2 on 3 November 1957 and the USA's EXPLORER-1 on 2 January 1958 confirmed that the Space Age had truly arrived. It was to lead to one of the most significant developments in the long history of meteorology: the WMO World Weather Watch Programme.

* See note on p. 8.

Satellite planning

Member countries of WMO were quick to realize that the further development of satellite technology gave promise of greatly expanding the volume of meteorological information covering the whole globe. It was envisaged that pictures of cloud systems and also data regarding the vertical distribution of wind and temperatures would increase the amount of data obtained from conventional networks and fill the important gaps in the areas, both land and sea, from which ground-based observations were not available. It was also recognized that the use of satellites would significantly help to improve the speed and efficiency of meteorological communications.

At its tenth session (Geneva, April/May 1958), the Executive Council* took the first opportunity to hold a general discussion on the various possibilities of using artificial satellites for meteorological purposes. Further consideration was given to the subject the following year when the Third WMO Congress was held in Geneva. As a result the Executive Council, at its meeting immediately after Congress, established a Panel of Experts on Artificial Satellites. The panel consisted of representatives from the two Member countries which were already launching satellites, Academician V. A. Bugaev (USSR) and Dr H. Wexler (USA). They were joined by Dr G. D. Robinson (UK), representing the WMO Commission for Aerology (now renamed the Commission for Atmospheric Sciences) and Dr W. J. Gibbs (Australia), on behalf of the Commission for Synoptic Meteorology, now the Commission for Basic Systems.

The panel held two meetings and by the first half of 1961 had submitted a report to the Executive Council* on the role of WMO in regard to activities in outer space and setting forth a number of proposals for the use of satellites in the international organization of meteorology and in fundamental research into the science of the atmosphere. However, consideration of action to be taken on the report was overtaken by events at the headquarters of the United Nations where the likely existence of ever-growing numbers of satellites in outer space was under anxious study from every foreseeable angle.

The United Nations, while usually welcoming new technology because of the benefits that may follow, is always concerned to ensure that its use should be confined to peaceful purposes. Just as, a decade or more earlier, UN had sought agreement among governments regarding the peaceful uses of atomic energy, so from 1958 onwards it recognized dangers as well as benefits as the likely outcome of more and more nations launching artificial satellites. UN appreciated that one of its Specialized Agencies, WMO, was a model of free and friendly co operation for the unrestricted exchanges of weather reports and forecasts among all nations and was glad to learn that WMO saw great advantages for national Meteorological Services in organized international arrangements for making full use of the new facilities available from satellites. Accordingly one of the earliest resolutions approved by UN for "international co-operation in the peaceful uses of outer space" laid heavy demands upon WMO. The resolution was adopted unanimously by the General Assembly of UN. It was abundantly clear that meteorologists in the various countries realized the favourable potentialities of satellites for the atmospheric sciences and were thus able to brief their governments effectively.

The resolution referred to, No. 1721 (XVI), was approved on 20 December 1961. It consisted of four parts, three of which concerned legal problems, an inventory of satellites and telecommunications matters. In section C of the resolution it was stated that the advent of satellites offered substantial opportunities for improvements in meteorological services and for progress in atmospheric research, thereby yielding world-wide benefits and illustrating the value of international co-operation. The text continued with a recommendation to:

> All Member States and to the World Meteorological Organization and other appropriate specialized agencies for the early and comprehensive study, in the light of developments in outer space, of measures:
>
> (*a*) To advance the state of atmospheric science and technology so as to provide greater knowledge of basic physical forces affecting climate and the possibility of large-scale weather modification;
>
> (*b*) To develop existing weather forecasting capabilities to help Member States make effective use of such capabilities through regional meteorological centres.

Finally, the resolution called upon WMO to carry out wide consultations with Unesco and other specialized agencies of UN, with governmental and non-governmental organizations, including the International Council of Scientific Unions (ICSU), and to submit a report on the organizational and financial arrangements required to complement the measures specified in the resolution. The report was to be addressed to Member governments of UN and to the Economic and Social Council (ECOSOC), and to be available by June 1962.

This resolution directed WMO to assume the leadership in a planning exercise of enormous scope and political importance. It was also of great value and encouragement to WMO since it was adopted by governments and in consequence carried far greater weight than would have been the case if a similar resolution had been carried in any forum other than the UN. Governments had already seen examples of meteorological satellites in action through the automatic picture transmission (APT) made by polar-orbiting vehicles. These pictures of pressure systems and of cloud structure could be intercepted by nations anywhere in the world using fairly simple and readily available receiving equipment. The transmissions provided an excellent example of "peaceful uses" of outer space and, in adopting the resolution, governments clearly expressed their determination to support atmospheric research programmes as well as efforts to improve the scientific and technical capabilities of national Meteorological Services.

It was clear that WMO had to act upon the resolution with the greatest urgency. It was also obvious that the EC Panel of Experts, referred to above, would not be able to cope with the General Assembly's directive because the conduct of business by correspondence with occasional meetings would be quite incapable of moving with the compelling urgency that was essential for providing the comprehensive study that had been requested. The Secretary-General of WMO, after consulting the President, discussed the problem with the Permanent Representatives of the two Member countries that were then operating satellites, namely the USA and the USSR. It was agreed that each country would send an expert to Geneva to work with the WMO Secretariat for as long as it would take to compile the required report. The experts were Academician V. A. Bugaev (USSR) and Dr H. Wexler (USA). The latter was the principal expert in the USA on meteorological satellites but he had other important commitments and was also in poor health. Another expert from the USA, Dr M. A. Alaka, was accordingly also sent to Geneva to assist Dr Wexler and to stand in for him whenever necessary. In the event Dr Wexler was able only to help lay down the governing principles for the task to be undertaken so the main work fell to Academician Bugaev, Dr Alaka and the Secretary-General of WMO and his staff in the Secretariat. In this way, a drafting group was formed. The preliminary work included the need, as laid down in the UN resolution, to consult all organizations which might have views to express or requirements to describe. Valuable contributions were received from numerous sources, notably from Unesco, the International Atomic Energy Agency (IAEA), the International Telecommunication Union (ITU) and the International Council of Scientific Unions (ICSU).

Dr H. Wexler (USA) and Academician V. A. Bugaev (USSR) discussing the World Weather Watch in the WMO Secretariat (1962)

There followed a spell of intensive activity, assembling all relevant material from within WMO and from elsewhere and, by critical examination, drawing up a clear and consistent document setting out the general objectives and suggesting ways of meeting them.

The group had the vision to appreciate that they were constructing a continuing programme, one which would be in a state of constant evolution where, from time to time, philosophies would be reviewed, principles revised as necessary and the tasks of implementation rewritten. The requirements for meteorology could not be set out unalterably and the face of advancing technology could not be accurately foreseen.

WWW—The first report

With commendable, even remarkable, speed the report requested by the UN General Assembly was completed and made available to ECOSOC in June 1962 and to the General Assembly which was due to hold its seventeenth session later in the same year. The report was also to be submitted for approval to the Fourth Session of the World Meteorological Congress, which was scheduled to meet in Geneva in the spring of 1963.

The report was entitled *"First report on the advancement of atmospheric sciences and their application in the light of developments in outer space"*. The term World Weather Watch (WWW), suggested by Dr Wexler, was introduced and the overall objective was described by Dr Bugaev as to develop WWW in such a way as to ensure that any country could receive meteorological information on any scale, including the global scale, for making weather forecasts and for taking measures both to improve its economy and to guard against natural disasters of meteorological origin. This description involved commitments not only concerning the data obtained from conventional surface-based observing systems as well as satellites but also for the development of adequate communications, using the most up-to-date computer-aided facilities and various automatic devices, so that meteorological services in all countries could, if so desired, function at the highest level of scientific and technical capability.

In conformity with the UN resolution, the report discussed a wide range of subjects including weather forecasting and climatology and their applications, the possibility of changes in the global climate (whether occurring through natural causes or as a result of man's industrial activities), small- or large-scale weather modification and the meteorological aspects of water resources. The report was realistic and practical and avoided any serious attempt to make proposals for a very long period ahead. Although some topics were treated in considerable detail and many of the views expressed still remain valid, the intention was to assess requirements in general terms for the next ten years and to suggest specific targets that might be implemented within the ensuing three years, e.g. by the end of 1966.

The report may also be described as creative, confident and containing promise of spectacular advances in the acquisition of meteorological data. A confident outlook was required since in the early days there were some shortcomings regarding the quality of satellite pictures of cloud systems and in determining the precise positions over the globe. The picture transmissions (APT) had dramatically demonstrated their usefulness by revealing the existence of a hurricane on 12 September 1961, two days earlier than it was detected on the routine weather maps. The teething troubles were soon largely eliminated and APT was seen to afford new benefits in showing the distribution of sea ice and snow cover. The drafting group also expressed confidence in the value of radiation measurements in different wavelengths for defining temperatures and cloud-top configurations, in their ability to delineate ocean currents and to determine the mean temperatures of atmospheric layers. Considerable stress was laid upon the benefits to be derived from the use of satellites in the field of telecommunications. One attribute of particular importance was cited as the ability of satellites to collect data from automatic weather stations located in remote areas or on the sea and to relay their observations to convenient collecting centres.

Most attention in the report was given to what might be described as operational or service aspects: observational data and the associated telecommunication system to permit rapid collection and dissemination, programmes for the improvement and extension of weather forecasts and the desirability of enabling countries to acquire as high a proportion as they wished of the total volume of available raw and processed data. Thus, a highly centralized organization was proposed consisting of world and regional centres with responsibilities for providing the information (data, analyses, actual and prognostic charts, rainfall forecasts and so on) required by the national Meteorological Services. The world centres were to be located in the USSR (Moscow), the USA (Washington, DC) and Australia (Melbourne) for the southern hemisphere, at the headquarters of the respective meteorological establishments. These centres, using the collected conventional and satellite data, would prepare analyses and prognoses on global and hemispherical scales and transmit the results to regional meteorological centres through which onward transmission to national Services could be arranged as required. Regional centres would use the products of the world centres and all available data from within their region and would provide the dependent national centres with detailed regional analyses and prognoses in chart form and also bulletins in plain language. In turn, each national Meteorological Service would supplement the guidance received from its regional centre and perhaps from one or more world centres with detailed local analyses in order to carry out its domestic and international responsibilities.

There were therefore three fields in which substantial advances were considered by the drafting group to be essential if the hopes that would be placed on the WWW were to be fulfilled:

- A very marked improvement in global observations to provide better and more complete data for analysis and prognosis (data from space were regarded as complementary and by no means replacing observations from surface stations);
- A data-processing system to permit the application of weather prediction techniques using high-speed computers;
- A world-wide co-ordinated telecommunication system to ensure reliable and rapid exchanges of observational data and analyses and forecasts.

Clearly, these requirements, while in line with the basic objectives defined by IMO, would, once approved, place a heavy planning load upon WMO. In particular, the existing arrangements for collection and dissemination of data, which had evolved over so many years and had never lost time in utilizing new technology, would have to be adapted to meet the challenges presented by almost overpowering volumes of satellite data.

It will be evident that as regards the operational aspects, the *First Report* gave a comprehensive treatment of the requests contained in the UN resolution and also provided clear guidance on the follow-up action that should be initiated. On the research side, however, the report was less forthcoming and was confined to some general remarks. The drafting group felt that the problems of atmospheric research were so complex and long-term that a larger forum would be needed in order to produce an adequate, penetrating discussion with recommendations as to the policies and programmes required to make significant progress over a vast field of research. It was therefore proposed that WMO, with the support of other organizations concerned, should establish an Advisory Committee composed of leading scientists to make proposals on what needed to be undertaken to fill important gaps in knowledge of the atmosphere so that projects could be planned with the aim of solving problems of weather prediction, climate changes, water resources and, possibly, large-scale weather modifications.

The *First Report* undoubtedly ranks as one of the major documents in the history of WMO. At the time it was written satellite meteorology was in its infancy and not all meteorologists shared the enthusiasm of Dr Wexler, Academician Bugaev and the Secretary-General of WMO as to the potential of space-based observations for weather forecasting and for atmospheric research. At the time, it was probably difficult to foresee the improvements and extensions that would be made in meteorological data from satellites and the rapid development in techniques that would occur. Nevertheless, as progress was made the report retained its consistency. It was clearly appreciated that in due time observations from polar-orbiting and geostationary satellites would be used in numerical weather prediction routines and would be valuable for estimates of wind and wind shear, for the recognition of stable and unstable air masses and for studies of other atmospheric properties.

Action by UN and WMO

The *First Report,* virtually in its final form, was submitted to the Fourteenth Session of the WMO Executive Council* in the spring of 1962. After some minor editorial changes, it was approved for transmission to the next session of the UN General Assembly and to other UN bodies including ECOSOC and the UN Committee on the Peaceful Uses of Outer Space. The report was, of course, prepared in response to a resolution of the General Assembly and was appropriately presented to that body but in its meteorological context it would also be submitted for approval to the Fourth Congress of WMO scheduled to assemble in Geneva in the first half of 1963.

The UN General Assembly held its seventeenth session in December 1962. It was very favourably impressed by the content of the report and by the sense of urgency which had clearly attended its preparation. The good will of governments was aroused in no small measure and the General Assembly proceeded to adopt a new resolution (Resolution 1802 (XVII)) as a sequel to the resolution passed at its preceding session in 1961. On the observational and other operational aspects of meteorological services, the General Assembly approved the report submitted by WMO and Member States to strengthen national Meteorological Services and to expand the organization of education and training in meteorology and associated disciplines. The resolution also placed special emphasis upon the increasing use of meteorological satellites.

With regard to research, the General Assembly gave further endorsement of the report submitted to it by WMO. However, it appreciated that WMO would probably be fully committed to implementing the World Weather Watch as an operational system designed to meet the requirements of national Meteorological Services. The General Assembly, therefore, in Resolution 1802, invited the International Council of Scientific Unions to develop, in conjunction with WMO and other relevant organizations, an expanded programme of scientific research into atmospheric processes and related phenomena. The declared intention was that the research programme and the wide-ranging activities organized by WMO for the development of the WWW would support one another and form a comprehensive unity.

When the Fourth Congress of WMO assembled in Geneva in April 1963 it considered the two UN General Assembly Resolutions, 1721 and 1802, together with a number of preparatory studies carried out by WMO including the *First report on advancement of atmospheric sciences and their application in the light of developments in outer space.* As already mentioned, this report was prompted by Resolution 1721 and had been given provisional approval by the WMO Executive Council.*

Fourth Congress unhesitatingly accepted the responsibilities placed upon WMO by the two resolutions of the UN General Assembly and confirmed the action taken by the Executive Council* in approving the First Report and in other related matters. The far-reaching decisions taken by Congress included:

- Approval of the World Weather Watch as an extension of long-established plans for facilities and services required by meteorologists;
- Approval of the proposed WMO Advisory Committee on Atmospheric Research;
- Establishment of a WMO Development Fund in support of the WWW;
- Establishment of a WWW planning unit in the WMO Secretariat.

Congress recognized that a great deal of planning had still to be done and accordingly requested the Executive Council* to arrange for a comprehensive study with special reference to:

(*a*) An analysis of the national requirements to be served by WWW and the advances in technology that should be utilized;

(*b*) An overall plan for observational methods and networks, communication systems, data-processing centres and other essential functions of the WWW.

The organization of this study placed an enormous task upon the new WWW planning unit. It had to collaborate with all parts of the WMO Secretariat and consult in meetings and by correspondence with Member countries of WMO, the Regional Associations and the Technical Commissions, the Advisory Committee on Atmospheric Research (when formed) and numerous external organizations which would be expected to use the WWW directly or indirectly. Inevitably, in due course the planning work expanded into one of the major activities of the WMO Secretariat.

As a result of UN Resolution 1802, which invited ICSU to develop a programme of research into the atmospheric sciences, WMO accepted that its Advisory Committee, which was also to be deeply concerned with atmospheric research, should seek to work in close collaboration or even in partnership with ICSU. WMO set up its Advisory Committee in 1963. It consisted of 12 scientists of high international reputation in meteorology and related disciplines, selected after consultations with Permanent Representatives of Members of WMO and with ICSU. The committee was given wide terms over the whole field of research and was also invited to advise on education and training in reference to the objectives laid down in the UN resolutions.

A year later, in 1964, ICSU formed its own Committee on Atmospheric Sciences. It held its first meeting in 1965 and the choice of location—Geneva—was a sure sign that a fruitful relationship between ICSU and WMO would soon materialize.

From Fourth Congress onward the intense activity associated with all aspects of the WWW proved the wisdom and necessity of the new WMO Development Fund. The objective was to finance essential WWW projects which could not be funded by the normal arrangements within WMO. Congress earmarked US $1.5 million for these purposes and it was envisaged that half the sum would be spent on the improvement of facilities (equipment, communications, data processing, etc.), about ten per cent on education and training and the balance on surveys and studies related to the planning of the World Weather Watch.

Main features of the WWW take shape (1963–1967)

Between Fourth Congress (1963) and Fifth Congress (1967) very considerable progress was made in the planning of the WWW. Member countries began to adapt their Meteorological Services to fit in with the structure of the WWW and also to reap the benefits of improved facilities and of expanded research programmes. Fourth Congress, acting upon the two resolutions of the UN General Assembly and guided by the *First Report* and the discussions which ensued, gave directives for the further development of the WWW, i.e. by means of planning an implementation. In an additional resolution, No. 1963 (XVIII), the UN General Assembly endorsed the decision of Fourth Congress. Additional advice was given by the Executive Council* at its annual meetings and thus the Secretary-General had a clear remit for carrying out the responsibilities of the WMO Secretariat to promote the WWW as vigorously as possible. The overall objective was to present detailed proposals on the WWW for adoption by Fifth Congress in 1967.

Although the research programme was generated by the same UN resolution as resulted in the WWW, it was not strictly a part of the WWW but complementary to it. The two were

* See note on p. 8.

dependent upon and supported each other. On the research aspects, therefore, it is sufficient to remark that discussions between ICSU and WMO continued and in 1967 the General Assembly of ICSU and the Fifth Congress of WMO approved the Global Atmospheric Research Programme (GARP) to be conducted by a Joint ICSU/WMO Organizing Committee. The subsequent progress of the GARP is described in Chapter X, which deals with WMO's research programme.

It was decided to proceed with implementation of the WWW during the period 1964 to 1967, that is, during the four-year interval between successive Congresses. A progress report would be submitted to Fifth Congress in 1967 together with planning proposals on the following lines:

(*a*) A plan for the WWW covering the steps to be taken toward its implementation in the four-year period 1968–1971;

(*b*) Proposals for further studies and investigations on new and promising techniques and technologies which might be incorporated in the WWW in subsequent plans.

The procedure adopted by the WMO Secretariat for the document submitted to Congress set a pattern for the future and, up to and including Ninth Congress (1983), successive four-year WWW Plan and Implementation Programme proposals were prepared. For Tenth Congress (1987) the Secretariat prepared long-term ten-year plans (1988–1997) for all major programmes of WMO, and WWW planning conformed to this arrangement.

The WWW Plan and Implementation Programme for the four-year period 1968–1971 was adopted by Fifth Congress and outlined the general features of the structure of the WWW as a dynamic, evolutionary system with world-wide ramifications. As capabilities of meteorological satellites increased, improving the quality of the data and adding new types of data, and as conventional networks improved and extended their coverage, adjustments were made to the initial Plan and the implementation targets became more extensive and urgent.

Structure and utilization of the WWW

From the outset, three components were regarded as indispensable parts of WWW. These were:

- The Global Observing System (GOS), consisting of equipment and other facilities for making observations at stations on land and at sea, from aircraft, meteorological satellites and other platforms such as buoys;
- The Global Data-processing System (GDPS), consisting of meteorological centres organized and equipped for the processing of observational data in real time or non-real time;
- The Global Telecommunication System (GTS), consisting of telecommunication facilities, including computers and various automatic devices, for the rapid and reliable collection and distribution of observational data and processed information according to agreed procedures.

As experience was acquired in monitoring the operation of the WWW and progress made toward further implementation, two additional elements were added to the WWW Plan beginning in 1984. These were:

- The Monitoring and Operational Information Service, consisting of arrangements for informing Member countries of the status of operation of WWW and its subsystems;
- The WWW Implementation Support, Activity consisting of arrangements for the exchange among Members of methodology, acquired experience and other information on the installation and operation of the WWW. This activity included a comprehensive training programme for personnel engaged in the planning, implementation and operation of the natural elements of WWW.

The utilization of the WWW has been largely confined to WMO Members in a three-tier structure consisting of World Meteorological Centres (WMCs), Regional/Specialized Meteorological Centres (RSMCs*) and National Meteorological Centres (NMCs), some of which are, of course, WMCs and RSMCs. However, a number of establishments with substantial interests in meteorology, such as certain university departments and the European Centre for Medium-range Weather Forecasts (ECMWF), have been enabled, by agreement with the relevant national Services and WMO, to have access to the GTS in order to receive the information they require.

The primary responsibility of WMCs and RSMCs is to provide a service—global or covering a hemisphere in the case of the former and regional in the case of the latter—to the national Meteorological Services. For their part, the national Services are primarily responsible for providing services to external users of meteorological information such as aviation, shipping, agriculture and so on.

The WWW is essentially an aggregation and co-ordination of all the facilities controlled by the Meteorological Services of the countries, in 1990 more than 160, which are Members of WMO. The organization and functioning of the WWW are governed by the WMO Technical Regulations, drawn up and agreed by the Members, and by annexes in the form of manuals on the GOS, GDPS, GTS and on the various codes which are

A French-built drifting buoy

*Formerly, Regional Meteorological Centres (RMCs)

GMS-4 at Tanegashima Space Centre, Japan (1989)

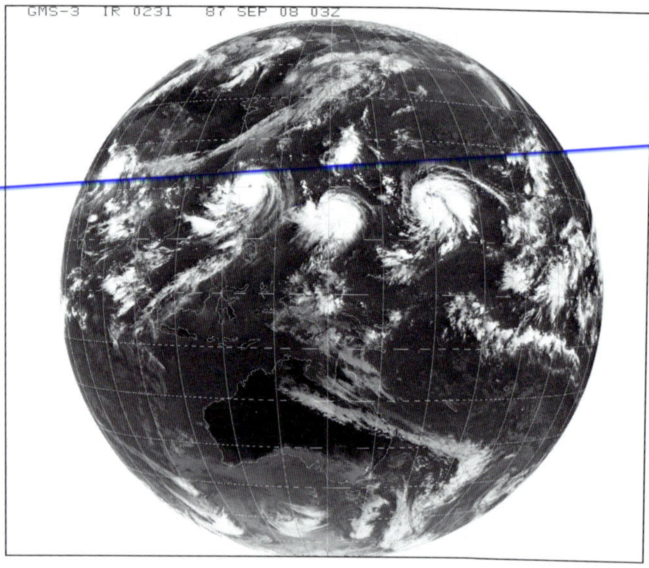

Infra-red image from GMS-3 in stationary orbit at 140°E, 0300 UTC, 8 September 1987

used for compiling meteorological messages. The overall or global aspects of the WWW are the responsibility of the WMO Commission for Basic Systems whilst the Regional Associations of WMO deal with the regional aspects. It would be difficult to exaggerate the importance of the Commission for Basic Systems in the planning, development and operation of the WWW; this Commission is the focal point for all aspects of the World Weather Watch.

The Global Observing System (GOS)

WMO, like its predecessor IMO, has constantly endeavoured to improve the world-wide network of surface-based observing stations providing high-quality data at fixed times, generally every three hours but in many cases at hourly or half-hourly intervals. Major efforts have been concentrated on the large data-sparse areas, usually uninhabited, of the land surface and on the extensive expanse of oceans which are remote from the recognized sea lanes. Attempts have also been maintained to recruit additional vessels to the fleet of Voluntary Observing Ships and to increase the number of commercial aircraft from which in-flight weather reports are obtained. WMO and its Members have always kept in touch with new technology with a view to possible applications to weather observing and important advances have been made in developing automatic weather stations on land, on ships or on buoys to operational status. Even so, in spite of all the determined efforts, important gaps have remained in the surface-based networks and much remains to be accomplished before the minimum data coverage set by the WMO Technical Regulations is attained.

When meteorological satellites first arrived on the scene they provided highly valuable data on cloud distribution during both day and night. Meteorological offices, suitably equipped, were able to receive cloud-cover information directly from satellites and this facility made an important contribution in fulfilling all types of forecasting requirements. During the four-year period 1975–1979, between Seventh and Eighth Congresses, special studies were carried out, taking into account the rapid development of meteorological satellites and other factors, and new specifications were agreed in respect of the Global Observing System for both operational and research (GARP) purposes. The data available from satellites became more extensive, improved sensors giving a whole range of important products such as radiation temperatures of clouds, land and ocean surfaces, vertical profiles of temperature and water vapour as well as the distribution of total outgoing short-wave radiation. In view of these developments it was decided to separate the GOS into two complementary sub-systems, one surface-based and the other space-based.

Throughout the two four-year periods 1980–1983 and 1984–1987, between Eighth and Tenth Congresses, the need to consider the scale of motions in relation to the spatial and temporal resolution of data was one of the important factors that resulted in further evolutionary changes in the GOS. Another factor was the launching of TIROS-N-type satellites, which gave a marked improvement in the global coverage of meteorological data. From time to time the degree of dependence on the space-based sub-system increased as local economic problems brought about temporary reductions in the amount of data provided by the surface-based sub-system. By 1986 the space-based sub-system consisted of polar-orbiting and geostationary satellites providing vast quantities of data directly to receiving stations established in nearly all countries.

The Global Data-processing System (GDPS)

The purpose of the GDPS is to make available to Meteorological Services the processed information they require for all their responsibilities, including weather forecasting (real-time) and climatology (non-real-time). The GDPS came into operation in 1968 and its organization has varied as the GOS has provided more data and as progress has been made in forecasting and other techniques. The main objective is that the GDPS should provide the best possible products using the most powerful computers available and taking full advantage of the results of atmospheric research. Each national Service has only to be connected to the Global Telecommunication System in order to

The Buenos Aires HRPT station, showing both the polar-orbiting and the geostationary satellite receiving antennas

enjoy full access to all the GDPS issues that it requires whether for its work for users of meteorological information or for its own research programmes.

The GDPS has been operating on a three-level system of World, Regional and National Meteorological Centres. WMCs have the task of providing data, information and advice for recipients to use for general forecasting of planetary and large-scale pressure and wind systems for various periods ahead, e.g. 24 hours, a few days and up to about a week. The RSMCs, some 26 in the network and well distributed over the globe, use WMC products for general guidance and also their own computerized analysis and forecasting procedures to furnish information and advice to national Services for forecasting on a large, meso- or small scale for varying periods ahead. NMCs are therefore able to receive a full measure of advice from the higher levels in the GDPS but they should also possess their own data-processing capability in order to cater to the special needs of all users of meteorological information.

The Global Weather Experiment (FGGE), which took place during 1979 and 1980, made an important contribution to the development of the GDPS. Experience showed the necessity of extending data-assimilation methods from three to four dimensions so that observations made at non-synoptic hours would be incorporated into analysis programmes. As a result of this ex-perience, a decision was made to carry out an Integrated WWW System Study (ISS) so that the acquisition and processing of data would ultimately become fully integrated and incorporate available technological and scientific advances.

The Global Telecommunication System (GTS)

The GTS is concerned with collection, exchange and distribution of observational data, mostly from the GOS, and of processed information made available by WMCs, RSMCs and other centres (see foldout diagram). The GTS is also used by the environmental programmes of other international organizations where data transmissions are required. WMO readily co-operates in such activities so long as spare capacity exists on the channels to be used. The GTS functions on a three-level basis:

- The Main Telecommunication Network and its branches (until 1984 called the Main Trunk Circuit);
- Regional Meteorological Telecommunication Networks;
- National Meteorological Telecommunication Networks.

The existence of a highly complex facility such as the GTS bears witness to the degree of co-operation that exists among national Meteorological Services, which readily support the co-ordination of their activities under the auspices of WMO. In most respects the GTS is the aggregate of the telecommunication functions of WMCs, RSMCs and NMCs. Junctions in the whole network are provided by Regional Telecommunication Hubs (RTHs), many of which are located with RSMCs. The form of the GTS and the Main Trunk Circuit resulted from prolonged meetings of experts who appreciated the fundamental importance of adequate communications for the efficient operation of the WWW. The Main Trunk Circuit was originally planned as a girdle round the globe but branches into Africa (RTHs at Nairobi and Cairo) and into South America (RTHs at Buenos Aires and Brasilia) were included to ensure the rapid collection of data from these continents. Ninth Congress (1983) reviewed the structure of the GTS when additional branches, Tokyo–Beijing and Beijing–Offenbach, were installed and the Main Trunk Circuit was seen to be something more than a girdle. It was decided that from 1984 the configuration of the Main Telecommunication Network should include three WMCs and 14 RTHs. In addition,

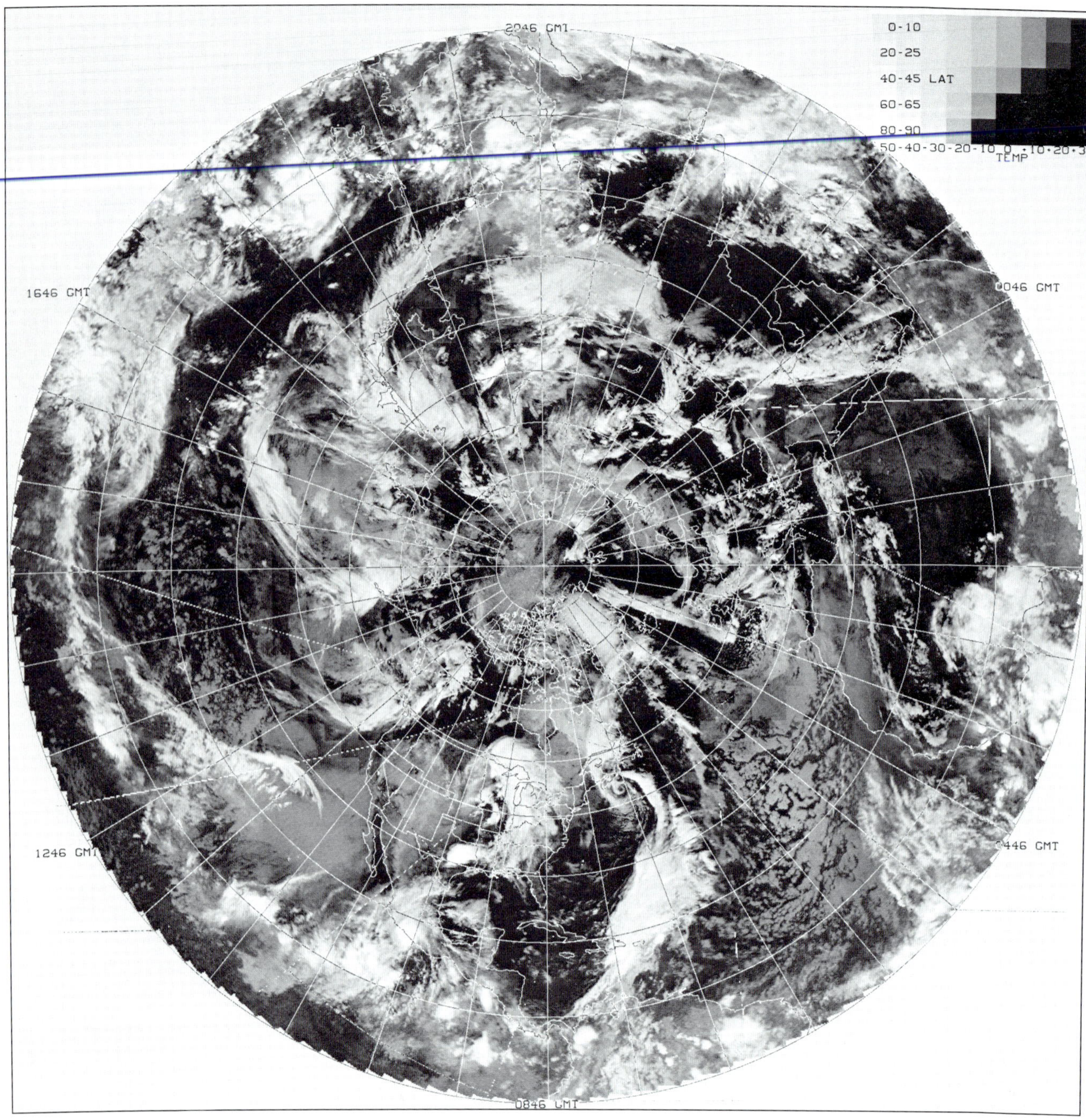

Composite cloud picture of the northern hemisphere derived from satellite

16 RTHs were designated to support Regional Meteorological Telecommunication Networks in the six Regions of WMO. The new, enlarged arrangements provided increased flexibility for the routeing of data and processed information with the result that outages could be dealt with more easily.

Some benefits of the WWW

When proposals for the WWW were approved by Fourth Congress (1963) a period of urgent planning began so that Fifth Congress (1967) was able to adopt a programme for the development of the WWW which would come into operation officially on 1 January 1968. It was realized that new technology, including satellites and electronic computers, would lead to enormous advances, using means not previously available, in the aspirations for world-wide coverage of weather data, which had motivated international meteorology since 1873. It does not belittle the WWW to view it as a continuation, or even a fulfilment, of long-established objectives but rather acknowledges the percipience of meteorologists who with rudimentary facilities maintained their efforts to promote the science of the atmosphere and realized the necessity of international world-wide co-operation. However, the changes brought about by the WWW Plans have been so dramatic that it seems permissible to view the WWW as a separate entity, a new beginning inspired by the dawn of the space age. The benefits that have accrued to meteorology as a result of the WWW are to be found mainly in the context of weather forecasting and atmospheric research. In assessing improvements it is not unusual to treat research and forecasting separately although it is well known that greater accuracy in forecasting is often the sign of successful research. In the case of the WWW, however, advances in forecasting since 1968 could have resulted from two other causes, either separately or in combination:

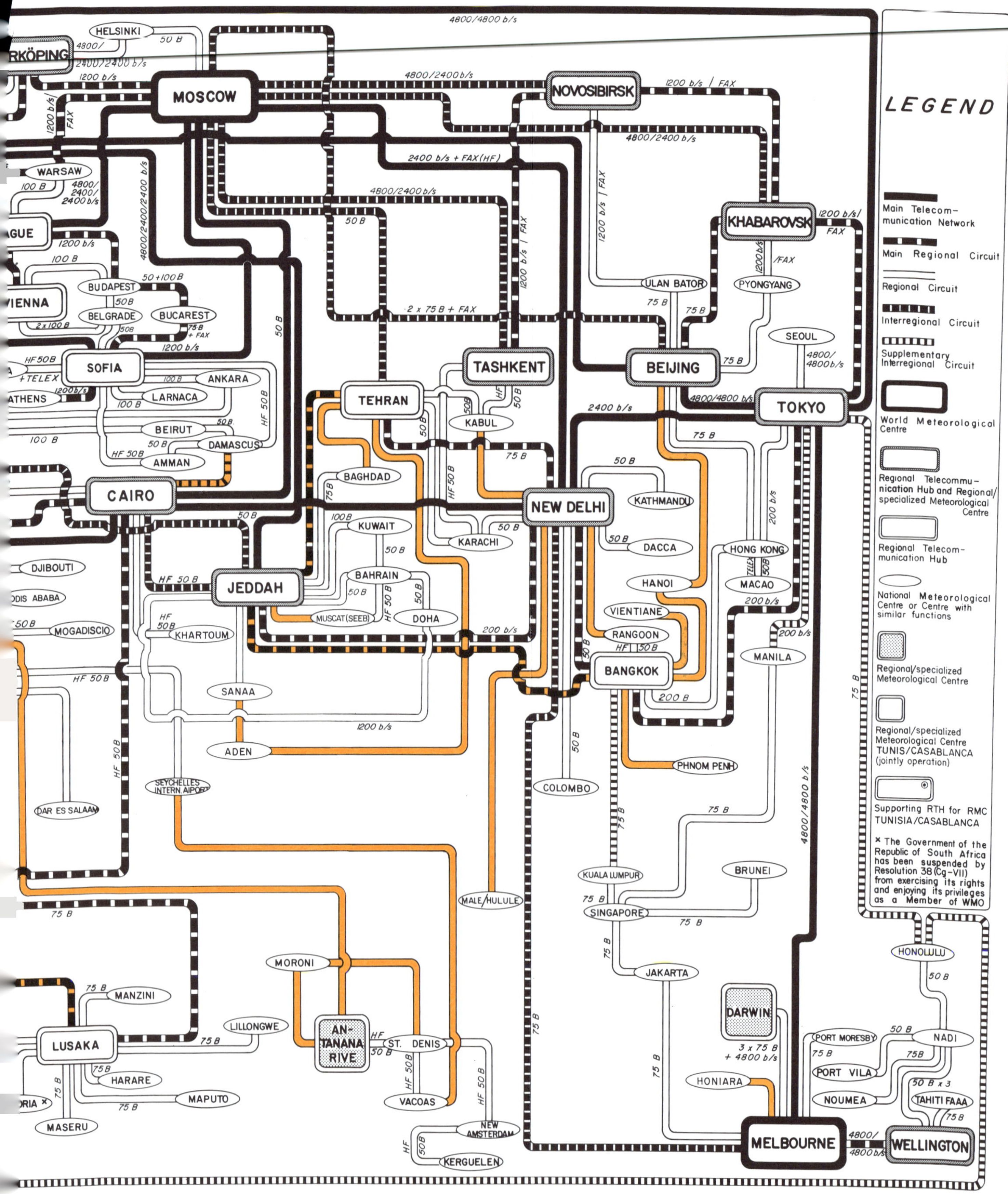

Las denominaciones empleadas en este mapa y la forma en que aparecen presentados los datos que contiene no implican, de parte de la Secretaría de la Organización Meteorológica Mundial, juicio alguno sobre la condición jurídica de países, territorios, ciudades o zonas, o de sus autoridades, ni respecto de la delimitación de sus fronteras o límites.

Употребляемые на этой карте обозначения и изложение материала не означают выражения со стороны Секретариата Всемирной Метеорологической Организации какого бы то ни было мнения относительно правового статуса страны, территории, города или района, или их властей, или относительно делимитации их границ.

IMPLEMENTATION OF THE GLOBAL TELECOMMUNICATION SYSTEM

MISE EN OEUVRE DU SYSTEME MONDIAL DE TELECOMMUNICATIONS

EJECUCION DEL SISTEMA MUNDIAL DE TELECOMUNICACION

ОСУЩЕСТВЛЕНИЕ ГЛОБАЛЬНОЙ СИСТЕМЫ ТЕЛЕСВЯЗИ

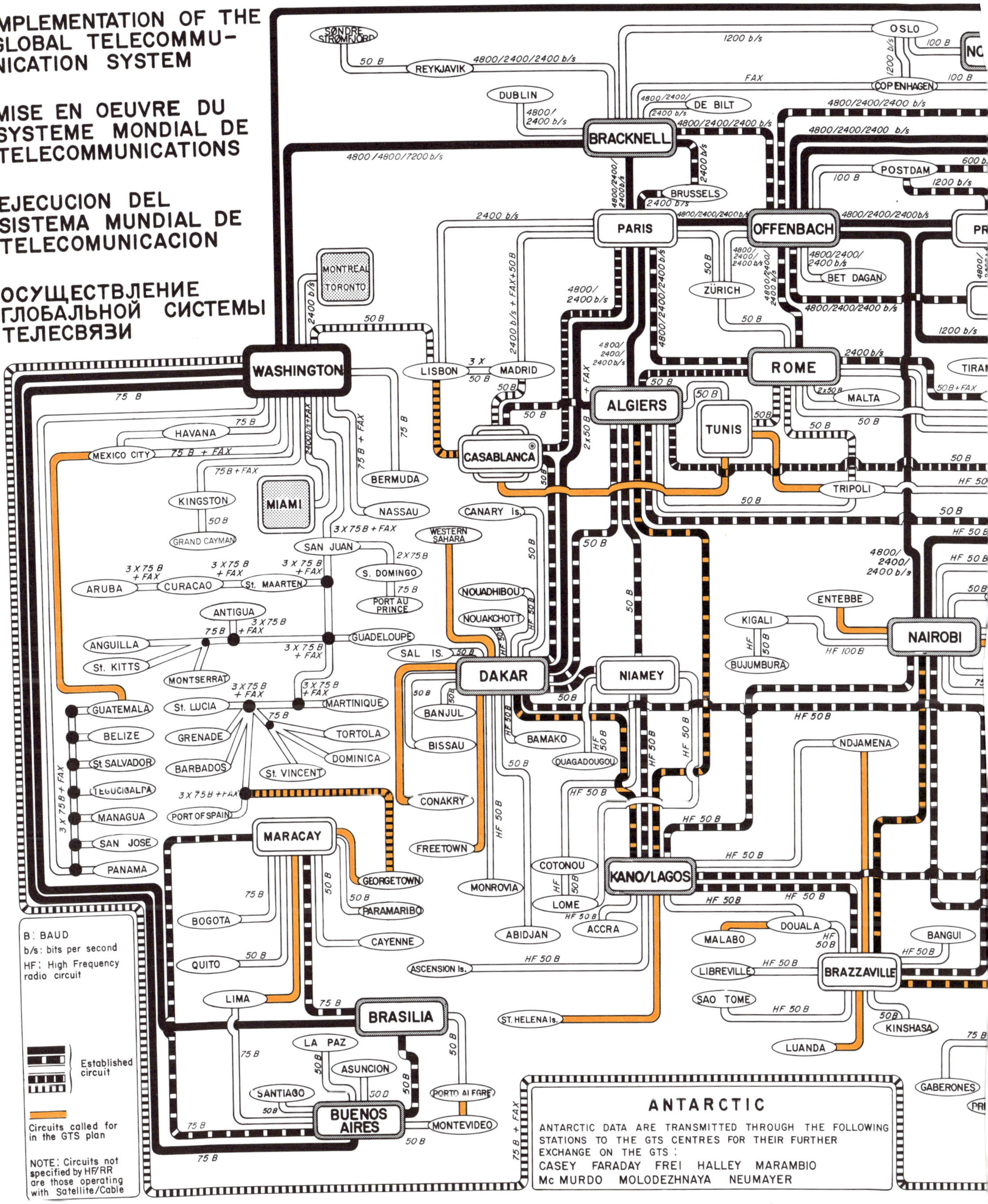

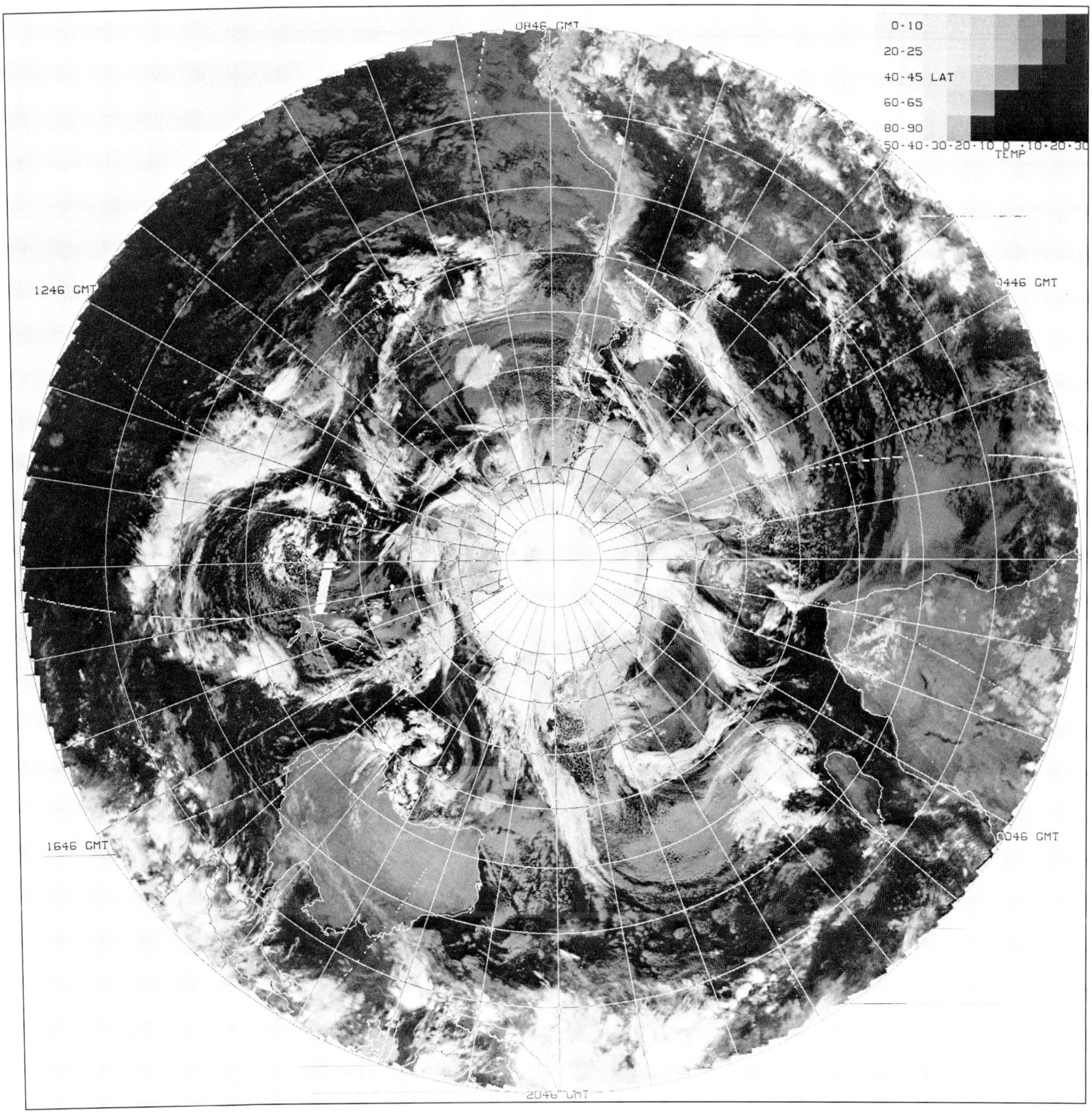

Composite cloud picture of southern hemisphere derived from satellites

(*a*) Much more complete coverage, due to satellites, of observational data within and surrounding the area for which forecasts have been prepared;

(*b*) Much more rapid reception, for which the GTS is responsible, of observational data and processed information at a forecast centre.

Most national Meteorological Services keep a statistical check on the accuracy of their routine forecasts and some of the progress that has been noted since WWW came into operation is given in the following list:

- Improvements in short- and medium-range forecasts for all purposes, e.g. for the public, aviation, agriculture, shipping, transportation, hydrology, industry and recreation;
- Improvements in extended-range forecasts for agricultural planning, water-resource management, etc.;
- Improvements in the timeliness and accuracy of warnings of adverse weather, e.g. gale warnings for shipping, warnings of tropical cyclones (or hurricanes or typhoons) which might cause natural disasters.

These benefits are available to all Members of WMO since the WWW is designed to allow each Meteorological Service access to all the data and processed information available within the system. Consistently accurate weather forecasts make a valuable contribution to almost every sector of a country's economy as various benefit-to-cost studies have convincingly demonstrated. The Members of WMO should therefore be in no doubt that their contribution to the WWW has been a good investment.

With regard to research, there is no question that the greatly improved data coverage has led to much more refined models of the general circulation of the atmosphere and in consequence forecasts up to five or more days ahead have proved more accurate than ever before and have been subject to fewer serious errors. As an example, recent studies have shown that the reliability of forecasts for five days ahead has attained that of forecasts for up to 72 hours ahead before the WWW was firmly established.

The success of the GARP and its subsidiary projects also pays tribute to the value of the WWW. These internationally organized research programmes have been involved, with the help of the WWW and additional facilities provided at the time, in studying the laws governing the formation of climate and its fluctuations under the influence of numerous factors. A striking feature of the WWW era is that meteorologists have considered it practical, realistic and not over-optimistic to try to construct mathematical models of the global climate so that it may eventually be possible to make quantitative assessments of the role of certain causes of climate change, whether these causes are of natural occurrence or the sequel to man's activities. Among the international groups researching into the global climate, climate variability and climate change—three aspects of the second objective of the GARP—discussions are apt to arise as to whether research may in the long term reveal practical ways of preventing the deterioration of regional climates or even of effecting improvements. These matters can be discussed in a practical way since the advent of the WWW with its facilities for observation, data collection and data processing with the aid of the most powerful computers and with its scientific programme, GARP, which has stimulated fundamental research on a global scale.

The importance of the WWW is very great, perhaps incalculable. It is not an end in itself but a vitally important means to numerous ends of social, economic and political consequence.

CHAPTER IV

INSTRUMENTS AND METHODS OF OBSERVATION PROGRAMME

Introduction

In the second half of the nineteenth century the development of meteorology as a scientific discipline was seen to be a practical proposition. There were compelling incentives to proceed because of the need to provide weather information primarily for the protection of shipping but other potential applications were also recognized. The scientists concerned had concluded that networks of observing stations would be required so that an assembly of meteorological data covering a wide area would become available for the construction of weather maps and for the study of the movement of weather systems and of the changes taking place. The next step, inevitably, was to adopt as governing principles that all meteorological observations, whether instrumental or visual, must conform to very high standards as to quality and that all the observations in a network of stations must be compatible with one another. These principles were accorded world-wide validity when, in 1873, IMO was established and set about the development of meteorology on an international scale.

Requirements for high-quality observational data and for compatibility involved a range of varied responsibilities—training, defining technical standards, intercomparisons of meteorological instruments and quality-control procedures. In addition, there had to be major concern for new and improved instruments and for constant oversight of observing techniques. Meteorology, from its foundation as a scientific discipline with practical applications, would clearly be in a continuous state of development and a progressive outlook would be indispensable. All these responsibilities, set out in appropriate detail, were assigned to the Commission for Instruments and Methods of Observation (CIMO), a basic Technical Commission and one of the first established by IMO. When WMO succeeded IMO and held its First Congress (Paris, 1951) CIMO was maintained in existence and, over the years, its responsibilities have been varied from time to time in order to take account of changing requirements or the introduction of new instruments.

Ninth Congress (Geneva, 1983) decided to organize the diverse activities within seven major scientific and technical programmes. One of these programmes, the World Weather Watch, was provided with a basic component in the form of an Instruments and Methods of Observation Programme (IMOP) for which CIMO was designated as the corresponding Technical Commission. The IMOP is clearly essential to the WWW and to the Commission for Basic Systems (CBS) but it also serves and co-operates in all other WMO programmes.

The main long-term objectives of the Instruments and Methods of Observation Programme may be summarized as follows:

- To prepare and keep up to date information and advice on instruments, observation systems and techniques and on methods of observation;
- In support of WMO programmes, to encourage new observing systems utilizing relevant advances in technology and taking into account cost-effective considerations;
- To promote the development of standardized methods of meteorological and related geophysical measurements;
- To establish rules and procedures for global and regional intercomparisons of meteorological instruments;
- To participate in the WMO Education and Training Programme as regards the preparation of training courses for instrument technicians and inspectors and the provision of support for WMO Regional Instrument Training Centres;
- To organize, in accordance with agreed rules and procedures, global and regional intercomparisons of meteorological instruments.

Since IMO was replaced in 1951 by the intergovernmental WMO, the interests and responsibilities of CIMO have expanded with the rapid and varying developments in technology. Instruments in operational use in the late 1940s have proved more versatile as, for example, when radar was at first employed mainly to locate rainclouds and was later found to be valuable in tracking the progress of tropical cyclones and also for the measurement of rainfall. Under WMO much effort has been applied to the development of automatic weather stations providing a fairly comprehensive range of surface measurements. The dawn of the space age, marked by the launching by the USSR of the artificial satellite SPUTNIK-1, was the starting point for intense activity in connection with instruments for remote sensing.

In the case of a Technical Commission such as CIMO and a programme such as the IMOP there are no easy solutions and few final ones. Some of the elements in the oldest observing routine, e.g. that of rainfall, still pose problems as regards measurement; some variables are difficult to define and in some cases (e.g. visibility) have to be defined according to the method of observation, which may be visual or instrumental. In different countries manufacturers of instruments exhibit a variety of designs in trying to meet a specification drawn up by CIMO. Thus standardization of equipment is a target of some complexity and intercomparisons have to be organized with stringent scientific and technical criteria.

Sessions of CIMO

Member countries of WMO are free to nominate experts to any or all of the Technical Commissions. A general, operationally aligned Commission, e.g. CBS, would expect to attract a large number of participants, representing a large majority of Members. CMM, on the other hand, a specialized Commission, might expect to draw most of its members from countries with a maritime tradition. CIMO is to some extent exceptional because its membership has not been large although it could not be described as a specialist Commission. Indeed, its field of responsibility covers all meteorological and related instruments. The membership of CIMO has steadily increased and by 1988 the Commission consisted of 225 experts nominated by 113 Members of WMO. Attendance at sessions of CIMO is shown in the table overleaf.

Sessions of the Commission for Instruments and Methods of Observation

Sessions	Place and date	WMO Member countries	Non-Member countries represented	Other international organizations	Total number of delegates
I	Toronto Aug./Sept., 1953	18	3	3	51
II	Paris June/July, 1957	38	4	4	87
III	NewDelhi Jan./Feb., 1962	31	-	5	70
IV	Tokyo October, 1965	25	-	5	67
V	Versailles September, 1969	49	-	6	110
VI	Helsinki August, 1973	48	-	5	109
VII	Hamburg August, 1977	56	2	8	123
VIII	Mexico City October, 1981	49	1	3	129
IX	Ottawa July, 1985	50	1	4	88
X	Brussels September, 1989	53	-	2	95

Technical Commissions usually hold a session between one Congress and the next. They report to each Congress on the work carried out in the four-year period since the preceding Congress and submit for approval plans for the activities to be undertaken in the ensuing four-year period and beyond. The plans are discussed by Congress, which then approves a work programme—in the case of CIMO, the IMOP—and authorizes the expenditure involved. The work of a Commission is largely carried out by working groups and rapporteurs in accordance with the guidance provided by Congress and the procedures agreed at the sessions of the Commission.

The agenda for a session of CIMO contains a number of items which figure permanently and this is regarded as inevitable. The *Guide to meteorological instruments and observing practices* must be kept under review so that chapters may be revised as necessary, new ones added and new editions prepared. In 1983 the fifth edition was published with the new title *Guide to meteorological instruments and methods of observation*. Technical Regulations have to be kept up to date and comparisons between instruments arranged, sometimes in a regional project but usually globally. In some quantities absolute standards are difficult to define and still more difficult to attain but progress must be a permanent target.

At the second session of CIMO (Paris, 1957) a detailed examination was made of the experimental comparisons of radiosondes, particularly of an extensive series of tests carried out in Switzerland the previous year. Other subjects considered included:

- An interim reference precipitation gauge;
- Intercomparison of barometers;
- Definition of optical range for meteorological purposes;
- Definitive psychrometric tables;
- Radiation measurements and atmospheric electricity.

At its fourth session (Tokyo, 1965) the Commission, in addition to its numerous routine items, discussed observations and instrumentation at aerodromes, where measurements of wind and temperatures along runways were being accorded increased importance. The Commission also gave considerable attention to the expanding uses of radar in meteorology and to the organization of specialized training in the field of instruments and methods of observation. At the request of Fourth Congress (Geneva, 1963) CIMO reviewed its responsibilities and terms of reference and prepared a report for submission to Fifth Congress (Geneva, 1967).

By the time of the fifth session of the Commission (Versailles, 1969), protection of the environment was beginning to increase in importance. The measurement of atmospheric ozone was placed on the CIMO agenda and studies were initiated with regard to instrumentation and measuring techniques. With the encouragement of Congress CIMO continued to give close support to the planning of training courses and their implementation. It was also making every effort to keep abreast of new technological developments since it was already clear that more refined instruments would be entering into routine use for the World Weather Watch and for such research programmes as the GARP.

The range of CIMO's interest widened steadily and from the 1970s onwards the Commission became involved in additional environmental measurements and in satellite instrumentation. Published reports and technical conferences were undertaken to keep national Meteorological Services fully informed of the development of instruments and methods of observation. In addition, the relevant sections of the WMO *Technical Regulations* were kept under permanent review and the authoritative *Guide to meteorological instruments and observing practices* was revised and expanded from time to time. In 1984, after sustained effort by working groups and individuals, a fifth edition of the *Guide* was published in English. The eighteen chapters of the previous edition had been thoroughly revised and the following seven new chapters were added:

- Meteorological rocket sensing;
- Lower troposphere soundings;
- Measurement of sunshine duration;
- Automatic meteorological stations;
- Soil moisture measurement;
- Quality control of observing practices and procedures;
- Techniques for sampling surface variables.

The *Guide* is the most influential of WMO's publications from the point of view of achieving standardization and uniformity in methods of observation throughout the world. The first edition, consisting of 12 chapters, was published in 1954 and the extension from 12 to 25 chapters in a period of 30 years demonstrates the growth of meteorology as a basic science and in a large number of applications.

The diagram opposite shows how the activities of CIMO have been organized in a three-pronged structure in pursuit of the desired improvements in the quality of basic data. CIMO itself has the governing role exercised through its sessions, working groups and rapporteurs. Further progress is achieved in essential ways by means of instrument comparisons, by technical conferences and by workshops and other training arrangements.

Intercomparisons of instruments

It is not possible to exaggerate the importance of testing and verifying the accuracy and robustness for operational use of the instruments used for making meteorological measurements. A

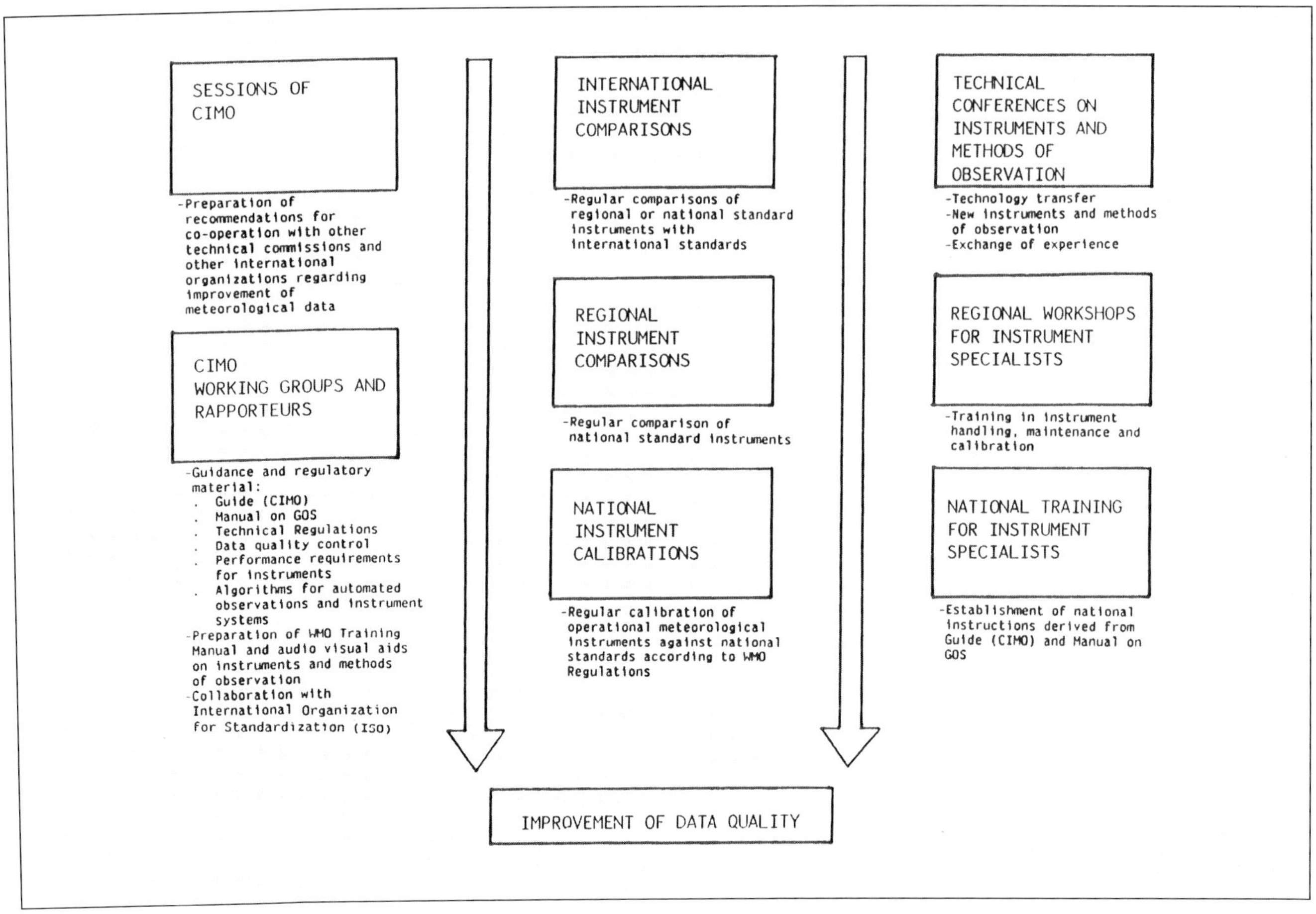

Hierarchy of the main activities of IMOP for improving meterological data quality

substantial proportion of the work of Meteorological Services is based on the data provided by network of observing stations; representativeness and comparability are therefore vital considerations. In the study of synoptic data plotted on weather charts, special value is attached to gradients, especially those of pressure and of temperature. Thus whilst the measurements recorded at each station may be highly accurate, it does not follow that the gradients derived from them would be equally so, unless an adequate testing programme had been carried out. Not all meteorological recordings are plotted on weather maps and used in forecasting. However, all data are used in climatology and in environmental control and therefore performance and quality standards must be applied to all instrumental equipment.

One of the principal activities of the IMOP is concerned with the intercomparisons of instruments. At any one time it is probable that a comparison is either at the project stage or actually in operation in accordance with a carefully organized plan. Intercomparisons are an important means for improving data quality. Some examples of intercomparisons are given below.

Upper-air measurements

In 1956 a successful radiosonde comparison programme was carried out in Payerne, Switzerland. Besides giving valuable results, it provided a number of suggestions as to the problems requiring consideration in any similar but more comprehensive exercise to be undertaken in future years. The CIMO Working Group on Comparison of Aerological Instruments carried out extensive studies in formulating guidance material on compatibility, reference sonde comparisons and laboratory tests. In due course work was completed as regards temperature reference sondes and the Commission turned its attention to the development of complete reference sondes, the improvement of techniques and the evaluation of radiosonde and satellite data. At the sixth session of CIMO (Helsinki, 1973) the Working Group on Radiowind Computations revealed that much of the current upper-wind measuring equipment was not capable of meeting the specified accuracy requirements. The use of automatic equipment and computer-based quality-control procedures was recommended and the Commission decided to establish a single Working Group on Upper-air Systems.

In the early 1980s, the Commission decided to arrange another intercomparison of radiosondes and an international organizing committee was formed to plan and implement the project. In reaching this decision CIMO had reviewed the results of earlier radiosonde comparisons and was greatly concerned about systematic inconsistencies of upper-air data. These were attributed to the variety of techniques used in the case of a large number of radiosondes in routine operation. The committee recognized that, for operational purposes, upper-wind data should be regarded as being equally important as data on temperature, humidity, pressure and geopotential. It was therefore considered necessary to include the different methods of wind-finding (radar, NAVAID) in the comparison programme. The committee further proposed that the prospects of carrying out a full international series of tests would be enhanced if the intercomparison were to be held in two phases—the first in the United Kingdom in 1984 and the second in the USA the following year.

At the end of the trials the committee considered that the project had been carried out most efficiently and with admirable co-operation from the host country for each phase. Preliminary results were presented to the WMO Technical Conference on Instruments and Methods of Observation, which was held in Ottawa in July 1985. The next photo shows the launch of five radiosondes—one each from Australia, Finland,

Launch of five radiosondes from the Goddard Space Flight Center Facility, Wallops Island

India, the United Kingdom and the USA—from the Wallops Flight Facility of the Goddard Space Flight Center of NASA, USA, during the second phase of the intercomparisons.

The final report of the international radiosonde comparison became available early in 1986 and was discussed in April of that year at a meeting of the CIMO Working Group on Upper-air Measurements which took place in Geneva. This comparison provided the largest uniform data set ever collected on the relative performance of operational radiosondes. Systematic discrepancies had been found in temperature, pressure and humidity measurements and it was possible to determine adjustment factors to improve the accuracy and comparability of the upper-air data received at analysis centres. The result was most valuable and the working group went on to discuss the possibilities of obtaining further improvements in the quality of radiosonde soundings, e.g. by developing algorithms for the production of data by an automated system.

Radiation measurements

CIMO's Working Group on Radiation Measurements has done valuable work in absolute radiometry and calibration procedures for standard and network instruments. A significant advance was made when self-calibrating (absolute) instruments were developed and as a consequence it became advisable to re-examine the International Pyrheliometer Scale (1956). The Commission has always stressed the need for standardization of radiation instruments and the associated calibration techniques. International pyrheliometer comparisons have thus been a recurring activity of CIMO. The fourth comparison was held in Davos at the World Radiation Centre in October 1975; it provides an example of the detailed planning necessary for such a complex exercise and of the thoroughness with which the work is accomplished.

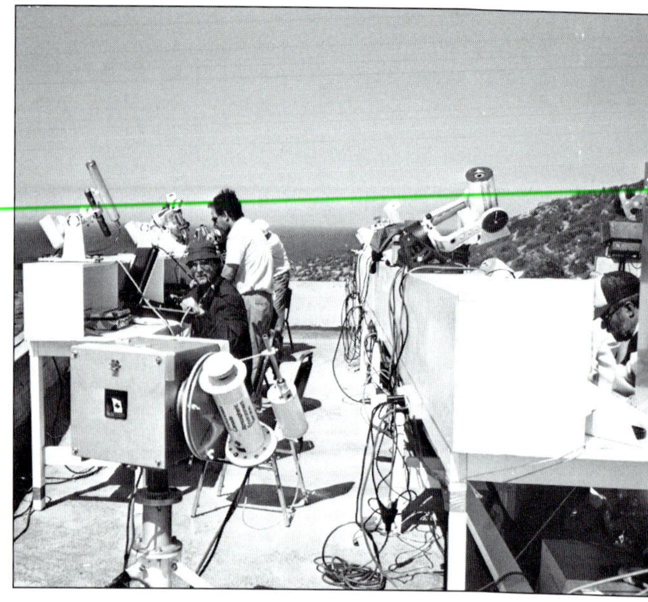

Regional Pyrheliometer Comparisons, Ensenada, Mexico (February 1989)

A large number of instruments were compared, including 33 Ångström pyrheliometers and one silver-disk model, one actinometer and 12 absolute radiometers of seven different designs. In view of the number of instruments involved, a computer-controlled data-acquisition system was developed in a form especially suitable for the simultaneous measurement of digital signals from a variety of sources. One of the main results of the comparisons was that the readings of all absolute radiometers were found to be consistent with the accuracy claimed for them. The basis was therefore provided for a change from the International Pyrheliometer Scale (1956) to an absolute scale, the World Radiometer Reference (WRR).

The sixth comparison in the series was held at Davos in October 1985, bringing together a total of 22 Member countries from the six WMO Regions. The primary purpose was to compare regional standard instruments with those in use at national reference centres. During the 1980s, regional pyrheliometer comparisons were carried out in all of the WMO Regions. One of them is illustrated in the photo above.

Ceilometers

Over many years instruments and techniques for measuring cloud base have been under development in many countries. In 1985 the Executive Council of WMO noted that no intercomparisons had hitherto been carried out and decided that one should be arranged. An international organizing committee was formed and arrangements were made to hold tests over a six-month period at the United Kingdom's experimental site at Beaufort Park near Bracknell. The planning was naturally complicated by the lack of a reference and of a widely accepted definition of cloud base. Nevertheless, seven countries took part in the trials with equipment using different measuring principles (light pulse, rotating beam and laser beam). In order to obtain independent reference measurements, provision was made for visual estimates of cloud base, for aircraft observations and for pilot balloon soundings. It was appreciated that this first intercomparison could not be conclusive but the participants agreed that it was instructive and well worth while. The analysis of the results involved a total of 75 million measurements, registered over a period of some four thousand hours.

Several other instrument comparisons were carried out in the 1970s and 1980s, such as for hygrometers, psychrometers, digital barometers, for measurements of visibility, sunshine duration and solid precipitation. The Executive Council decided at its forty-first session that an international ozonesonde comparison should be arranged as a matter of high priority.

Technical conferences and technology transfer

Technical conferences have been a notable feature of WMO's activities in the field of instruments and methods of observation. Such conferences offer an excellent opportunity for experts to exchange ideas and experience concerning the effective use of instruments and the continuing aim of seeking improvements in observational data. The presentation of reports, discussions and question-and-answer sessions, supported by exhibitions of technical equipment, serve a valuable purpose in supplementing the scientific literature in instrumental developments. These conferences have served as an indispensable adjunct to CIMO in its fundamental role as organizer for the planning, co-ordination, evaluation, technology transfer and exchange of information on the efforts of national Meteorological Services in all matters relating to weather observations.

A WMO Technical Conference on Automated Meteorological Systems took place in Washington, DC, in February 1975. Some five years later, in September 1980, a follow-up WMO Conference on the Evolution and Standardization of Observing Techniques in the Light of Automation was held in Sweden. These two conferences helped draw the attention of WMO Members to the potentialities of automation but also provided warning signals which emphasized that the pace of development would be slow, particularly as regards sensors, and that the automation of data acquisition in the case of upper-air observations was presenting major problems. In view of the economic benefits to be derived from automation, Members were informed that they would need to develop their own systems or commission them, since commercial firms would not be able to foresee any significant profits in going ahead on their own.

In September 1984 WMO held a further Technical Conference on Meteorological Instruments and Cost-effective Meteorological Observations. The conference was held in the Netherlands, TECO-88 was held in Leipzig in 1989 and TECIMO-IV in Brussels in September 1989, immediately prior to the tenth session of CIMO.

These technical conferences attract an attendance of over 100 instrument specialists and a good number of manufacturers—usually more than 30—participate in an exhibition of the latest equipment for meteorological use. Papers are presented on surface and upper-air systems; remote-sensing techniques; sensor development; the deployment of automatic weather stations on land, at sea and in the air; the installation and maintenance of instruments; and methods of calibration and comparison.

Apart from the *Guide* and the reports on instrument comparisons, CIMO published a great variety of guidance material in the *Instruments and Observing Methods Reports* series. From about 1980 as many as 45 reports were issued and served as an important contribution to technology transfer. Over many years another noteworthy activity of CIMO in promoting technology transfer has taken the form of workshops for instrument specialists in several of the WMO Regions. These workshops have had the primary aims of improving the operational use, maintenance and calibration of meteorological instruments.

Meteorological instrumentation: Problems and challenges

Towards the end of 1982, under the presidency of Dr Seppo Huovila, of the Finnish Meteorological Institute, the Commission considered a selection of topics, chosen mainly for their urgency, in the domain of meteorological instruments.

Taking an anemometer reading at a demonstration centre for fish farming in the Philippines

CIMO was particularly concerned about the future trends of instrumentation in an age of rapidly advancing technology. At the time, in the early 1980s, it was noticeable that progress was somewhat slow in replacing manual observing stations by automatic ones. Experience had also shown that more had been expected of satellites and powerful computers than such equipment was capable of providing at that stage of their development. In view of the high cost of up-to-date technology, there were difficult problems of allocating priorities and deciding whether available funds should be used for computers, telecommunications or meteorological instruments, including automatic weather stations. Four important aspects were noted: automation, service and maintenance, instrument manufacture and responsibility for observations.

Automation

World-wide expenditure on surface-based automatic weather stations had been modest in the extreme. The reasons appeared to be twofold: first, it was necessary to accept the limits of available technology in regard to some meteorological parameters, e.g. recognition of cloud types; second, the change to automatic systems would involve significant modifications to the current meteorological codes, and the development of software for the necessary microprocessors would be a long and arduous task. However, it was suggested that the automated radiosonde systems (which were able to record and evaluate the radiosonde-radiowind sounding, select significant levels and compile the coded message) demonstrated that similar systems could be devised for surface synoptic stations to be converted to automation.

It was, of course, envisaged that, where observing stations existed for a particular purpose (e.g. agriculture, aviation, pollution), an automatic station would be required to conform to specified standards to be laid down by the appropriate Technical Commission of WMO. Overall, it appeared that an economic approach would be achieved by a combination of man and machine where a human observer is needed in many cases for all or part of the time.

Service and maintenance

It was noted that in many national Meteorological Services the maintenance of instruments and telecommunications was partly or wholly contracted out to commercial firms. Costs had therefore assumed increased importance and trends towards automation had to be preceded by cost-benefit analyses of maintenance costs and the reliability required. Whereas a manned station could offer a serviceability of 100 per cent, the same standard for an automatic station could be prohibitively expensive. The numerous practical problems were understood; the Meteorological Services had to find a balance within their resources.

Instrument manufacture

Meteorological equipment formed a very small, almost negligible, percentage of the total world sales of electronics. Companies would not see much commercial attraction in developing meteorological instruments and it was therefore essential that national Meteorological Services and research institutes should accept responsibility for developing and testing sensors and for assessing instruments and methods of observation.

Responsibility for observations

It had to be recognized that, in the chain of real-time operations, beginning at the observing site and ending in the office of the data user, potentially weak links existed because some of the components were contracted out. Meteorological Services had to accept overall responsibility for their data-producing organization and ensure that effective control and surveillance were exercised.

Looking ahead

WMO's Second Long-term Plan, 1988–1997, contains a full and demanding programme for CIMO. The responsibilities vested in the Commission for Instruments and Methods of Observation and undertaken under the programme known as IMOP cover one of the fundamental purposes of WMO as set down in the WMO Convention. It is to be expected that technology will be the pace-setter; it is therefore incumbent upon CIMO to be receptive of new developments that might be exploited by meteorology and hydrology.

CHAPTER V

THE TROPICAL CYCLONE PROGRAMME

Introduction

One of the most important activities of WMO, from the humanitarian and economic points of view, is to contribute to the unending struggle to reduce the loss of human life and the enormous damage caused by tropical cyclones. Some fifty countries threatened by these storms are to be found mainly around South-east Asia, the Bay of Bengal, the Caribbean and the South-west Indian Ocean. The northern part of the South Pacific is another area prone to tropical cyclones (see figure on page 37). The responsibilities of WMO in regard to tropical cyclones can be traced back over many years. The international organization of meteorology, which began more than a century ago, sprang from the need to provide weather forecasts and gale warnings for shipping. Since tropical cyclones form over the open sea, one of the earliest priorities of meteorologists in the development of synoptic and marine meteorology was to acquire basic data on these storms and to improve the techniques for forecasting their formation and subsequent development and movement.

The immediate threat of a tropical cyclone is to shipping and fishing fleets but if the storm approaches close to land, and especially if it crosses the coastline, the effects, measured by loss of life and material damage, can assume the proportions of a disaster. In the case of a developing country, a single tropical cyclone can cause losses on such a scale as to annihilate the social and economic progress of the preceding 12 months.

Such storms have of course been of great concern to the meteorologist from the earliest time. Initially, however, there was little that could be done in giving warning of their existence and future movements. With the improvement of meteorological techniques, especially the use of satellites, highly significant progress has been made possible in this field in recent years. A high degree of co-operation with the United Nations and the Specialized Agencies has also been developed in this same field. In the following paragraphs the factors that contribute to such disasters are first reviewed. This is followed by an explanation of the steps taken by WMO to improve the warning systems throughout the world—which, in turn, leads to the steps taken recently to establish the WMO Tropical Cyclone Programme and the subsequent progress.

Factors that contribute to disasters

The intense depressions which occur in the Tropics are called tropical cyclones, typhoons or hurricanes, depending on the regions in which they occur. Characteristic of these storms are winds and turbulence of extreme violence, heavy and prolonged rainfall, river flooding and storm surges. Large areas may be affected and the destructive potential is so great that tropical cyclones are often described as the most dangerous of all natural phenomena.

In regard to the effects of wind, there is a tendency to underrate the contributions they make to the total devastation resulting from a tropical cyclone. In many instances, the extreme winds operate some hours before the onset of flooding and storm surge and, besides killing people and destroying buildings, establish the conditions which turn flooding and storm surge into disaster. The pattern of wind damage is usually widespread and the impact of heavy objects moving at speed through the air may be too much even for stoutly designed buildings. Once a structure has been breached, the subsequent wind damage can be extreme. On land, wind and flying debris can lead to serious loss of life but generally on a smaller scale than the deaths from drowning resulting from storm surge and widespread flooding. At sea, however, the total loss of life may be almost entirely due to the effects of winds and waves on shipping and particularly on small craft.

Storm surge during the passage of Typhoon 8412, Pusan, Republic of Korea (1988)

GOES geostationary satellite image of Hurricane Allen *over the Gulf of Mexico in August 1980 and some of the damage caused by the same hurricane on the north coast of Jamaica*

Flooding seems to be the inevitable outcome of the excessive rainfall of a tropical cyclone. Contributory factors, in addition to the rainfall, are the layout of towns and countryside and the inadequacy of drainage. Experience has shown that in most major storms, the greatest loss of life and property damage can be attributed to floods. When large numbers of human lives are lost, the primary cause is probably the flash floods produced by rainfall of very high intensity.

Storm surges have accounted for some of the world's worst human disasters. As a tropical cyclone approaches a coastal region, the onshore wind causes a piling up of sea water and the level may rise several metres. The effects are most pronounced in wide and shallow bays exposed to the tracks of tropical cyclones—as, for example, in the northern part of the Bay of Bengal. A particularly dangerous feature of a storm surge is that it may occur several hours in advance of the landfall of a tropical cyclone and cause severe flooding at a time when the evacuation of people is still in progress.

Disaster prevention and preparedness

The loss of life and the damage to property that might be caused by a tropical cyclone can be reduced to a minimum if a vulnerable country has a well-planned programme for disaster prevention and an effective organization for community preparedness which includes a reliable forecasting and warning service. Over many years, a number of countries with a calamitous experience of tropical storms seemed to be unable, or perhaps could not afford, to take measures to prevent disasters, e.g. by the construction of dams and coastal defences, or to form an emergency organization that would go into action as soon as warning of an approaching tropical storm was given by the Meteorological Service. During recent years, the weather forecasts have become more accurate and the timeliness of warnings has constantly improved. When meteorological satellites were launched, it became possible to detect tropical cyclones as soon as they came into existence and to keep their development, direction and speed of movement under almost continuous surveillance. During these years also, Hydrological Services have extended their flood-forecasting services and knowledge of the measures necessary to guard against floods in general and flash floods in particular.

In a number of countries affected by tropical cyclones arrangements for disaster prevention and preparedness were given high priority from the early years of the present century. Before long, the whole organization reached a high standard of efficiency and a series of subsequent reviews showed the way to improvements that were implemented, again with high priority. For disaster prevention, extensive long-term measures were carried out including construction works to reduce the risk of floods; also, land use and building development were made subject to planning controls. Disaster preparedness took the form of well-trained and well-equipped teams under central direction and brought into operation on the issue of a tropical cyclone warning. The co-operation of the general public was ensured by the issue of printed instructions and by practical exercises. Past experience was a useful motivating factor for everyone.

The countries mentioned above and some others demonstrated that a programme for disaster prevention and preparedness supported by the whole community and triggered into action by an effective warning system can substantially reduce the death toll from tropical cyclones. The figure on the next page, taken from an official publication of the USA, illustrates the value of protective measures in terms of lives saved. It shows as well that continued building development in vulnerable areas leads to increased damage. In the 30-year period 1900 to 1929, hurricanes, as tropical cyclones are called in the region, killed more than 10 000 people. But, in the 30 years from 1945 to 1974, after continuing improvements in prevention, preparedness and warning systems, fewer than 2 000 lost their lives as a consequence of hurricanes. This reduction was achieved in spite of the growth of population in the areas concerned. By contrast, as the figure shows, the average annual damage, after allowing for inflation, was nine times as great in the decade 1965 to 1974 as in the decade 1915 to 1924. Of course, in the years from 1915 to 1965 there was considerable economic development with large increases in the number of buildings and other installations and so the total damage was likely to be that much greater.

Significantly, it is in countries where sustained efforts are applied to the struggle against tropical cyclones that there is a strong awareness that the time never comes when everything possible has been done. In such countries, every attack by a tropical cyclone is followed by an examination of the emergency arrangements and of the public response to the warnings and instructions that were issued. It may be stated that over many years, all vulnerable countries have made strenuous efforts to do everything within their power to provide safeguards against tropical storms. However, in some of the developing countries, the problems have presented special difficulties apart from the shortage of finance and other resources. For example, in a highly developed country such as the USA, an order to the threatened population to leave their homes can be carried out speedily and efficiently because of the availability of numerous roads and adequate transportation. In developing countries, the situation could hardly have been more different.

The problem of helping developing countries in tropical storm regions was a continuing subject for study and discussion within the UN family and elsewhere. Some countries, e.g. the Philippines,

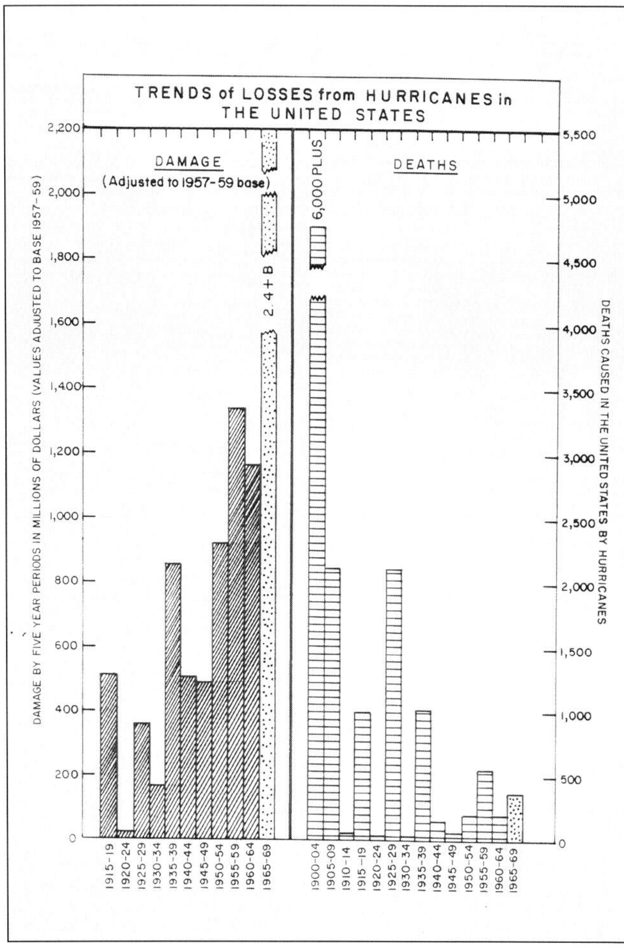

Flooding in the Philippines caused by Typhoon Olga *(May 1976)*

were accustomed to experiencing more than one storm in the tropical cyclone season. With the aid of meteorological satellites, so much was known about the location and approach of a tropical cyclone that it was frustrating to find that efforts to protect the people and to keep damage to a minimum were meeting with such limited success. At length, in the early part of the 1960s, one of the UN regional commissions, namely the Economic Commission for Asia and the Far East (ECAFE)*, concluded that the vulnerable countries operating separately were probably making less progress than they would achieve if they came together and made a co-operative attack on the complex problems that each had to endure. In the ESCAP region, where tropical cyclones are called typhoons, it was considered that Meteorological Services operated efficiently although there were urgent requirements for additional weather radars and improved telecommunications. However, ESCAP was greatly concerned about the damaging floods that seemed to occur at each typhoon visitation and therefore approached WMO, as a world-wide international organization and as a Specialized Agency of UN, raising the possibility of bringing the countries together for a concerted attack upon the problems presented by typhoons. This far-reaching initiative on the part of ESCAP was to lead to the formation of the Typhoon Committee and then to the WMO Tropical Cyclone Project which in due course was upgraded to one of WMO's programmes under the responsibility of the World Weather Watch and the Hydrology and Water Resource Programmes.

The Typhoon Committee

The initial discussions between ESCAP and WMO resulted in the arranging of a meeting of experts which was held in Manila in 1965. After all aspects of the typhoon question had been reviewed, a mission was set up to visit the countries concerned in order to discuss their individual problems and to assess the possibility of collaborative activities. The mission was headed by Dr F. W. Reichelderfer (who had served as first President of WMO for the four-year term 1952–1955) and included a meteorologist and a hydrologist. Countries were visited during 1966 and 1967. The mission's report contained constructive recommendations, the most important of which was that there was a compelling need for a co-ordinated regional programme aimed at developing an integrated warning system capable of helping the attainment of maximum protection against typhoons. The report also stressed that, within the overall regional context, individual countries must retain responsibility for the effective operation of their meteorological, hydrological and community-preparedness services.

The report was favourably received by both ESCAP and WMO and it was decided that both organizations should jointly sponsor the establishment of a Typhoon Committee as an intergovernmental body. The terms of reference were widely defined and covered disaster prevention and preparedness as well as meteorological and hydrological matters, including the provision of flood-forecasting services. The committee was officially approved in 1968 by ESCAP in April and by the WMO Executive Council** at its twentieth session, in June. Membership initially consisted of China, Japan, Hong Kong, the Republic of Korea, the Lao People's Democratic Republic, the Philippines and Thailand. The Khmer Republic (now Cambodia), Malaysia and the Socialist Republic of Viet Nam joined in 1972, 1976 and 1979, respectively. In addition to WMO and ESCAP, international organizations which assisted in the work of the committee included the UN Development Programme (UNDP), the Office of the UN Disaster Relief Co-ordinator (UNDRO) from 1971, the UN Environment Programme (UNEP) and the League of Red Cross Societies (LRCS).

The Typhoon Committee, very ably supported by a highly qualified secretariat, was making encouraging progress, meeting regularly and also exchanging ideas by correspondence. Then, within little more than two years of its existence, the participating countries and the supporting international organizations received a number of reminders—totally unnecessary reminders—of the vital need to take precautions against typhoons. In the autumn of 1970, in the space of a few weeks, a series of typhoons ravaged the Philippines, causing heavy loss of life and enormous damage over large areas. In November of the same year, a tropical cyclone, generally considered to have been the worst ever experienced, struck the low-lying coastal areas of Bangladesh and the number of deaths, mostly from drowning as a result of storm surge, was assessed at more than 200 000.

In both the Philippines and Bangladesh, the loss of life and the material damage were on such a scale that it was quickly realized that, in the developing regions, single countries or small groups of countries (as in the Typhoon Committee) did not possess the resources or the technological skills to prevent, or even mitigate to any great extent, the havoc wrought by tropical cyclones. These calamities had demonstrated that international action, comprehensively organized, was required in support of the efforts of developing countries are vulnerable to tropical cyclones. The Typhoon Committee, as an intergovernmental body, therefore exercised its right to appeal directly to the United Nations, asking urgently for international assistance. The UN General Assembly gave immediate attention to the appeal and adopted Resolution 2733. This called upon WMO to mobilize scientists, technologists and other appropriate resources in international efforts to discover ways and means to reduce the harmful effects of tropical cyclones and remove or minimize their destructive potential.

The Tropical Cyclone Project

The UN Resolution prompted a decision by the Sixth Congress of WMO (Geneva, 1971) to establish a WMO Tropical Cyclone Project and thus international attention became focused upon the need for combined efforts over a wide range of activities to counter the threats presented by tropical cyclones. In December 1971, another resolution of the UN General Assembly dealt specifically with the provision of aid to countries affected by natural and other disasters and also appointed a United Nations Disaster Relief Co-ordinator with responsibilities in the various aspects of prevention, preparedness, relief and rehabilitation. Other international agencies mentioned above were also involved in carrying out the UN directives and the Typhoon Committee was able to proceed with its plan of action with strengthened determination.

In setting up the Tropical Cyclone Project, Sixth Congress laid down the following primary objectives:

- Improving the present capabilities of detection, tracking and forecasting of tropical cyclones;
- Developing and making generally available techniques for quantitative storm-surge forecasting;
- Improving flood forecasting capabilities;
- Improving tropical cyclone warning systems;
- Supporting disaster prevention and community preparedness and related activities;
- Providing data, for planning and associated purposes, on the risk of loss by cyclone wind, storm-surge flood and river flood.

* Now Economic and Social Commission for Asia and the Pacific (ESCAP)

** See note on p. 8.

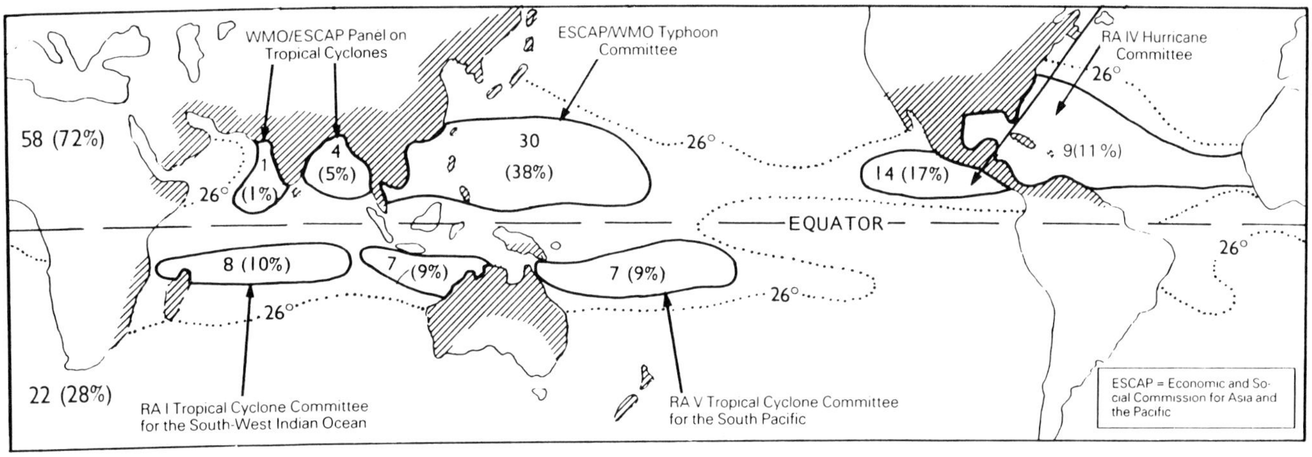

The global occurrence (in numbers and percentages) of tropical cyclones. Cyclones form only where the sea temperatures exceed 26° (dotted line); the land areas affected are shown hatched (After W. M. Gray, 1975). *Tropical cyclone bodies and the basins covered by their programmes are also indicated*

The encouragement implied in the resolution of the UN General Assembly combined with the support of Congress resulted in other regions following the example of the Typhoon Committee and establishing the machinery for dealing in an organized manner with their own problems. In due course, therefore, there were five regional cyclone bodies functioning in accordance with the guidance issued by Sixth Congress. They were:

- ESCAP/WMO Typhoon Committee;
- WMO/ESCAP Panel on Tropical Cyclones in the Bay of Bengal and the Arabian Sea;
- WMO (Regional Association for Africa) Cyclone Committee for the South-west Indian Ocean;
- WMO (Regional Association for North and Central America) Hurricane Committee;
- WMO (Regional Association for the South-west Pacific) Cyclone Committee for the South Pacific.

The programme was structured in two main divisions: a regional component and a general or planning and development component. Each of the above committees or panels prepared its own long-term technical plan concerned with the development of meteorological and hydrological networks and with the various activities involved in disaster prevention and preparedness. The countries themselves made praiseworthy efforts to implement the regional programmes but, as the UN General Assembly appreciated, there was a compelling need for external aid in the form of international support from such sources as the UNDP and from the Voluntary Assistance Programme of WMO and also from donor countries which would undertake to provide experts and finance for one or more items of the technical plan. Although these external sources had to consider many applications for financial and material assistance, the various tropical storm committees received generous support which supplemented in a most helpful way the individual and combined efforts of Members. Quite remarkable progress was made in each region.

The second of the two main divisions of the project was essentially a planning and development component concerned largely with broad objectives that would be applicable to the countries in each of the five tropical storm committees. A plan of action contained a number of sub-projects concerned with various aspects of instrumentation, the use of data from geostationary satellites, and meteorological and hydrological forecasting techniques. A prominent place was also given to education and training. There was also in the course of meetings considerable encouragement for exchanges of ideas that might lead to special studies.

The various sub-projects have been undertaken by appropriate WMO Technical Commissions or by the Secretariat in collaboration with other international organizations. For example, a handbook prepared by a group drawn from ESCAP, WMO and LRCS, entitled *Guidelines for disaster prevention and preparedness in tropical cyclone areas*, was published in 1977. This comprehensive publication drew extensively upon official publications issued by countries such as Australia, Japan and the USA, which had developed disaster safeguards, including warning systems, to a high level of efficiency over many years. The guidelines were of course aimed primarily at the developing countries and have proved of great use. Another publication, entitled *The quantitative evaluation of the risk of disaster from tropical cyclones* (WMO-No. 455), was distributed to the various countries in 1977; it consisted of two parts, meteorological aspects and hydrological aspects. At about the same time, a scientific paper on storm-surge prediction was completed by a team of experts from India, Japan and the USA, filling an important gap in the technical procedures of the forecasting and warning system.

From the institution of the tropical cyclone project, symposia and seminars have been a recurring feature of the activities in each region. Countries with vast experience in coping successfully with threats of natural disasters have provided specialists to conduct many of these courses and to impart their experience.

Review of tropical cyclone activities

At the time of WMO's Eighth Congress (Geneva, 1979) the Tropical Cyclone Project had been in existence for about eight years as a result of its establishment by Sixth Congress in 1971. During the intervening period between the two Congresses, the committee in each of the five regions prone to tropical storms had been very active and had made encouraging progress over the whole field of disaster prevention and community preparedness. Responsibilities in each component of organization against disaster were better understood and the various countries had achieved closer integration of their arrangements. Furthermore, there was improved co-ordination among the Member countries in each committee and exchanges of information among the committees had been placed on a continuing basis. The WMO Regional Associations and the countries participating in the work of the committees gave all possible support.

As work progressed in the regional plans of action, new ideas emerged and were often discussed in the symposia and seminars that took place. Shortly before Eighth Congress assembled in Geneva, the Typhoon Committee had begun feasibility studies for a Typhoon Operational Experiment, to be known as TOPEX (see below). This was an example of the lively attitude taken by all involved in this work of such high importance. At Eighth Congress, the Member countries of WMO showed their interest in the Tropical Cyclone Project and their desire that it should be given all possible encouragement and material aid. Congress appreciated the total effort involved over the five regions and the benefits to be gained by adequate co-ordination. It was therefore decided to upgrade the project, to be known in the future as the Tropical Cyclone Programme (TCP), a higher level in the hierarchical programme structure of WMO activities.

Congress made no change in the administrative arrangements under which the Tropical Cyclone Project formed part of the World Weather Watch Programme. However, shortly afterwards it was considered that there would be mutual benefit if the Tropical Cyclone Programme had closer links with the Tropical Meteorology Research Programme. Accordingly, the TCP was transferred to the Research and Development Department. It was certainly useful to bring research scientists into contact with the day-to-day problems which forecasters and hydrologists had to contend with in connection with the system for issuing warnings against wind, rain, flood and storm surge. The transfer was maintained for about three years and then it was decided that, in view of the almost entirely operational nature of the Tropical Cyclone Programme and the wide range of its activities, including disaster prevention and preparedness with no obvious relationship to research in tropical meteorology, a return to the World Weather Watch Department was the logical step. By this decision it was acknowledged that the TCP, although organized and co-ordinated by WMO, was virtually a joint programme involving, besides WMO, UNDP, LRCS and the appropriate UN commission for the region.

The Typhoon Operation Experiment

At the eleventh session of the Typhoon Committee (November 1978) there was some preliminary discussion of a suggestion that an operational trial should be arranged to test emergency procedures for guarding against disaster from tropical storms. In July 1979, a feasibility meeting was held in Tokyo and a decision was reached to recommend that a Typhoon Operational Experiment (TOPEX) would be practicable and would provide valuable data and other information. The proposed plan envisaged an operational study of the various systems involved in analysis, forecasting and warning, using actual typhoons as they occurred. The experiment would have three components—meteorological, hydrological and the dissemination of warnings and associated information. The proposals also made provision for a TOPEX centre to be established in Tokyo and for each participating country to set up a sub-centre.

The proposals for TOPEX were approved, subject to some small adjustments. The operational part began with a three-week period during the 1981 typhoon season and continued in each of the years 1982 and 1983 with an 11-week trial period. Altogether, nine typhoons were experienced during this three-year operational phase and every opportunity was taken to exact the maximum benefit. A meeting was convened in Tokyo in 1984 in order to evaluate the results of the experiment to the extent possible although some aspects had to be left for long-term studies.

It was widely agreed among the participating countries that TOPEX had been well worth while and had yielded benefits of both national and international value. Significant improvements were observed in forecasting ability and in the operation of measures for disaster preparedness and mitigation. Important gaps in the Global Observing System and in the Global Telecommunication System of the World Weather Watch were identified and action was undertaken for improvements. The experiment had also inspired a great deal of support for training and additional resources became available for strengthening the infrastructure of the organization against disaster. Another valuable product was the closer and more effective co-operation among the countries forming the Typhoon Committee.

Conclusion

WMO programmes are invariably stimulating and conducted with enthusiasm. This noteworthy characteristic is explained partly by the fact that Member countries nominate as participants in the various programmes experts prepared to contribute to joint efforts as part of a team and anxious to achieve results. The Tropical Cyclone Programme bears comparison with any other of WMO's activities. Its central objectives—to save lives and to reduce damage—are themselves sufficient to demand absolute dedication and it is clear that this has always been present in the programme as a whole and in the regional bodies where the vital efforts are carried out.

It should also be mentioned that, in the absence of a single body dealing with the overall aspects of the WMO Tropical Cyclone Programme, the Secretariat, in addition to playing its normal fundamental role, assumed responsibility for the management and planning of this important programme.

Like the World Weather Watch Programme, the Tropical Cyclone Programme is a continuing one. It is not possible to foresee the day when WWW is judged to be complete or when countries are fully satisfied that no improvements are required in their defence and emergency measures against tropical cyclones. The scope of the TCP is very wide in meteorology and hydrology and also in the related aspects of prevention and preparedness. The organization in each country is inevitably of great complexity, involving a variety of services and skills that must all be blended together so that in time of emergency everything will function smoothly and meet its objectives. Between the various components of the organization there is a high degree of interdependence and, if one part of the system were to fail, the other elements would be seriously handicapped. The tropical cyclone warning service is from many points of view the most vital element of the emergency organization because, if operated efficiently and conscientiously, it can provide time: time for the different operational services to be brought to a full state of readiness and to implement pre-disaster measures; time for the removal of the people from threatened areas; time to protect installations and to close down industrial activity.

Consideration of the benefits to be gained shows that the Tropical Cyclone Programme must have a prominent place among the many challenging programmes of WMO. The responsibilities are great and to be welcomed since the UN General Assembly has placed WMO with other concerned international organizations in the forefront of the struggle against tropical cyclone disasters.

CHAPTER VI

MARINE METEOROLOGY AND ASSOCIATED OCEANOGRAPHIC ACTIVITIES PROGRAMME

Introduction

Applications of meteorology are to be found in nearly all the social, economic and industrial activities of a country and also in the numerous ways in which a nation carries out its international commitments. Meteorology can thus be seen as a connecting thread in so much that helps to define the life of a nation or many nations working together. Professional meteorology finds its fulfilment in the applications of the science. These applications cover a very wide range and are of very considerable variety.

To the meteorologist all applications of his scientific calling are worth the effort and there is no tendency to discriminate on grounds of importance between one or other user of meteorological information. However, it is possible to recognize a category of user whose requirements demand a substantial effort on the part of WMO and its Members and also, in their variety, bring to bear many of the resources and techniques of applied meteorology. This category is fairly large and, as examples, includes marine meteorology and associated oceanographic activities, aeronautical meteorology and agricultural meteorology. These three represent major centres of applied meteorology and Ninth Congress (Geneva, 1983) decided to group them together in an Applications of Meteorology Programme, forming one of the principal Programmes of WMO. The three spheres of activity—marine and oceanographic aspects, aeronautical meteorology and agricultural meteorology—are respectively the subjects of the present and the two succeeding chapters.

Growth of marine meteorology

From the earliest times when enterprising sailors made voyages in search of trade and wealth, the weather in coastal waters and over the oceans has naturally been a subject of vital importance to mariners. Accumulated experience gave valuable indications as to ocean trade routes offering the least danger but, even so, disasters occurred as a result of such phenomena as tropical storms, often accompanied by tidal waves, and gales, mainly associated with temperate-latitude depressions. Experience was clearly not enough but it was the primary source of guidance until towards the middle of the nineteenth century when it became possible to consider as a practical prospect the scientific study of the atmosphere, based on observational data from networks of stations equipped with instruments and integrated with the help of telecommunications. Progress was bound to be slow but mariners lacked neither vision nor determination. In 1853 in Brussels delegates from ten seafaring nations began the task of drawing up agreed procedures for the meteorological logs to be compiled by the masters of ships sailing on the oceans. The initial, somewhat rudimentary steps were thereby taken in the development of meteorology as an international science providing a range of services and with the efforts of all countries co-ordinated by a world organization of increasing reputation and influence.

Since these early days meteorology has prospered under the stimulus of users' requirements. Shipping first and then aviation smoothed the way for all nations to establish meteorological services with observing networks forming a world-wide system under the leadership of IMO and, in due course, of its successor, the intergovernmental WMO. As a result, many other interests were able to benefit from these developments in meteorology. The organization and the networks required to satisfy such major users as shipping and aviation were able to provide the weather forecasts and climatological information that served the needs of agriculture, various sectors of industry, the general public and many others. More recently, since about 1979, the data, assembled and statistically analysed, have been making a valued contribution to such important undertakings as the World Climate Programme (see Chapter XII).

The developing role of marine meteorology

The traditional role of meteorology in the marine environment has evolved from being a modest contribution to the protection of shipping against adverse weather to the provision of many and varied meteorological services such as weather bulletins which are broadcast by a number of National Meteorological Centres around the world and certain actual weather reports, analyses and forecasts for the high seas and coastal waters. The WMO body mainly responsible for this evolution is the Commission for Marine Meteorology (prior to 1972 entitled the Commission for Maritime Meteorology). Throughout its history the Commission has owed much to the willing co-operation of other international bodies concerned with marine affairs, e.g. the Inter-governmental Maritime Consultative Organization (IMCO) which, in 1982, was renamed the International Maritime Organization (IMO), the International Chamber of Shipping (ICS) and the International Council for the Exploration of the Seas (ICES). Special tribute should be paid to the fleet of Voluntary Observing Ships which began in a small way more than a century ago and, with the necessary support of mariners, have provided highly valued meteorological data from the oceans of the world. After World War II, from 1947, the efforts of the Voluntary Observing Ships were augmented by the more frequent and more comprehensive observations made by weather ships in the North Atlantic and other oceans. Unfortunately the substantial costs of weather ships have led since the 1970s to a steady decline in their numbers.

Since the First Congress of WMO in 1951, advances in meteorology have enabled national Meteorological Services to increase the variety and widen the scope of the information and advice they are able to offer in forecasting and climatology. A notable feature has been that customers of many years' standing have intensified their demands with regard to existing services and have introduced fresh requirements in the pursuit of their own operational efficiency. Important new tasks have arisen in the marine environment. An interesting example is the weather routeing of ships, a service aimed at helping mariners to navigate clear of the worst weather conditions, thus protecting the ship and its cargo from damage and usually saving time on the voyage from departure port to destination. Offshore drilling operations in search of oil and gas gave rise to a new range of services covering marine climatology as well as day-to-day weather forecasting. It was quickly realized that the requirements were wider than meteorology and the assistance of experts in physical oceanography had to be enlisted. Marine pollution was another problem which called for joint efforts by oceanographers and marine meteorologists. From about 1960 onwards, incidents involving tankers and resulting in the release of oil slicks on the open sea appeared to become more frequent and a great deal of public concern was aroused. This concern increased and has become ever watchful because oil pollution constitutes a very serious threat to inhabitants of the sea and to bird life. In addition, it has become widely known that the oceans have a major role in the stability of the global climate and any phenomenon such as contamination of the seas that might adversely affect this role is to be avoided.

As the importance of meteorology in the marine environment steadily expanded as regards both services and research, the presence or impact of oceanography—in particular, physical oceanography—became more marked. From the earliest days of the Voluntary Observing Ships, ships' weather reports have included measurements or estimates of such elements as sea-surface temperature, height of waves and the nature of swell. More recently, in the World Weather Watch Programme, certain meteorological centres have provided routine forecasts of mixed layer depth. It may seem odd that these variables are present in weather forecasts for mariners and in ships' observations but their inclusion illustrates that the atmosphere cannot be regarded as a closed system. Indeed, the meteorologist has always been aware of the intermingling of atmospheric and oceanic processes and has constantly stressed the importance of oceanic data if progress is to be made in the science of meteorology. It is well known, of course, that these oceanic quantities are controlled to a major extent by the weather conditions, in particular the wind speed and direction. However, perhaps the main point is that oceanographic services, making routine observations and issuing messages of existing conclusions and anticipated changes, have not been established in any manner comparable and analogous to meteorological services. In the field of research there have for many decades been joint projects with oceanographers and meteorologists working together but, for good reasons, the oceanographers up to the 1960s had not attempted much progress in establishing permanent observing networks. The situation changed when the General Assembly of the United Nations drew attention to the need for greater knowledge of the oceans and for wider use of their resources.

The resources of the sea

Just as in 1961 and in the following years a series of resolutions adopted by the UN General Assembly spurred WMO and its Member countries to the far-reaching development known as World Weather Watch, so in 1966 another UN resolution, the first of a series, focused attention upon the oceans and requested nations and concerned international organizations, including WMO, to make effective arrangements for an expansion of activities in marine science and technology. The subjects covered by the UN resolutions included:

- International co-operation in matters related to the oceans;
- Development of a long-term and expanded programme of oceanic exploration and research (given the acronym LEPOR);
- Measures for the prevention and control of marine pollution;
- Peaceful uses of sea beds;
- Exploitation and conservation of living marine resources.

It was evident from the first of the UN directives, which was entitled "Resources of the sea", that existing observing networks would have to serve wider purposes than previously and that new observing systems leading to denser networks would have to be established.

The first resolution produced a strong increase in activity among nations with maritime interests and among international organizations exercising marine functions and responsibilities on behalf of their Member countries. The resolutions which followed over the next decade maintained the momentum. There were three international organizations that might be described as being very substantially involved in

Fishing boats off the coast of Peru. Weather information is of vital importance to fishing fleets

Marion Dufresne—*a ship used by the NAOS network (see also Chapter VII)*

taking all possible action on the UN resolutions. These organizations were:

- Unesco, through its Intergovernmental Oceanographic Commission (IOC);
- FAO, through its interests in fishing and marine biology;
- WMO, through its Commission for Marine Meteorology.

The above organizations were primarily concerned with the implications for science and technology and the requirements for associated services arising from the resolutions. Other organizations, such as the International Maritime Organization, had major interests in such aspects as administration, the safety of life at sea and the Law of the Sea.

Action involving WMO

At the time of the UN resolution on resources of the sea (1966), there was considerable interest shown by governments and in international circles about commercial possibilities in the oceans and on the sea bed. There was also wide discussion concerning the rights of exploration. It was very clear that the whole subject brought to the forefront by the United Nations contained elements of a highly political nature. However, the three international organizations mentioned in the resolutions had much to consider in their own special fields and had also to discuss how co-operation might best be achieved.

The Executive Council* of WMO held a full discussion on the UN initiative and concluded that the responsibilities likely to fall upon the organization would probably be much wider than the terms of reference of the Commission for Marine Meteorology. It also viewed the possibility that urgent questions might arise and call for high-level decisions or advice. It was therefore considered that the Council should set up a group to advise it in all matters arising directly or indirectly from the UN resolution and any subsequent ones regarding the oceans and their resources. Accordingly, the Executive Council decided to set up an EC Panel on Meteorological Aspects of Ocean Affairs and invited Member countries that wished to do so to nominate suitably qualified experts to assist in the tasks to be undertaken by the panel.

*See note on p. 8.

The first task of the panel was to hold a meeting in order to discuss its responsibilities and to make plans for action that might be taken and for any preliminary studies that seemed to be required. The meeting also decided that urgent arrangements should be made for a partnership to be developed with Unesco's Intergovernmental Oceanographic Commission. It was accepted that close relations would also be required with the Fishery and Marine Biology division of FAO but IOC was given greater priority since WMO's scientific and technical interests in the oceans were mostly concerned with physical oceanography.

The WMO panel and IOC arranged to hold joint sessions once or twice a year, according to the work in progress, and a joint IOC/WMO Operational Programme was set in motion for an Integrated Global Ocean Station System (IGOSS). Eventually, it was agreed that a joint IOC/WMO working committee should be established to carry out the implementation of IGOSS and related matters.

The three organizations—FAO, Unesco and WMO—reacted to the UN General Assembly's demands with energy and determination. After a series of informal discussions, involving also the Scientific Committee on Oceanic Research (SCOR) of ICSU, it was arranged that a joint working party should meet to consider the scientific content of a comprehensive programme of international co-operation as required by UN. The meeting took place on the island of Panza in 1969 with the following defined purpose:

> To increase knowledge of the ocean, its contents and the contents of its sub-soil and its interfaces with the land, the atmosphere and the ocean floor and to improve understanding of processes operating in or affecting the marine environment with the goal of enhanced utilization of the ocean and its resources for the benefit of mankind.

The joint working party conducted its studies under four main headings:

- The ocean circulation and ocean-atmosphere interactions;
- Marine biological production;
- Marine pollution;
- Dynamics of the ocean floor.

In all the above subdivisions meteorology was involved with regard to the provision of services covering both forecasting and climatology. In two of the subdivisions, namely, ocean

A supply ship beset by ice in the Arctic

circulation and ocean-atmosphere interactions and marine pollution, the co-operation of meteorologists was considered to be in the field of research. The findings of the working party and the recommendations were published by each of the four participating organizations under the title *Global ocean research*. The report, which came to be known as the *Panza Report*, aroused considerable interest and the suggestions regarding targets to be aimed at in, for example, world fish catch, offshore oil production and mineral extraction were frequently referred to in the ensuing years.

In addition to the activities of the EC panel and of the Commission for Marine Meteorology, WMO played a large part in a number of research projects including the Indian Ocean expeditions and others carried out under its Research and Development Programme. These projects were generally organized independently but made full use of the facilities and services established by WWW and by the CMM Marine Meteorological Services System.

In the early 1970s interest and activity in ocean affairs had greatly increased and a number of international organizations were closely involved. Co-ordination clearly became a matter of major importance and it was therefore agreed to establish an Inter-Secretariat Committee on Scientific Problems relating to Oceanography (ICSPRO). Unesco, FAO, WMO and the International Maritime Organization took part in the work of this committee. It was also found necessary to assemble another working party, the Group of Experts on the Scientific Aspects of Marine Pollution (GESAMP), with FAO, Unesco, IMO and WMO as participants. In due course, these committees were joined by representatives of the UN Environment Programme (UNEP), an organization created by the UN Conference on the Human Environment (Stockholm, 1972).

Progress to report

From the time of the First WMO Congress, in 1951, the Commission for Marine Meteorology has maintained efforts to develop and co-ordinate a world-wide system for the provision of weather bulletins and warnings for shipping. However, an area south of latitude 60°S has been the main responsibility of an EC Working Group on Antarctic Meteorology with full arrangements for liaison with CMM and with the WMO/IOC Working Committee for IGOSS. All areas with regular commercial shipping lanes were served with frequent weather bulletins as a result of plans drawn up by CMM and implemented through the agency of WMO's Regional Associations. Special arrangements have always been made for fishing operations in the southern oceans and for expeditions in Antarctic waters.

The system devised by CMM has been subjected to a form of market research and has proved to give satisfactory service to the great majority of those receiving the weather bulletins and acting upon the advice contained in them. The arrangements are expected to become more effective and comprehensive when it is practicable to transmit all weather forecasts and warnings on INMARSAT, the advanced communication system being organized by IMO for maritime safety. This new development, combined with satellite communication facilities and the establishment of Regional/Specialized Meteorological Centres and with the advent of more exacting shipping requirements for safety and operational efficiency, may lead to an overall review by CMM of marine meteorological services before the end of the century.

The collection of weather reports from ships by means of the HF/MF Maritime Mobile Service encountered many problems during the 1960s and 1970s because reductions were taking place in the number of radio officers carried on ships. Eventually the majority of ships supplied weather reports only during the internationally agreed watch hours. Data coverage at some synoptic hours was adversely affected and a further difficulty arose as a result of traffic congestion on HF coastal radio stations. However, early in the 1980s the decline in the availability of weather reports from some sea areas was reversed by the use of INMARSAT as a collecting agency. By 1987 about 15 per cent of ships' reports were collected in this way and the percentage is expected to be doubled by 1990. The use of satellites is also expected to reduce delays in data collection. Weather reports from the oceans have always made a vital contribution to the whole range of forecasting services and CMM has to maintain its efforts both in recruiting Voluntary Observing Ships and in ensuring that weather reports reach the designated meteorological centres without delay. The strength of the fleet in 1988 was more than 7 300 ships and in the preceding year the average number of reports collected each day was around 3 300.

The Marine Climatological Summaries Scheme, developed by CMM in the 1960s, provides for the extraction of climatological data from ships' log books and has already yielded a large data base for use in a variety of research programmes.

A balloon being ejected by an automatic launcher aboard a Japanese freighter in the North Pacific Ocean

IGOSS, the Integrated Global Ocean Services System, the joint responsibility of WMO and IOC, has been and is being developed in parallel with the World Weather Watch—i.e., the essential elements of IGOSS are an observing system, a data-processing and service system and a telecommunications organization making extensive use of the WWW Global Telecommunication System. In the plans for IGOSS provision has been made for monitoring operations, for training and for research and development. Progress in IGOSS has been somewhat slower than was hoped for in 1966 when the first UN resolution on the sea and its resources was passed by the General Assembly. Practical difficulties in deriving data from sea areas have been very formidable and every effort is to be made to exploit advanced satellite techniques for obtaining data for sea-surface temperature, wind stress and wave state.

Concluding remarks

There has been a noticeable difference in outlook between the activities of CMM and the activities arising out of the UN directives on the oceans. Although CMM has some long-term studies in its programme, its major effort is operational, aimed at securing as quickly as possible practicable improvements in data coverage and in the efficiency of collecting weather reports and disseminating them. Each progressive step is reflected in slightly more accurate and timely weather forecasts and warnings for shipping and, on occasion, in the service becoming more comprehensive.

On the other hand, the UN resolutions were concerned with objectives and were therefore essentially of a very long-term nature. The difficulties are immense and the costs enormous. From 1966, the date of the first UN resolution, the planning was of overriding importance because it set out to provide the framework within which advances could be made whenever the opportunity arose. The first requirement, in line with the work of CMM, was to keep traditional facilities at their current strength and to maintain efforts to enrol more ships in the Voluntary Observing Fleet. The next step, side by side with the first, was to plan the deployment of buoys with automated measuring and recording equipment and to make use of the potential of satellites, all this in order to extend the coverage and scope of the organization co-ordinated by CMM. Every small step, such as additional bathythermograph observations, represented valuable progress.

The differences in outlook referred to above were not, of course, incompatible. What can be achieved rests with the Member countries of UN and WMO which receive consistent advice on the marine environment from their own meteorologists and oceanographers, whether the meetings are held by UN or by WMO. A global observing system is primarily the integration, in terms of organization and technical facilities, of all the material efforts which may exist for each country's domestic requirements and also, in some cases, for a nation's international responsibilities. The international organizations mostly concerned have been studying the deficiencies and encouraging nations to greater efforts for the common good. For this purpose and to clarify ideas, some separate monitoring systems have been suggested, reasonably consistent with the *Panza Report:*

- A health-monitoring system;
- A monitoring system to describe the Earth's living and non-living resources;
- A monitoring system for the total physical environment, including pollutants of a physical nature;
- A monitoring system for marine biology and fisheries.

In all these classifications meteorologists are involved, wholly involved, in the monitoring of the physical environment. An advantage of considering separate systems is that individual requirements can be investigated as to network spacing, frequency and regularity of data acquisition, real-time and non-real-time and so on. But each observing platform would have a comprehensive role rather than a specialized one.

The role of the Marine Meteorology and Associated Oceanographic Activities Programme is inevitably an expanding one. The programme is one of the most challenging confronting WMO.

CHAPTER VII

AERONAUTICAL METEOROLOGY PROGRAMME

Introduction

The advent of aviation in the early part of the century and its spectacular growth gave an enormous impetus to the development of Meteorological Services and to the already-created IMO. International civil aviation wanted observing stations at airports and forecasting offices at the more important ones. Another essential need was for a telecommunications system to permit the rapid interchange of reports between one airport and another, including special priority for reports of sudden deteriorations. All these requirements were so compelling that civil aviation authorities throughout the world were prepared to help in the financing of surface and upper-air observing stations at important locations, including airports, in the world-wide planning of air routes.

The rapid expansion of civil aviation resulted in existing national Meteorological Services improving their observing networks and new national Meteorological Services being created, some of them by the civil aviation department in their country and so with responsibilities exclusively for civil aviation. Reports from all these observing stations were, of course, available for other purposes including research and the provision of weather forecasts for shipping, agriculture, the general public and so on.

During the 1930s the growth of international civil aviation was substantial and intercontinental air routes came into existence. The staging points on such routes were generally less than 1 000 kilometres apart but in 1939 successful trials were carried out to demonstrate the feasibility of operating air routes over the North Atlantic Ocean, the distance from take-off to landing being of the order of 3 500 kilometres. It is well to remember those days because some of the aircraft in service could be flown safely only in good, preferably cloudless weather, and that when better-equipped aircraft which were not restricted in this way appeared, they were liable to encounter ice-forming clouds, turbulence and strong winds. In that period of civil aviation development the responsibilities of Meteorological Services to contribute to the safety and efficient operation of air routes were fully recognized.

WMO and ICAO

At WMO's First Congress (Geneva, 1951), the list of Technical Commissions that were set up included one for aeronautical meteorology (CAeM), thus officially continuing the work of the corresponding Commission of IMO. One of the earliest undertakings of WMO was to enter into negotiations with the International Civil Aviation Organization (ICAO) with a view to including a working arrangement between the two organizations. This is a common practice among international organizations, especially those which are Specialized Agencies of the United Nations, if an area of common interest exists. This was clearly the case with WMO and ICAO since the latter, in view of the great importance of meteorology to civil aviation, had within its organization a MET division. Working arrangements were agreed and approved by both organizations during 1953 for entry into force on 1 January 1954. The arrangements were designed to secure close co-operation and to establish a machinery regarding the allocation to one organization or the other of primary responsibility for certain sectors within the field of common interest. With ICAO's profound concern with meteorology it was inevitable that some difficult problems would arise but the presence of goodwill on both sides allowed agreements to be reached.

As a result of the working arrangement, CAeM has held many joint sessions with one of the components of ICAO, e.g. with the Air Navigation Conference of ICAO or with the MET Division. The eighth session of CAeM (Geneva, 1986) was only the second occasion on which the Commission met on its own.

An important subject of great concern to WMO was the North Atlantic Ocean Station (NAOS) scheme. This scheme was set up in 1947 by means of a joint financing Agreement between countries with airlines which operated, or planned to operate, on air routes over the North Atlantic. These routes bringing North America and Europe closer together were regarded in terms of commercial interests as the most important in the pattern of civil aviation throughout the world. Administrative responsibility for NAOS was eventually transferred from ICAO to WMO, an important change described in a separate section below.

The first simultaneous session

For CAeM its first such session (1954) was a formative one defining a high proportion of its activities for the years ahead and also crystallizing its responsibilities. For example, proposals for special meteorological codes for aviation purposes would be referred to the Commission for Synoptic Meteorology; climatology requirements for aeronautical use would be developed by the Commission for Climatology.

The work of the session fell broadly into four main parts:

- Facilities;
- Standards and associated procedures;
- Technical matters;
- Climatological aspects.

With regard to facilities, the session discussed the complex question of network density for upper-air observations and developed a number of principles related to the aeronautical requirement. Recommendations were also made concerning the potential value of sferic networks for providing warnings to aircraft about thunderstorm areas and the further development of automatic weather stations and radar for the assistance of aviation.

Standards and recommended practices and procedures for air navigation services are of major operational importance; they require considerable study and discussion and finally have to be officially endorsed by both organizations. All regulations of WMO and ICAO concerning the provision of meteorological services to aviation are identically worded and have the same status in the two organizations.

On the subject of technical matters the first session devoted considerable attention to the need to develop appropriate forecasting facilities and techniques for high-level operations. The Comet, the first of the civil jet aircraft, was already in operation at the time. The requirements drawn up for this type of aircraft, cruising at much greater altitudes than were typical of current airline operations, had been tested in proving flights and on passenger carrying schedules and seemed to provide a useful guide to some of the meteorological demands of the future. The session, looking to the future, appreciated that during the 1950s the role of the piston engine in transport aircraft would be reduced. The introduction of jet propulsion was heralding a major breakthrough and the major airlines were already planning the widespread introduction of jet-propelled aircraft on all routes, usually operating well above the clouds and weather.

The joint session, in addition, examined a number of measures intended to assist progress in various questions affecting the safety or economy of aircraft operations. Aircraft icing, turbulence and gusts, and (atmospheric) jet streams were among the questions on which recommendations were made. With regard to aeronautical requirements for climatology, these were specified for aerodrome conditions and for world maps of mean wind and temperature distribution up to about 200 hPa over the globe. Another very valuable feature of the meeting was the series of scientific lectures arranged on topics related to air operations. Mountain waves, the application of radar observations, and numerical forecasting were among the subjects of the lectures.

Thus, at its first session, CAeM clearly realized that the initial long and slow development of airline operations was coming to an end. New operations based on the latest technology would need to be found and these presented new and more complex challenges.

Wind-measuring instruments at Hong Kong Airport

Special requirements for turbine-powered aircraft

After the first session of CAeM, in 1954, a problem very much in the minds of the delegates as the meeting drew to a close was the imminent arrival on the operational scene of long-haul, turbine-powered aircraft. The Executive Council* of WMO was apprised of these developments and formed a Panel of Experts on High-level Analysis and Forecasting Techniques. The panel met in Brussels in March 1959, six months before CAeM was scheduled to hold its second session in a simultaneous meeting with the MET division of ICAO, which would be holding its fifth session.

The panel already had available in draft form a WMO Technical Note on high-level forecasting techniques. Its task was therefore to review this draft and to bring it to completion. In addition the panel had to consider and, if necessary, recommend action to be taken by Members of WMO and by the constituent bodies of WMO.

The panel emphasized three important requirements calling for action by Members:

- Checking the accuracy of upper-air temperature and wind forecasts;
- Reducing the effects of systematic errors in radiosonde observations. The panel drew attention to differences which could arise from calibration methods as well as computation techniques, especially in the case of upper winds;
- Recognizing the importance of the analysis and forecasting of tropopause data and instituting accuracy tests.

*See note on p. 8.

The panel also considered that the analysis and forecasting of jet streams was one of the major difficulties encountered by aeronautical forecasting services. Further studies were urgently called for. The panel's report provided valuable guidance for CAeM for one of the main items of its agenda at its second session.

A useful sequel to the work of the panel was a decision taken jointly by ICAO and WMO to hold seminars on analysis and forecasting for turbine-powered aircraft operations. Two seminars were held in Cairo and in Nicosia towards the end of 1961 and a third was held a year later in Bangkok. All these seminars were a resounding success.

Problems of aeronautical meteorology

By the time of the third simultaneous session (Paris, 1964), CAeM had retained all its traditional requirements for operations carried out by piston-engined aircraft, mainly because such aircraft would continue in use for some time to come, especially for general aviation, and had come to terms with the requirements for turbine-powered aircraft with thus much greater operating heights. This does not imply that all problems were solved. Indeed, all kinds of aviation presented both old and new problems for the meteorologist working in the aeronautical sphere. The meeting in Paris began to consider the requirements for supersonic transport (SST) aircraft and how best these could be met.

This simultaneous session involved two Divisions (Operations and MET) of ICAO and CAeM for WMO. The theme of the session with its long agenda may be summarized as follows: how far and by what processes of refinement can meteorology help further to increase safety, reliability and efficiency in international civil aviation in the supersonic jet age? The problems that faced CAeM were of three kinds: organizational, scientific and technical, or a combination of these three categories.

In the context of organization the session examined the principles governing the centralization of aeronautical forecasting services and the measures to be taken to encourage such a development. The problem arose from the expansion of air traffic which had resulted in selected offices in some areas providing route forecasts to a number of other offices which had not sufficient staff to meet all their requirements independently. Discussion of the subject was based on a report by the CAeM Working Group on the Meteorological Aspects of Area Forecast Systems. The meeting found the report very helpful as a guide to the formulation of policy and stressed the special value of an area forecast system in fulfilling existing long-range operational requirements and those anticipated for operations at supersonic speeds.

Scientific and technical problems were discussed for each phase of an operation—take-off, cruising and landing. Runway Visual Range, Slant Visual Range and the reporting of wind shear in the lower layers on the approach to landing were among the problems where precise measurement was of prime importance. With regard to SST operations a requirement was stated for exchanges of raw and processed data for levels up to 10 hPa.

A highly important discussion arose in the separate session of CAeM. Many of the delegates felt that meeting in joint sessions with an operationally oriented body like ICAO resulted in a concentration on administrative and purely technical matters so that important scientific problems were only superficially considered—if at all. It was agreed that the president of CAeM would consider this question urgently and investigate the possibility of holding a scientific and technical conference on the many problems of aeronautical meteorology. The conference was duly arranged.

Conference on Aeronautical Meteorology

The WMO Scientific and Technical Conference on Aeronautical Meteorology was held in London from 18 to 29 March 1968. The conference was designed as a forum for the presentation of new scientific and technical knowledge resulting from experience, experimentation and research, for the theoreticians to present the latest scientific thought to the practitioners and for the latter to highlight the operational meteorological problems calling for scientific research and guidance. Over 50 countries and six international organizations, in addition to WMO, sent representatives to the conference which, since its announcement, had aroused widespread interest.

The conference was divided into three symposia held consecutively and entitled: Meteorological Conditions in the Vicinity of Aerodromes; Meteorological Conditions in the Troposphere and Lower Stratosphere; and Meteorological Information for Supersonic Transport Operations. Each symposium was divided into seven or eight subject items which, in general, represented the most important problems confronting the aeronautical meteorologist. Under each item one or more leading experts invited prior to the conference presented papers which were followed by a general discussion. In all some 40 papers were presented.

The first Scientific and Technical Conference on Aeronautical Meteorology was undoubtedly a great success. The discussions were stimulating and a notable feature was the interchange of ideas between research and operations.

The first separate CAeM session

At its Fifth Session (Geneva, 1971) CAeM for the first time departed from a provision of the working arrangements agreed between ICAO and WMO and held a separate meeting. However, ICAO, IATA and the International Federation of Airline Pilots Associations (IFALPA) sent representatives to the session of CAeM.

The separate session provided a long-delayed and much-needed opportunity to put purely operational aspects into the background and to concentrate on aeronautical meteorology. It was an opportunity to consider the position of aeronautical meteorology within the whole field of meteorology; at one time it had been the principal field but other applications had already appeared and were steadily growing in prominence.

CAeM took note of a number of new factors entering into aviation in the 1970s. Pressures were being exerted for more rapid turn-around, for greater economy combined with safety, for improved efficiency and also for less noise and less pollution of the environment. The Commission therefore assessed the anticipated developments and took steps to undertake long-term planning of facilities and services and to commission or carry out the necessary research and investigations. Important topics referred to and discussed were the detection and

Meteorological briefing at Antananarivo Airport

forecasting of clear-air turbulence, Slant Visual Range and vertical wind shear in the lowest 60 metres of the atmosphere. It was also recognized that all-weather operations would put greater emphasis on the representativeness of observations and the accuracy forecasts for take-off and landing.

In the end the session provided a valuable period for reflection which seemed to be effective in producing a great upsurge of interest in and enthusiasm for the problems of aeronautical meteorology. The main foundations for the work of CAeM were laid for many years ahead.

Later sessions had much to do in revising and bringing up to date the regulatory documents published by both ICAO and WMO. Another organizational project was to bring into operation the Area Forecast System (AFS) and then the World Area Forecast System (WAFS).

A new NAOS Agreement promoted by WMO

On 30 June 1975 the North Atlantic Ocean Station (NAOS) Agreement administered by ICAO came to an end and was replaced on the following day by a new joint financing Agreement under the auspices of WMO. The transfer of responsibility from ICAO to WMO has a special significance because, in promoting the new Agreement, WMO for the first time in its history convened a Conference of Plenipotentiary Delegations (at which the principal delegate of a participating country is vested by his government with full negotiating powers).

When World War II ended, international civil aviation expanded at an increasing rate. It had long been foreseen that the operation of transatlantic air routes between North America and Europe would require a network of weather ships covering the North Atlantic Ocean. The responsibility for such a network could not be undertaken by a single nation so ICAO brought together representatives of countries whose airlines intended to fly on North Atlantic air routes. The result was the first NAOS Agreement, which came into existence in 1948 and had as its primary purpose the provision of meteorological data to assist in the safe, regular and economic operation of airline services over the North Atlantic region. However, from the inception of the network its meteorological data were always widely disseminated on the WMO meteorological telecommunications circuits and thus contributed not only to aviation requirements but also to all other meteorological purposes. In the course of time aviation made impressive advances and cruising altitudes rose to levels above the prevailing weather. The value of the network to aviation therefore diminished steadily whereas its importance to meteorology, which itself had made substantial scientific progress, continuously increased for forecasting, climatology and research. It therefore gradually became increasingly clear that any long-term future for the NAOS network would rest with WMO rather than ICAO.

The NAOS scheme was not long in operation before it was beset with vital questions concerning finance and the availability of suitable ocean-going vessels. ICAO held a succession of meetings to consider these important practical questions and in 1954 the original NAOS Agreement was replaced by a new joint financing scheme which, remarkably, remained in force until 1975 although periodic meetings had indicated that the problems of maintaining the network as an aeronautical facility were becoming insurmountable. Eventually the seventh ICAO Conference on NAOS (Paris, 1972) decided that the system under the administration of ICAO should be terminated on 30 June 1975. It was also recommended that if the network remained in existence after that date, there should be a new Agreement developed by WMO and entering into force on 1 July 1975.

With the support of more than 20 countries WMO entered into a period of intensive efforts aimed at prolonging the life of the NAOS scheme from mid-1975 onwards. Already in 1972 the scheme was in jeopardy. Plans were being made to reduce the network from nine ocean stations to seven owing to a shortage of weather ships when suddenly the USA announced a phased withdrawal of its three ocean stations for budgetary reasons. It was therefore clear that any new NAOS Agreement would be concerned with a smaller number of ocean stations than in the past.

There was no doubt that a strong consensus of meteorological opinion considered that the continuation of a NAOS network was essential but the practical and financial problems seemed to predominate. These were examined in detail in a series of preparatory meetings arranged by WMO. The work of the Secretariat in assembling and distributing information received great encouragement from the WMO Executive Council*, which adopted resolutions stressing the need for a network of ocean weather ships to provide adequate meteorological services for the North Atlantic, Europe and surrounding areas.

The time factor had to be kept in mind in the interests of continuity and, in the latter half of 1973, the Secretary-General of WMO, after holding consultations with ICAO, convened a Conference of Plenipotentiary Delegations to conclude a new joint financing Agreement on North Atlantic Ocean Stations. Arrangements were made for the conference, which was co-sponsored by ICAO, to take place in the WMO Headquarters in Geneva in February 1974 and the draft of a new Agreement was included in the documentation.

The conference was attended by delegations from 26 countries and a number of international organizations. However, although the continuation of a NAOS network was supported unanimously and good progress was made on certain aspects of a new Agreement, some of the major problems could not be resolved in the time available. The conference therefore adjourned so that additional information could be gathered and national positions re-examined. After much work by the Secretariat and several information meetings, the Secretary-General reconvened the conference for the period 4–15 November 1974 in the WMO Headquarters. The second part of the conference was attended by delegations from 22 countries as Participating States, by observers from two other countries and by observers from various international organizations, including the Inter-Governmental Maritime Consultative Organization and the Intergovernmental Oceanographic Commission. The conference adopted the text of a joint financing Agreement and also produced a Final Act and a report. At a signing ceremony on 15 November presided over by the Secretary-General there were 21 signatures to the Final Act and four to the Agreement. The Agreement remained open for signature until 31 May 1975 and after that date remained open for accession.

The conference had a most unenviable task in finding a balancing point between the meteorological requirements and the practical considerations. Finally the required network was specified as follows:

Ocean station	Location	Operating party
C	52°45'N, 35°30'W	USSR
L	57°00'N, 20°00'W	UK
M	66°00'N, 02°00'E	Netherlands, Norway, Sweden
R	47°00'N, 17°00'W	France

* See note on p. 8.

Dr D. A. Davies, Secretary-General, addressing the Conference of Plenipotentiary Delegates (Geneva, 1974)

The Agreement entered into force on the appointed date but the long-term future of ocean stations deployed on the North Atlantic has always seemed insecure. To non-meteorologists it has generally appeared that satellite data have proved comprehensive in global coverage to the extent that expensive observing sub-systems such as weather ship networks were probably superfluous. However, meteorologists (whether engaged in forecasting, in climatology or in research) have accepted that the data provided by ocean stations—hourly surface observations, routine measurements of wind, temperature and humidity in the upper air, oceanographic observations, etc.—are essential and cannot be provided efficiently and economically by other means.

The transfer of NAOS from ICAO to WMO marked the end of the system as an essential aeronautical requirement and the beginning, under WMO, of a new phase as a general meteorological facility within the framework of the World Weather Watch. It is interesting to note that WMO, mainly through the Commission for Basic Systems (CBS), spent several years, ending in December 1988, in a detailed study of the operational WWW system required for the North Atlantic. The network design assessments resulted in specifications for a composite observing system for the North Atlantic. The range of sub-systems included three NAOS vessels as well as polar-orbiting and geostationary satellites, drifting buoys and so on.

Conclusion

The activities of CAeM are characterized by liveliness and enthusiasm, which are equally evident in the user of its products, civil aviation, with a common determination to be near the frontiers of technology and to make full use of any advances that give promise of improvements in safety, economy and efficiency of airline operations.

The Tenth Congress of WMO (Geneva, 1987) reviewed the Aeronautical Meteorology Programme and issued the following long-term objectives which are reflected in the Second Long-term Plan, 1988–1997:

(i) To provide meteorological services required by air navigation, and enhance, in co-operation with ICAO, guidance and assistance to achieve this objective;

(ii) To promote the further development of meteorological support to aviation, incorporating the advances in atmospheric sciences and new technology;

(iii) To initiate specific studies aimed at improving the accuracy of forecasts of meteorological parameters affecting the safety and economy of air navigation;

(iv) To assist in the development of observing and information distribution systems at aerodromes;

(v) To provide, in co-ordination with ICAO, specialized training in aeronautical meteorological forecasting and briefing practices to ensure a high level of standard services for aviation.

A promising start has been made in implementing this Plan.

CHAPTER VIII

AGRICULTURAL METEOROLOGY PROGRAMME

Introduction

The close relationship between agricultural production and weather and climate is too evident to need stressing. The need to apply meteorological knowledge and skills to assist agricultural activities has therefore been recognized at all times as being of high importance. Indeed, the various aspects of agricultural activities require a wide range of meteorological information to be made available: from micrometeorology, with its emphasis on instrumental observation, through weather forecasting for short periods ahead, to extended ranges such as monthly and seasonal forecasts, and finally, towards the end of the spectrum, to climate, climate variability and climate change. Before Meteorological Services came into existence, workers in agriculture, in common with sailors, had no choice but to acquire, consciously or unconsciously, as wide a knowledge as possible of weather and its variations, including seasonal changes, and of climate. As is well known, weather lore, which reflects the experience accumulated over generations, has originated largely from farmers, shepherds and mariners.

From the formation of IMO in 1873 and, more particularly, since its replacement by WMO as an intergovernmental body and a Specialized Agency of the United Nations, agricultural meteorology has developed along scientific and technological lines, leaving weather lore well behind. There was never any question that almost every aspect of agriculture—from strategic decisions in long-range planning affecting land use and selection of crops to the various tactical decisions that have to be taken in day-to-day work—is dependent upon climate and weather. Large areas previously considered barren have been made productive as a wider understanding has emerged of the significant meteorological factors involved in plant growth and as more has been learned about the controlled use of available water resources. However, as well as the positive or promotional side, there is a negative aspect, and meteorologists are also deeply involved in the struggle against disasters. Such weather events as droughts, floods and frost may cause crop failures and lead to famine in many parts of the world. Notable examples are the effects of recent drought in parts of western and eastern Africa and of floods in Asia.

Since about 1960, the United Nations and the Specialized Agencies concerned with the natural environment have regarded world food production as one of the most pressing problems confronting mankind. The whole subject was discussed and plans for action were prepared at the UN Conference on World Food held in Rome in 1974.

WMO had already several times* chosen subjects connected with agriculture as themes for World Meteorological Day, an annual event which takes place on 23 March, the anniversary of the establishment of WMO, and is designed to draw worldwide attention to a subject of major importance. In 1984 the theme selected for the occasion was *"Meteorology aids food production"*, which was highly appropriate because by that time it was being widely recognized that the progressive development of agriculture was dependent upon the application of modern scientific methods. Furthermore, although the role of agricultural meteorology had expanded, and was continuing to expand, it would always be worth while to emphasize that if agrometeorological information and advice were made available to planners and farmers, and properly taken into account, serious losses could be avoided and agricultural production increased significantly.

Meteorologists engaged in applying the science of the atmosphere to agriculture very quickly realized that this whole

* *"Weather and agriculture"* (1968); *"Weather and food"* (1976); *"Meteorology aids food production"* (1984).

Technicians in meteorology at the Regional Training Centre, Niamey, removing soil samples to test for water content and assessing the establishment of seedlings and of crop response to various agrometeorological factors such as soil temperature and water content of topsoil

activity was multidisciplinary. Meteorologists, specialists in their own field, found themselves working in combined undertakings with experts not only in agriculture but also in forestry, livestock husbandry, hydrology, rural development, economics and the social sciences. Agrometeorological departments of national Meteorological Services played a key role in these interdisciplinary projects whilst WMO, through its Commission for Agricultural Meteorology, promoted the transfer of knowledge in agrometeorology to all countries. WMO has carried out this work within the framework of its Agricultural Meteorology Programme and its various training activities as well as through its Technical Co-operation Programme, which provides essential resources to developing countries.

A notable feature of WMO's activities in agrometeorology is the co-operation that has developed with other organizations such as the UN Food and Agriculture Organization (FAO)—like WMO a Specialized Agency of UN—and various international agricultural research institutes, national agricultural establishments and individual specialists. Some of this co-operation has been related to specific problems in jointly operated projects but much of it is officially based and comprehensive. For example, as early as 1954, FAO, in conference with all its Member countries, adopted a resolution noting the work undertaken by WMO in the application of meteorology to agriculture and food production throughout the world, and called upon governments to arrange co-ordination between their agricultural and meteorological services in order to ensure the practical implementation of their joint programmes.

When the Ninth WMO Congress (Geneva, 1983) decided to establish *inter alia* an Applications of Meteorology Programme, as one of the major programmes of WMO, the inclusion of the Agricultural Meteorology Programme was inevitable in view of the range, scientific scope and importance of its activities.

The Commission for Agricultural Meteorology (CAgM)

CAgM was one of eight Technical Commissions created by the First Congress of WMO, held in Paris in 1951. The terms of reference laid down by Congress included the following subjects of particular importance:

- Weather forecasts to meet the requirements of agriculture;
- Weather and climate in relation to animals and plants;
- Weather and climate in relation to soils and vegetation;
- Meteorological phenomena, including condensation and evaporation, relating to the bare soil and cultivated areas;
- Artificial influences on weather conditions in relation to agriculture.

Successive Congresses reviewed the terms of reference of CAgM and made necessary amendments and also gave guidance as to priorities. Third Congress (Geneva, 1959) noted the wide range of studies which the Commission had undertaken and which were being carried out by working groups and rapporteurs. Congress was also informed that FAO was planning to organize a "Free the world from hunger" campaign in 1963 and requested CAgM as well as the Secretariat to give full co-operation to FAO in this programme. In addition to these matters, Congress gave directions on three subjects of general application but of special interest to activities in agriculture, namely:

- Arid land research and development;
- Meteorological work in the humid tropics;
- Research into tropical meteorology.

The Agricultural Meteorology programme, as it existed in the late 1980s, was adopted in broad terms by Fifth Congress (Geneva, 1967) after its review of the activities that had been carried on by CAgM since the formation of WMO as an intergovernmental body. The programme approved by Congress was in three parts: basic requirements, including data and fundamental and applied research; local requirements, including operational systems and provision for warnings of very adverse conditions; and national and international requirements, including development planning and education and training. Congress also confirmed its agreement to the establishment of a panel consisting of representatives from FAO, Unesco, UNDP and WMO. The responsibility placed on the panel was to develop and implement an agrometeorological programme in aid of food production and to ensure that collaboration took place in any project involving experts from several disciplines.

CAgM held its seventh session in 1974 with the knowledge that later that year UN would be holding a World Food Conference (see p. 53). The Commission made arrangements for WMO to participate fully in the conference and set up a Working Group on Weather and Climate to advise Congress and the Secretary-General on the measures required for WMO to co-operate in all relevant recommendations approved by the conference.

Seventh Congress (Geneva, 1975) endorsed all the commitments arising out of the recommendations of the World Food Conference and also decided that more emphasis should be placed on training in efforts to assist national Meteorological Services to co-operate fully and effectively in all programmes involving agrometeorology. Approval was therefore given for short- and medium-term expert missions to developing countries and the concept of roving seminars was brought into the Agricultural Meteorological Programme. Eighth Congress (Geneva, 1979) requested CAgM to give special attention to the problems of desertification and established the World Climate Programme, comprising four components: the World Climate Data Programme, the World Climate Applications Programme, the World Climate Impact Programme and the World Climate Research Programme. The succeeding Congress in 1983 adopted a "Food" component of the World Climate Applications Programme as one of the priority areas to be closely co-ordinated with CAgM.

A Sudanese meteorologist using an Assman psychrometer to make microclimatic observations in a cotton field

From 1974 onwards a number of studies undertaken by CAgM came to fruition and the resulting reports were given wide distribution in WMO publications. The subjects treated included meteorological aspects of plant and animal production, plant pathology, biological aspects, forest fire services, wind breaks and shelter belts, forecasting requirements for aircraft spraying and numerous topics in agroclimatology.

An important section of the Agricultural Meteorology Programme has been devoted to activities in an Inter-agency Group on Agricultural Biometeorology. International discussions of problems of the biosphere were initiated in 1968 and since that time the relevance of the subject to all human welfare has been increasingly accepted. Founder members of the inter-agency group were FAO, Unesco, UNDP and WMO. An additional member of the group was the UN Environment Programme (UNEP), which was established by the UN Conference on the Human Environment (Stockholm, 1972) and was subsequently invited to join the group. Over the years the group has carried out valuable agroclimatological surveys, some examples of which are as follows:

- Semi-arid and arid zones of the Near East;
- Semi-arid zone in Africa south of the Sahara;
- The highlands of East Africa;
- The Andean zone;
- South-east Asia;
- Humid tropical lowlands of South America.

Tenth Congress (Geneva, 1987) carried out a comprehensive review of the work of the Organization in the field of agriculture and provided guidance on long-term objectives for the Agricultural Meteorological Programme which may be summarized as follows:

- To help increase agricultural production and improve its quality;
- To keep users informed concerning up-to-date knowledge of relations between meteorological factors and agricultural, forestry and livestock production;
- To assist in the establishment of agrometeorological data banks for application to problems in all sectors of agriculture;
- To assist national Meteorological Services to develop, in co-operation with agricultural services, scientifically devised operational techniques;
- To promote agrometeorological research;
- To co-operate in the monitoring of drought and desertification.

Services performed by CAgM

Agrometeorology is relatively young as an applied science but it already has many achievements to its credit. Its practitioners have been highly successful in explaining to actual and potential users the economic benefits of agrometeorological services. Compared with some other applications of meteorology, agricultural workers and those engaged in research and development needed little persuasion about the need of information and advice on weather and climate. The growth of agrometeorological services since about 1960 has been remarkable, particularly in regard to the production and protection of crops and animals. The nature, extent and supply of agrometeorological information have varied regionally and nationally and have depended to a great extent upon the level of economic development of the countries concerned. The services required have been controlled not only by climatological, meteorological and hydrological conditions but also by long-established agricultural practices, although these have tended to evolve as greater advantage has been taken of prevailing conditions.

Rice terraces in the Philippines

CAgM has been very active in spreading knowledge by means of the preparation and distribution of reports drawn up by its working groups and rapporteurs. Member countries of WMO have also been encouraged to compile national progress reports on the applications and value of meteorological information. All this material has covered the agrometeorology of farming practices in every climate region. In addition to the production of crops and animals, such topics as plant pests and diseases, drought assessment and monitoring, the struggle against desertification, forestry, and the effects of pollution and atmospheric gases on crop production have been covered.

UN Conference on World Food

The United Nations convened a Conference on World Food which took place in Rome in November 1974. WMO played a full part in the preparations for the conference and in the proceedings. Aspects which WMO highlighted were that the variability of weather and climate caused important fluctuations in food production and that more comprehensive use of meteorological information in all phases of agriculture was essential.

At the time of the conference there were serious food shortages in a number of countries but arrangements, mainly bilateral rather than international, were in progress to alleviate the conditions to the extent possible. The conference therefore gave most of its attention to the longer-term problems of finding ways and means of increasing world food production and of keeping track of the global food situation through the establishment of a world food security system. In a resolution dealing with a global information and early warning system for food and agriculture, WMO was invited to co-operate with FAO in setting up a scheme to utilize meteorological data to assess crop yields in key areas of the world. WMO was also requested to undertake urgent research in crop/weather relationships so that the early warning system could be in a constant state of development.

Another resolution called upon WMO to co-operate with other international organizations in assessing on a world-wide basis the potential for additional land areas to be used for agriculture. In this task the main problem for CAgM was to study the climatic limitations of any areas selected for consideration. Scientific water management formed the subject of a separate resolution in which FAO and WMO were urged to carry out desert-encroachment surveys, promote irrigation schemes and designate areas where flood control systems were required.

More than half of the many resolutions adopted by the conference had a bearing on the activities of CAgM. Seventh Congress (Geneva, 1975) considered the matter and the items relevant to the work of the Organization were incorporated with slight adaptation into the Agricultural Meteorological Programme.

At this Conference on World Food the UN Specialized Agency most deeply involved was, of course, the Food and Agricultural Organization. After the conference, FAO adopted at its nineteenth session a resolution on co-operation with WMO and gave guidance to its Director-General in the following terms:

(*a*) To ensure that the meteorological aspects of FAO projects be given due consideration, in consultation with WMO, with a view to ensuring that, in the case of those FAO projects in which weather is seen to be an important factor, a meteorological component be included in the planning and implementation stages of the projects;

(*b*) To seek the collaboration and advice of WMO in matters related to the meteorological input to agricultural projects;

(*c*) To continue and, where necessary, reinforce the close relations with WMO in matters of mutual interest in the field of agriculture;

(*d*) To continue, within the limits of budgetary provisions, the practice of jointly convening and financing FAO/WMO Conferences on Agriculture and Meteorology and to invite WMO co-sponsorship of FAO meetings in which there is a meteorological component.

Locust control

In the period 1955–1960 WMO carried out a project in East Africa studying the relationship between meteorological factors and the breeding of locusts and their transport by the wind. A report on this project was published by WMO as a *Technical Note* under the title "Meteorology and the migration of desert locusts". The experience thus gained by WMO proved very valuable when in 1960 it was invited to collaborate with FAO in a Desert Locust Project which was set up and funded by the United Nations. The project involved various aspects of meteorology, including climatology, special observing procedures and the preparation of weather maps. WMO arranged for ships and observing stations in areas of major concern to provide weather reports and also to notify any sightings of locust swarms. Training was a major component of the project and, in association with FAO, WMO held a seminar in Tehran in 1963.

The menace of locusts remains a matter of major concern and the combat continues. Over the years WMO has provided climatological and current weather information for control operations in areas such as the Red Sea, the Horn of Africa, western Africa and western Asia.

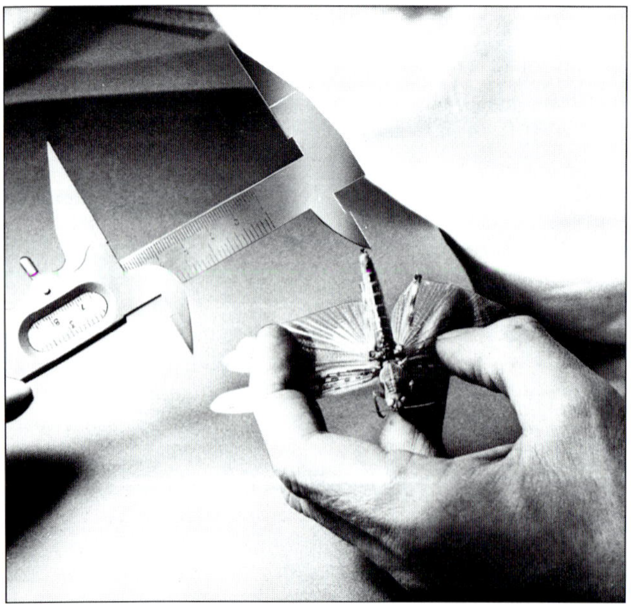

Examination of a locust

UN Conference on Desertification

Desertification is understood to indicate an expansion, in a region, of conditions which tend to a reduction in the capacity of the land to support livestock and crops. The subject aroused great anxiety in the United Nations, mainly about developments in certain parts of Africa although areas in other continents were also affected. The United Nations organized a World Conference on Desertification, which was held in Nairobi in August/September 1979. The subject concerned several of WMO's major programmes and the corresponding Technical Commissions, notably the Commissions for Climatology, Hydrology and Agricultural Meteorology. Land management and the planning of land use were of particular concern to the latter.

For a long time it had been assumed that climate change was the cause of desertification and that little in the way of preventive action would be worth while. However, a series of studies, including case histories of areas where desertification presented acute problems, demonstrated that although climate was an important factor, climate change was not the main element; it became evident that over-exploitation of the land, whether in use for agricultural or pastoral purposes, in an area where the ecological balance was known to be delicately poised, was a primary cause. The case studies showed that desertification could be halted and waste land reclaimed provided land use was carefully planned, account taken of weather and climate, and efficient control exercised over pastures, livestock and afforestation.

Many of the recommendations made by the conference were therefore naturally directed at WMO. The plan of action adopted by the conference contained a number of features which were easily fitted into the WMO Agricultural Meteorology Programme and others. Eighth Congress accordingly approved a WMO Plan of Action on the meteorological and hydrological aspects of the combat against desertification.

Economic benefits

The importance of meteorology to agriculture has never been in question. Nevertheless, a number of national Meteorological Services in conjunction with agricultural interests have carried out cost/benefit studies and have found that meteorology, applied efficiently over all relevant sectors of agriculture and supported by a strong research effort, can yield quite remarkable advantages expressed in monetary terms. These results have encouraged governments to strengthen the agrometeorological departments of their national Services. Not every project is measured in commercial terms; in certain examples, such as the encroaching deserts and recurring droughts, a compelling

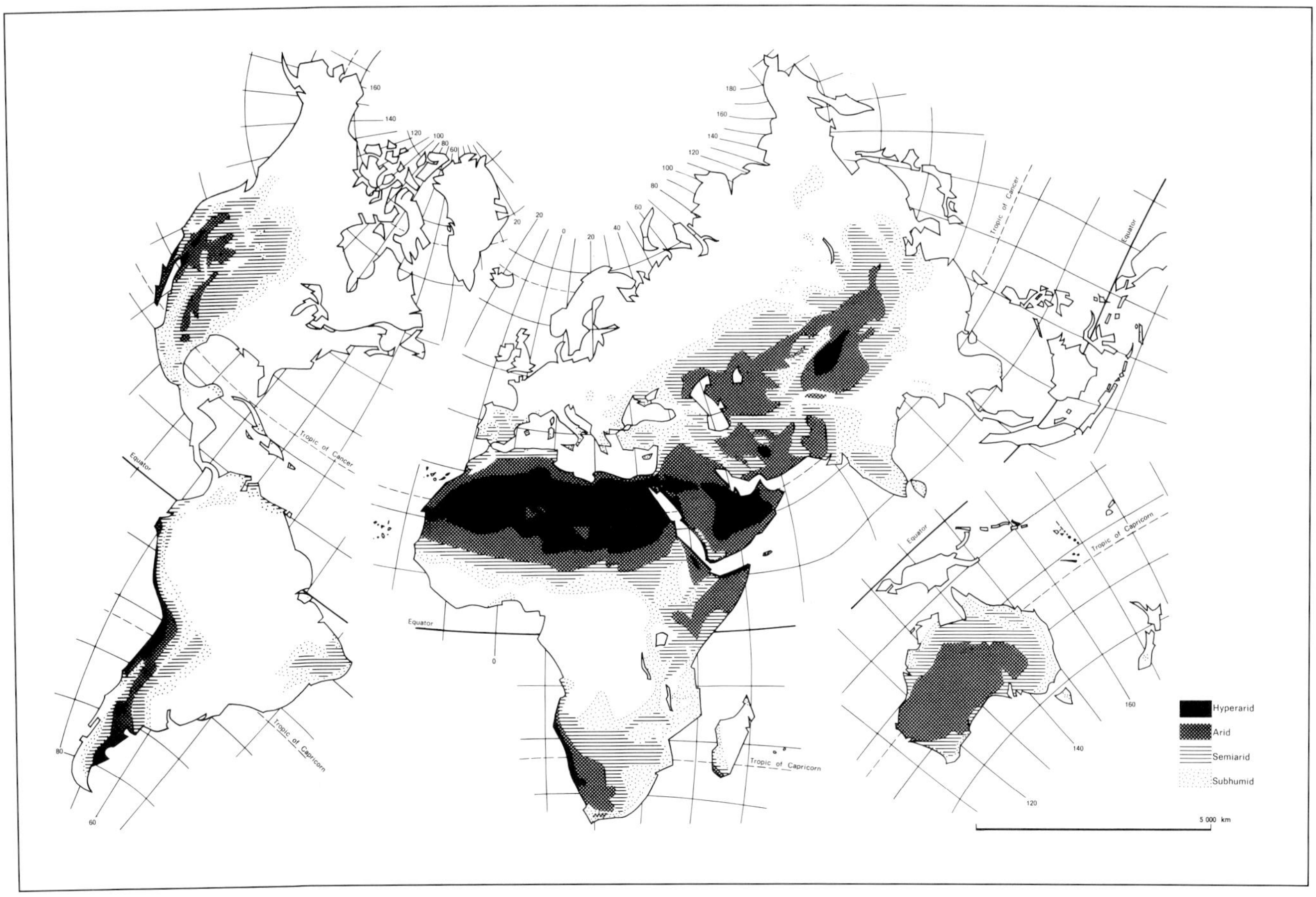

World distribution of arid regions

urgency exists to save lives and this represents the driving force which, if successful, is of immeasurable value.

One of the largest projects managed by WMO has been known by the acronym AGRHYMET, which signifies the presence of agriculture, hydrology and meteorology. The project was a regionally co-ordinated programme aimed at providing the Sahel countries with agrometeorological data and advice to strengthen food production and to reduce the adverse impact of droughts. The programme was supported by a number of donor countries contributing funds and expert advisers, and the Sahel countries have testified to the value of the results achieved by the programme. AGRHYMET has indeed become a quasi-permanent organization with a centralized or regional component at Niamey (Niger) and national components in all Sahel countries (Burkina Faso, Cape Verde, Chad, the Gambia, Guinea Bissau, Mali, Mauritania, Niger, Senegal). In the mid-1980s other regional co-operative projects, inspired by the success of AGRHYMET, were initiated, particularly in the form of drought-monitoring centres located in Nairobi (Kenya) and Harare (Zimbabwe).

Of wider purpose but potentially of very great value to agriculture in Africa is a project covering the whole of that continent for the establishment of the African Centre of Meteorological Applications for Development (ACMAD). This project is the outcome of studies made by the United Nations Economic Commission for Africa (ECA) of the widespread droughts of the 1970s and 1980s which affected some 35 African countries. ECA held a ministerial conference in April 1983 which adopted a resolution to convene a scientific symposium in order to examine causes of droughts and their effects on African national economies and to propose measures to deal with the problem. ECA held consultations with WMO and with many other organizations of the United Nations system and eventually a draft was prepared of a regional plan to combat the impacts of drought. The draft plan was considered at a meeting in February 1984 when experts from 25 African countries reported on the droughts experienced by their countries and the consequences.

The regional action plan was adopted by the ECA Conference of Ministers at which the Secretary-General of WMO gave a discourse on the scientific aspects of drought problems in Africa. The ministers also requested the Executive Secretary of ECA to consult WMO and other UN Agencies concerned about the feasibility of establishing an advanced centre for meteorology in Africa. The results of the consultative process were presented to the ECA Conference of Ministers at their meeting in April 1985 and it was then decided that ACMAD should be established. ECA will be the responsible authority for the centre whilst WMO is the main associated Agency; the Meteorological Services of all countries in Africa are also participating.

The decision to set up ACMAD must be regarded as imaginative and far-reaching. Agriculture will clearly have a prominent place but numerous other sectors of each country's economy will require meteorological support in a variety of forms: data, weather forecasts and climatology.

Conclusion

Problems for the agrometeorologist are so numerous that it would require a large effort to list each and every one. The whole task of the agrometeorologist is most challenging and the prime necessity is to be forward-looking because the users of meteorological information—plant breeders, experts in pests and diseases, farmers and many others—are constantly making significant progress, with the result that new meteorological requirements emerge and long-standing ones become more specific and demanding.

CHAPTER IX

HYDROLOGY AND WATER RESOURCES PROGRAMME

Introduction

The formal extension of WMO's responsibilities into the field of hydrology was accomplished after lengthy and at times contentious debate spread over many years from about 1946. IMO frequently discussed, without apparently seeking a firm conclusion, whether certain branches of hydrology, in particular those activities falling into the wide sector known as operational hydrology, should be regarded as lying within the boundaries of WMO's functions or just outside. A decision on the question would have had limited force since IMO did not enjoy government status. In practice WMO, as well as its predecessor IMO, has been deeply involved in hydrology, as certain components of the hydrological cycle (see diagram below) consist of meteorological elements—namely, precipitation in its various forms and evaporation. As scientific and applied disciplines, meteorology and hydrology each have numerous branches. Some branches of meteorology are closely related to some in hydrology. The data, forecasts and advice which the meteorologist provides over a wide range of meteorological services with special reference to rainfall and evaporation are indispensable to the hydrologist over almost the whole field of his activities, whether these be directed towards basic research or to the applications of hydrology in so many sectors of everyday life.

Both meteorologists and hydrologists provide data and associated information and advice for use in operational management and in planning and research over a wide range of activities. Co-operation has existed from time immemorial and has been to mutual benefit. The role of the meteorologist in providing information has been valued by the hydrologist for his application of that information in hydrological forecasting, water resources assessment and a number of other purposes. This close co-operation has engendered a team spirit between hydrologists as customers and meteorologists as providers. The term *hydro-meteorologist* came into use referring either to a hydrologist with an adequate knowledge of meteorology, but still a hydrologist, or to a meteorologist who, without departing from his primary discipline, had made a close study of the manner in which the hydrologist requires and applies meteorological data. The question of giving official recognition to the interrelationship between meteorology and hydrology was to remain a lively one until a satisfactory decision was reached and implemented.

In 1946, shortly after the end of World War II, the directors of national Meteorological Services, using IMO as the focus,

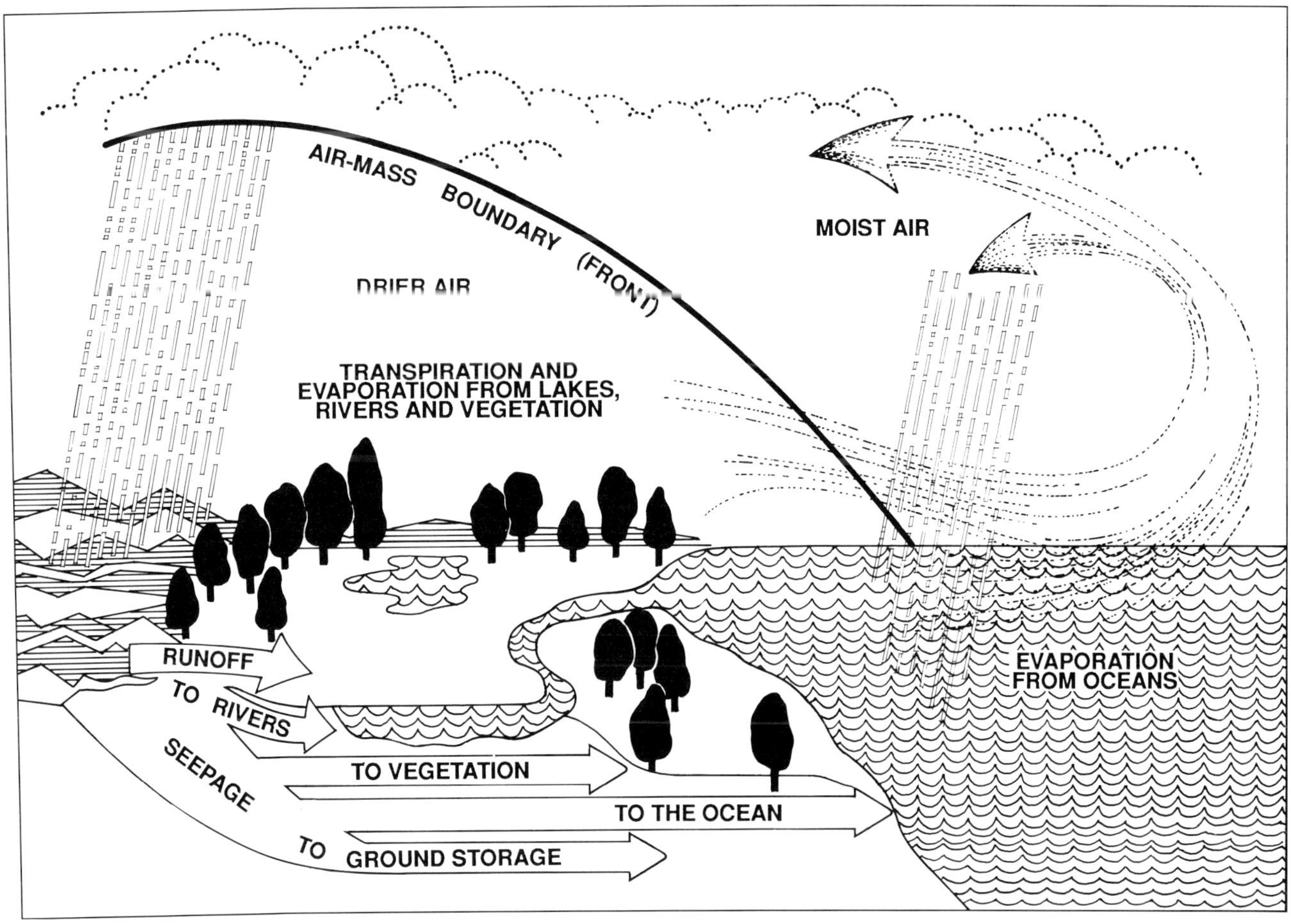

Pictorial representation of the hydrological cycle

embarked upon the revival of international co-operation in meteorology. The structure of IMO was restored and, where desirable, amended. A Hydrological Commission was established as one of the constituent bodies. However, when WMO replaced IMO a few years later it did not include among its Technical Commissions a separate one for hydrology and instead distributed its interests in hydrology among designated Commissions such as those for agriculture and climatology.

After this lapse of time it seems difficult to understand why WMO in setting up its new organization did not maintain the Hydrological Commission established a few years earlier by IMO in its declining years. To try to assess the reasons would merely be an exercise in speculation and it may be sufficient to mention two factors. First, WMO was established as an intergovernmental body and as a result the delegates to WMO were accredited representatives of their governments, working to official briefs. (This was a marked change from their position when they assembled in IMO as an informal gathering of directors of national Meteorological Services.) Second, in the late 1940s and during the 1950s international bodies were reforming after the enforced suspension of activity caused by the outbreak of World War II in 1939. Moreover, a number of new bodies, either intergovernmental or non-governmental, were being set up and were in the processes of growth and identifying their roles on the international scene. WMO would not have wished to forestall any developing plans for the organization of hydrology. Whatever the reason, the lack of a Commission for Hydrology in WMO's new structure presented the meteorological and hydrological communities with a topic for continuing and lively discussion since there was no questioning either the existence of or the necessity for the closest collaboration between the two disciplines. From the date of the establishment of WMO, therefore, there was strong pressure for hydrology to be accorded recognition by WMO through the setting up of a Technical Commission that would be responsible for all WMO activities in hydrology instead of having this important work distributed among other commissions.

There was, however, a difficulty to be overcome. Formal action on the part of WMO to incorporate hydrology into its field of responsibility would require amendments to the WMO Convention. This Convention, which has the status of a treaty between governments, defines the international objectives of WMO and outlines the ways and means of pursuing them. Conventions of this nature change, if at all, in an evolutionary manner and are not subject to amendment unless proposals have been examined and discussed at considerable length over a period of time.

Despite these factors, from the perspective provided by forty years, the decision of First Congress not to set up a Commission for Hydrology looks rather strange. For if this decision had been reached, hydrologists would probably have worked successfully within WMO to the degree that a change in the Convention to encompass hydrology might not have been considered necessary. However, the lack of action by First Congress (and several subsequent ones) on this matter resulted in an accumulation of pressure for hydrology to feature in an amended Convention. This change, which was eventually made in 1975, brought hydrology and its applications—such as water resources management—formally into the field of activity of WMO, to the great benefit of the Organization, the science and those involved in it. It must be left to speculation whether an earlier change in the Convention would have provided even greater benefit.

The foregoing explanation is perhaps open to question because it has been suggested that if First Congress had set up a Commission for Hydrology, the delegates would have deserved praise for their farsightedness, a complete contrast to what has been said above. At the time the issues were numerous, conflicting and complex. Taking one consideration alone, however, it does appear that the delegates were justified in avoiding any step that might have prevented, directly or indirectly, the establishment of a separate international organization for hydrology.

Caution at First Congress (1951) and Second Congress (1955)

As has already been indicated, when the transition from IMO to WMO occurred, the new organization at its First Congress (Paris, 1951) retained a number of IMO's Technical Commissions but left in abeyance for further study the possible requirement for a Commission for Hydrology. In subsequent Congresses a number of delegates expressed the view that the abolition of the IMO Hydrological Commission seemed to have been a mistake. Nevertheless, as a small step forward, Congress passed a series of resolutions stressing the importance of co-operation between Meteorological and Hydrological Services and drawing attention to various areas where joint activities would be beneficial.

There were many reasons to justify WMO delaying such an important development as the incorporation into its official responsibilities, which were already very wide, the task of organizing international co-operation in hydrology. One consideration, discussed in a lively manner over several years, was the possibility of creating a World Hydrological Organization which, like WMO, would become a Specialized Agency of the United Nations. The question eventually lapsed partly because two such agencies already carried out substantial work in the field of hydrology. These were WMO, mainly in operational hydrology, and the United Nations Educational, Scientific and Cultural Organization (Unesco), dealing mainly with hydrological research. Research in hydrology was also the concern of the non-governmental International Association of Scientific Hydrology (IASH) which in 1971 changed its name to the International Association of Hydrological Sciences (IAHS).

Other problems arose because of the organization of meteorology and hydrology in various countries. Although some countries had established combined Meteorological and Hydrological Services, often referred to as Hydrometeorological Services, there were a number of countries where both meteorology and hydrology were split among several government departments. For example, separate meteorological units were sometimes to be found in government departments concerned with civil aviation, agriculture and defence. Departments also had their own staffs dealing with relevant aspects of hydrology, e.g. water resources, flood control and so on. In such countries the directors of Meteorological Services, who were acutely conscious of the urgent need for progress in the science and applications of meteorology, were naturally anxious about WMO expanding officially into new fields with the risk that its limited funds would become so thinly spread that progress in high-priority areas of meteorology would be slowed or even halted. From the end of the Second World War, countries were determined to restore their economic and industrial strength and in consequence the demands upon directors of national Meteorological Services were extremely exacting and extended to numerous sectors of each country's activities. It was perhaps only to be expected therefore that the WMO Convention, adopted at First Congress, made no specific reference to hydrology; but that subject was clearly implicit in such phrases in the Convention as "geophysical observations related to meteorology" and applications of meteorology to "human activities". Furthermore, the Commission for

Climatology established at First Congress was given terms of reference which provided for "the arrangement of climatological data in order to meet the needs of hydrology".

Thus First Congress left undecided the high-policy question of WMO's official involvement in hydrology. At the same time, it need hardly be said, at working levels meteorologists and hydrologists in every country were collaborating to the extent which each required. However, the benefits of international co-ordination were still to be harvested and exploited in such important matters as observing standards, hydrological forecasting techniques, network design and training. The reservations of Congress were a reflection of both IMO's and WMO's preference for proceeding only when a wide measure of agreement had been reached rather than by a vote. If the hydrological question had been put to a vote at First Congress, the result might well have been close, leaving a substantial minority dissatisfied and perhaps sceptical of the eventual outcome.

After First Congress the arguments for WMO to become much more active in hydrological activities began to gather strength and, in the long run, were to prove irresistible. The Regional Associations of WMO were greatly concerned and at their meetings took the opportunity to pass resolutions on such topics as hydrological networks, the collection and exchange of data and the need for co-ordination between Meteorological and Hydrological Services. One of the Associations pointed to the importance of providing guidance to all countries on professional and technical standards to be observed in the various procedures, and recommended that WMO should urgently consider the establishment of a Commission for Hydrology. In addition to the Regional Associations, the Commission for Climatology discussed a number of hydrological questions although well aware that many aspects of operational hydrology could not be dealt with satisfactorily in a forum where the primary interest was climatology.

External pressures were also being applied to WMO. In 1954 the influential Economic and Social Council (ECOSOC) of the United Nations drew the attention of Specialized Agencies of UN to the importance of hydrology and suggested they should collaborate with governments to organize and strengthen international co-operation in the field of water resources management and utilization. It was further proposed that one of the Specialized Agencies—namely WMO—should undertake the responsibility for the collection and dissemination of hydrological data, working closely with national Hydrological Services and with IASH, the International Association of Scientific Hydrology. This proposal was widely supported at a meeting arranged by ECOSOC in 1954 for the discussion of water resources development. It was seen to be logical to persuade WMO to undertake the role proposed because it was well known that the Organization had vast experience in the assembly and broadcasting of meteorological data and associated information. It was also considered that the world-wide telecommunication facilities used for exchanges of meteorological reports might also be used to convey operational hydrological information to all users.

During this period, quite early in the existence of WMO, it was becoming clear that, as a Specialized Agency, WMO would be more and more involved with UN affairs, notably with the activities of regional economic commissions and also with the expanding programme of technical assistance to which the UN attached major importance. A feature of this involvement was that WMO received appeals for assistance in hydrological matters as well as in the field of meteorology.

These events with their mounting pressures were clearly highly significant for the future of WMO and accordingly the Secretary-General prepared a report for the Executive Council.* The report was of wide scope but laid special emphasis upon the discussions in ECOSOC and the resulting proposals for WMO to play a leading role in hydrology. The time was approaching for WMO's Second Congress, arranged for the spring of 1955, and the Executive Council concluded its consideration of the Secretary-General's report by requesting him to prepare a document for Second Congress on the policy, principles and future programme of WMO in regard to water-resource development. The paper prepared by the Secretary-General discussed in detail the various aspects of hydrology in which WMO might become involved and submitted the following recommendations:

(*a*) WMO should accept the responsibility of being the UN Specialized Agency dealing with the meteorological and, to a lesser extent, the hydrological aspects of the water-resource development programme of the United Nations;

(*b*) On the hydrological side, WMO should be prepared to meet specific requests made by UN or by other members of the UN family;

(*c*) The WMO programme in meteorology and hydrology should include for both disciplines promotion of the collection of data, the establishment of international standards, the sponsoring of technical conferences and the provision of technical assistance.

These recommendations would have gone far to meet the pressures, building up within WMO and being applied by external sources of considerable standing, for WMO to undertake the leading role in what may be described as the operational aspects of hydrology. However, Second Congress decided to move cautiously and its decisions fell some way short of the Secretary-General's proposals. It accepted that WMO should become the Specialized Agency responsible for those aspects of the UN's water-resource development programme "which fall into the common ground between meteorology and hydrology". Congress therefore requested the Executive Council* to construct a WMO programme, in line with the policy directive and aimed at meeting the needs of the UN family and providing the assistance and guidance required by the Members of WMO.

Panel on Water Resources Development

Immediately after Second Congress the Executive Council* of WMO held a meeting—its seventh session—for the purpose of planning its work programme for the implementation of the decisions reached by Congress. In regard to hydrology, the Council, strongly supported by the delegations of the USA and the USSR, decided to form a panel of experts to prepare a draft programme for WMO's participation in water resources development and to advise the Secretary-General on all matters related to the Organization's activities in the field of hydrology. Looking back to 1955 and considering the history of international hydrology since that time, it seems appropriate to commend the Executive Council in the warmest terms for establishing this panel and for its composition. The six experts—two appointed by the Executive Council, two representing the Commission for Climatology, one representing the Commission for Agriculture and a representative of IASH—were widely known and of the highest standing. They set about their important and complex task with a sense of urgency and determined to produce a constructive and realistic report. They had few, if any, of the inhibitions or misgivings that were so

* See note on p. 8.

noticeable among some of the Permanent Representatives of Member countries with WMO. The panel was united in its opinions as to what was required for the international co-ordination of hydrology and also as to the very prominent part which WMO should undertake. In effect, the panel defined a policy objective in the following terms: "WMO should assume responsibilities in the field of hydrology similar to its present responsibilities in the field of meteorology".

In the four years of its existence the panel held two meetings, in July 1956 and in March 1958, and at other times maintained a high level of activity by correspondence. There was full discussion and agreement on the amendments to the WMO Convention that would be required if Hydrological Services were to be placed on an equal footing with Meteorological Services. The principal amendment proposed by the panel was that WMO should be renamed the "World Meteorological and Hydrological Organization". The panel also decided to press strongly for WMO to form a Commission for Hydrology and to provide machinery for close collaboration with IASH.

After each of its two sessions the panel submitted a report to the Executive Council.* The reception was somewhat mixed. The first report was strongly endorsed by the Executive Council, which asked the Secretary-General to distribute copies to Member countries so as to give them an opportunity to express their views. Thirty-eight Members replied to the Secretary-General's letter and of these 25 took a favourable attitude to the panel's proposals whilst 13 Members were opposed in varying degrees from "not quite satisfied" to fairly firm disagreement. At the tenth session of the Executive Council (1958) the role of WMO in hydrology was again considered in the light of the comments received from the 38 Members and a fresh report from the panel which had recently held a second meeting. This time the Executive Council was divided, some members being reluctant to accept the panel's recommendation that the WMO Convention should be amended in such a manner as to place hydrology on an equal basis with meteorology. The majority of the Council were prepared to see WMO advance carefully and steadily over a wide range of hydrological activities but some members had retreated to the position that WMO's formal incursion into hydrology should be strictly limited to that discipline's common ground with meteorology. After considerable discussion the Executive Council* agreed unanimously to propose to Third Congress that the Organization should accept responsibilities in all aspects of hydrology "which involve meteorological considerations". This proposal was vague and must have been discouraging to the clear-sighted and forward-looking panel. Nevertheless, the door to progress along the lines desired by the panel remained open—if not widely at least ajar—and it also appeared that the Council would invite Third Congress to establish a Commission for Hydrology. The panel appreciated that, although these developments were headed in a positive direction, no lasting changes would occur unless and until the reasons for WMO to take over international hydrology became so strong as to justify changes in the Convention. On this aspect the Executive Council decided that any recommendation would be premature.

The panel's efforts to persuade WMO to take hydrology under its wing, therefore, suffered a setback by forces within WMO itself, the Executive Council* feeling unable to move decisively until all interested Member countries showed themselves ready to reach a fairly complete measure of agreement. The internal WMO picture was that progress towards hydrology was delayed, not abandoned. External forces then came forward to sustain the efforts of the panel. In July 1958, shortly after the tenth session of the Executive Council, a meeting of the UN Specialized Agencies on the development of water resources agreed that WMO should be responsible not only for rainfall and evaporation in the hydrological cycle but also for elements related to surface water. (These elements comprise river stage and river flow, lake level, lake and river ice, glaciers, sediment transport, flow of springs, water temperature and water quality.) Evidently the other Specialized Agencies were concerned to encourage WMO to play an active role in these important aspects of hydrology. There appeared to be good prospects that Members of WMO would agree.

In due course the report of the inter-agency meeting came before the UN Administrative Committee on Co-ordination and the proposals relating to WMO were approved. This was a step of high significance. Somewhat earlier the Council of the non-governmental IASH recommended that WMO was the most appropriate UN agency to accept responsibility for hydrology. The recommendation was submitted to the parent body of IASH, namely, the International Union of Geodesy and Geophysics, and was approved by the Union's General Assembly.

Commission for Hydrological Meteorology (CHM)

By the time of the WMO Third Congress (April, 1959) it appeared, not only to the panel but to many others within WMO as well as to perhaps all hydrological interests external to WMO, that the establishment of a Commission for Hydrology could not be delayed much longer. As previously noted, WMO's activities in hydrology were centred rather inappropriately in the Commission for Climatology but mounting concern was being expressed in the Regional Associations and the Commission for Agricultural Meteorology. The whole arrangement could not provide an efficient machinery for international co-ordination of hydrology or for the promotion of high scientific and technical standards in a subject that was becoming ever more important in all the United Nations development programmes. However, with so much support coming from outside bodies of substantial influence, it seemed that the portents could not have been more favourable for WMO to accept important responsibilities in international hydrology.

In the event Congress, the policy-governing body of WMO, held back. The pressures applied from outside could not prevail against the indecision within. Decisions at Third Congress were conflicting and, as is customary when important subjects are debated at high levels, the greatest vigour was shown by the minority. This minority was preoccupied with keeping WMO strictly to its existing commitments and with safeguarding national Meteorological Services from too deep an involvement in hydrology. After this lapse of time and with international hydrology prospering as an integral part of WMO, the opposing arguments seem trivial and were, in fact, contrary to the evidence already existing. For example, the assertion that a Meteorological Service could not be adapted successfully to take on responsibilities in hydrology seemed to ignore the fact that a number of Member countries of WMO had long since combined their Hydrological and Meteorological Services and were very satisfied with the resulting arrangement. An interesting argument, but one that was far from new, was that operational hydrology consisted of a series of independent closed systems—i.e., an island's problems would not be affected by problems elsewhere; a continent might give rise to a need for co-ordination if several countries were drawing water from a common river, but such a problem would be regional rather than global. However, this type of argument ignored the benefits that would inevitably follow for all countries if high standards were adopted universally for training, instrumentation and observing techniques. Furthermore, exchanges of information on such

* See note on p. 8.

complex subjects as hydrological forecasting would be extremely valuable to all countries and provide a stimulus to further research.

In addition to some national Meteorological Services opposed to the further expansion of WMO into hydrology, a number of national Hydrological Services also had reasons of their own for wanting to keep their field separate from meteorology except, of course, for the obvious common ground. There were still some lingering hopes that a Specialized Agency for the development of water resources might be created and it was feared that this possibility, however remote, would be jeopardized if hydrology were to occupy a central position in WMO alongside meteorology. Some directors of Hydrological Services could foresee serious problems if, on moving from the national to the international scene, they became subordinate to the directors of Meteorological Services. It was, however, largely accepted that this well-founded objection would become of less importance if the WMO Convention were suitably amended, giving official status to hydrology as well as to meteorology in international affairs.

Third Congress, perhaps a little confused by the wide-ranging arguments, failed to move decisively in the most logical and most favoured direction and instead took a small step leaving open the possibility of choosing among numerous directions for any subsequent move. Congress merely agreed that WMO would co-ordinate activities in "hydrological meteorology" and in associated aspects of hydrology and water resources. The Commission for Hydrological Meteorology (CHM) was established and the Secretary-General was authorized to set up a section in the Secretariat to provide support for the developing work related to hydrology.

The decision was unclear as to the responsibilities to be undertaken. The term "hydrological meteorology" was quite unfamiliar to practising meteorologists and hydrologists and so its meaning was loosely understood rather than defined. The terms of reference of the new Commission referred to "the measurement and study of those parameters in the hydrological cycle which involve meteorological considerations" and to "the application of meteorology to hydrology in such problems as river-stage forecasting, flood forecasting and the study of seiches".

However, if a decision is vague in some respects, the vagueness itself can be turned to advantage. It was soon appreciated that the creation of a separate commission to look after WMO's interests in hydrology was a highly significant and most important step forward. As a result activities in hydrology became integrated after being somewhat diffuse and were straightaway amenable to planning by specialists. Furthermore, the existence of the Commission and its developing programme imparted additional momentum to the movement for WMO's official involvement in hydrology. However, this momentum, together with the various internal and external pressures guiding WMO further into the field of hydrology, could well have been ineffective without the vision, dedication and leadership of those who recognized in good time that meteorology and operational hydrology had to become closely linked under one international organization. To that end they worked tirelessly and, it may be added, selflessly for the efficient development of water resources in the service of mankind.

At this point in this historical account it is appropriate to mention some of the personalities who greatly contributed to the advancement of international hydrology and who perceived that this major objective would be best served by bringing hydrology into close partnership with meteorology within WMO. The appearance of names might also breathe a spark of life into a narrative that may be in danger of failing to measure up to the dedication and enthusiasm that attended the hydrological activities of the Organization.

Some leading contributors

Without question the first person to be mentioned must be Mr Max Kohler, a hydrologist from the USA who had had a long association with the meteorological requirements for hydrology and, in the international field, with WMO in the manner in which its various activities bordered on or overlapped into hydrology. Mr Kohler was eminently qualified, and an obvious choice, to be chairman of the EC Panel of Experts on Water Resource Development set up in August 1955. A few years later, in 1959, when the preparatory work for the formation of CHM had been completed, it was entirely appropriate that he should be elected first president of the new Commission, an office he held for two four-year terms from 1960 to 1968.

At this critical time the presence and availability of Mr Kohler were especially important for hydrology and for the relevant activities of the United Nations family, including WMO. He had the vision to see what was required for the greater efficiency of Hydrological Services and he also had the capacity to plan the way ahead and to enlist the co-operation of enthusiastic colleagues to deal with diverse and complex problems. Mr Kohler's leadership was imaginative and restrained, as it had to be, because there was a wide range of conflicting opinion and the main chance of ultimate unity lay in persuasion.

An outstanding colleague of Mr Kohler in the development of plans for a separate commission for hydrological matters was Professor Evgueni Popov of the USSR, where Meteorological and Hydrological Services were already integrated into a single organization. In due course Professor Popov succeeded Mr Kohler as president of the Commission and also served two four-year terms from 1968 to 1976.

It became a tradition for the Commission to elect as its president a hydrologist of high standing who would be both energetic and effective in conducting the affairs of the Commission in order to ensure progress throughout the broad range of its activities. Thus the third president, Mr R. H. Clark, a hydrologist from Canada, was a logical choice and he maintained the practice of serving for two four-year terms, namely from 1976 to 1984. Mr Clark, a specialist in water resources engineering, had been—like his predecessors as president—a delegate to the first session of the Commission for Hydrological Meteorology and over the years had acquired a thorough knowledge of the objectives being pursued by WMO in operational hydrology.

The staffing of the hydrology section in the Secretariat was a matter of major importance. As soon as the Commission was approved, work began on the organization of activities and on the plans to be submitted for its first session. The work of the Commission covered a wide range and preparations had to be made for an increasing load to be undertaken in the Secretariat. The duties of the section comprised the technical and administrative aspects of the policies and programmes formulated by the Commission. During the presidency of Mr Kohler the unit dealing with hydrology in the WMO Secretariat made valuable contributions at a time when the Commission was developing and firm foundations were being laid. In due course the work in the Secretariat expanded considerably and eventually the Hydrology Section responded to this expansion and became first a Division and then a Department—i.e. one of the major components of the WMO Secretariat. The staff of the Department carried out their duties to the evident satisfaction of the Commission and indeed originated many of the ideas that were adopted by the Commission in its policy-making.

The development of CHM

When Third Congress (April 1959) ended, the Executive Council* met and, among other matters, considered implementation of Congress decisions setting up the Commission for Hydrological Meteorology (CHM). As a preliminary step the Panel of Experts on Water Resource Development was invited to continue in operation and to deal with all questions falling within the terms of reference of the new Commission. The Secretary-General also wrote to the Permanent Representatives of Member countries, drawing their attention to the action required for CHM to come into existence and inviting them to nominate appropriate experts to assist in the activities of CHM. Within a short period 30 Members had provided nominations and the Secretary-General made arrangements for the election of a president. The chairman of the Panel of Experts on Water Resource Development, Mr Kohler, was elected president and immediately set in motion the work of organizing the activities of CHM and of planning the first session. Mr Kohler took up his appointment as president towards the end of 1959 and in the following year CHM held its first session in Washington, DC, USA.

Discharge measuring using the moving-boat method on the River Exe, United Kingdom

At long last, therefore, WMO had a separate Technical Commission for hydrological matters. The name given to the commission, the Commission for Hydrological Meteorology, did not please everybody but hydrologists were greatly encouraged by having their own forum for discussion of their activities and no longer had to be content to meet in other Technical Commissions—namely, those for climatology and for agricultural meteorology—where hydrological questions could not always be given the necessary attention.

At its first session CHM gave urgent attention to clarifying its terms of reference since there was a divergence of opinion as to the scope of the Commission's responsibilities, which would, of course, help to define its work programme. A small group of delegates wished to confine the Commission within fairly narrow boundaries. A larger group wished to avoid arbitrary or artificial limitations and argued that a wide variety of topics, of interest mainly to hydrologists, should be regarded as being within the Commission's terms of reference. With such an interpretation, questions related to river stage and river flow, lake level, lake and river ice, glaciers, sediment transfer, flow of springs, water temperature and water quality would fall within the area of responsibility of the Commission. Some of these subjects may not be closely related to day-to-day work in meteorology, but it is now realized that most of them have a bearing on the global climate and all are of importance to operational activities in hydrology.

The interpretation of terms of reference is apt to result in endless discussion and an uneasy compromise. The Commission therefore turned its attention to considering the responsibilities which WMO had accepted, with the support of other agencies, in the UN programme of technical assistance. In addition, delegates were able to set out the requirements of countries for advice on such questions as technical standards and observational networks for those meteorological and hydrological elements that are common to many of the basic problems requiring attention. In the light of these considerations, relating to external as well as internal responsibilities, the Commission arrived at practical solutions offering a sensible work programme. By this procedure a practice that has become a tradition for consultative and working groups was confirmed. In other words, the best way forward is to set aside terms of reference, draw up a programme of action to cover foreseen requirements and in the end it is found that all that is proposed accords closely with the terms of reference and solves the problems of interpretation that arose the outset.

CHM devoted its first session to a comprehensive review of the subjects that needed attention and to the formulation of a strategy that would be appropriate for a Commission which would be recognized as the leading body in the UN family on hydrology and water resources. An ambitious programme was prepared with tasks delegated to working groups whose progress would be considered when the Commission met for its second session. In all, seven working groups were set up and assigned responsibility for the following topics:

- Hydrological forecasting;
- Network design;
- Terminology;
- Instruments and methods of observation;
- Hydrological design;
- Guide on hydrological meteorology;
- Publication and exchange of data for hydrological meteorology.

---

* See note on p. 8.

The Commission also discussed various other questions and recorded in a series of recommendations its concern for: climatological maps for hydrological purposes, the study of seiches, radar measurement of precipitation, and the effects of land use.

For what was achieved, the first session of the Commission was widely perceived to have been most successful and to have totally justified WMO's decision to establish a separate commission for hydrology.

The International Hydrological Decade—WMO and Unesco

In 1961, the year of CHM's first session, the General Assembly of the United Nations designated the current decade, 1961 to 1970, as "The United Nations Development Decade" and called upon Member States and their peoples to intensify efforts to mobilize and support all measures required by developed and developing countries to accelerate progress towards economic growth and social advancement of each individual nation. In the following year ECOSOC adopted a resolution requesting that "proposals for a priority programme of co-ordinated action in the water resources field within the framework of the UN Development Decade be drawn up by the UN Water Resources Development Centre as soon as possible, with the co-operation of the various organs concerned".

Arising from the UN initiative on a development decade, action in the field of hydrology was discussed in 1962 at a meeting of the General Assembly of IASH. Proposals for an International Hydrological Decade (IHD) were communicated to WMO and to Unesco, the two Specialized Agencies principally concerned. The General Conference of Unesco, meeting in 1963, supported the proposals and, in conformity with the UN development decade, decided to enlist the help of governments to plan and implement a programme generally on the lines proposed by IASH. This was clearly a highly important development in international hydrology and one that was of obvious interest to WMO. In its plan Unesco proposed that the IHD should include seven basic components as follows:

- Appraisal of the level of existing knowledge and the identification of major deficiencies;
- Standardization of instruments, observations, techniques and terminologies;
- Establishment of basic networks and improvement of existing networks to provide basic data;
- Research on hydrological systems (representative basins);
- Research on specific hydrological problems;
- Theoretical and practical training in hydrology and related subjects;
- Systematic exchange of information.

Reference to the preceding section shows that there was the closest imaginable correspondence between the programme put forward by the Commission for Hydrological Meteorology at its first session (1961) and many of Unesco's proposals for the IHD. Participants in CHM derived great encouragement from this upsurge of interest in international hydrology and felt that wide governmental support of the IHD would be of lasting benefit to CHM's working programme and general strategy. Furthermore, it was generally accepted throughout UN that the IHD programme would be heavily dependent upon the whole-hearted involvement of WMO. The observational stations forming WMO's Global Observing System all produced data of hydrometeorological relevance and, of importance to any operational applications that might arise, exchanges of collected reports were made in real time over the Global Telecommunication System co-ordinated by WMO.

At Fourth Congress WMO, which was fully apprised of the two decade programmes, discussed their implications and took a most constructive view. Congress decided that WMO should collaborate with UN, the Specialized Agencies and other international organizations in supporting the UN Development Decade and in the implementation of the International Hydrological Decade. Taking these and other related matters into account, Fourth Congress enlarged the role of CHM in order to ensure that it would be prepared to meet meteorological and associated hydrological requirements in the field of water resources. Furthermore, a change was made in the name of the Commission, which for the next few years was known as the Commission for Hydrometeorology (CHy). This new title gave little satisfaction to members of the Commission. Under the strong and enlightened leadership of Mr Kohler, the Commission's president, hydrology had achieved an identity in WMO but still had to be content with a hybrid label.

In accordance with custom the WMO Executive Council* met immediately after Congress to organize the administrative and operational consequences of the decisions and recommendations made by Congress. It was recognized that careful planning of WMO's participation in the IHD would be essential. The Executive Council therefore appointed a Panel of Experts for the IHD and requested CHy to take adequate steps to ensure the maximum contribution towards the successful conduct of the IHD. The panel, under the chairmanship of Mr Kohler, was in effect a small inner cabinet of CHy and was thus in an excellent position to plan and implement WMO's support for the hydrological decade. Meetings of the panel were held each year throughout the decade and were, of course, greatly helped by advice from other members of the Commission and by contacts with other organizations taking part in the IHD.

The Commission held its second session in Warsaw in the autumn of 1964. Much of the business of the session was inevitably concerned with the IHD, since both the decade and the Commission were involved in hydrology in its wide context, and it was appreciated that efforts to crown the IHD with success would be of value in furthering the work of CHy.

One of the most significant and lasting outputs of the second session was the approval for publication of the WMO *Guide to hydrometeorological practices*, published in 1965.

There were, however, a number of problems facing the Commission in relation to its participation in the IHD. These stemmed, at least in part, from the reluctance of WMO at successive Congresses to take the bold step into hydrology that was so widely desired. The term "hydrometeorology" was regarded as restrictive and other organizations involved in the IHD tended to infer that WMO's contribution would be on a limited scale of narrow scope and possibly somewhat half-hearted.

With these organizations various misunderstandings had to be cleared up. It was, for example, disconcerting on all sides when, as a result of serious underestimates of WMO's support, plans were drafted independently of WMO for the use of observing networks and associated facilities which were controlled by WMO Members and co-ordinated by WMO.

On the whole these problems were susceptible of explanation and so were fairly short-lived but it was unfortunate that they should have occurred. There was perhaps one redeeming feature: the Commission recognized that some radical steps were essential in the interests of clarification of its role and in order to safeguard its smooth functioning over the long term. It appeared

---

* See note on p. 8.

that the Commission was moving towards recommending that its title should be changed to Commission for Hydrology and that its terms of reference should more accurately reflect the wide range of its activities within and beyond the elements of the hydrological cycle. The Commission also considered that national Hydrological Services would have to be brought closely and officially within the executive bodies of WMO if the views of these Services were to be adequately represented and, an important proviso, were seen to be adequately represented.

The business of the second session might appear to have been largely dictated by the plans of Unesco to organize an International Hydrological Decade. Certainly nearly all points on the Commission's agenda had a bearing on the IHD but it would also be true to say that the agenda for a meeting on the IHD would have contained few aspects that were not included in the work of the Commission. There were, in addition, matters of long-term importance that claimed the attention of the Commission, which was still in a state of development, and its members felt confident that its range of activities was in no danger of contracting and indeed offered every prospect of significant expansion. The work of the Commission at its second session can perhaps best be summarized by noting that the requirement for working groups to pursue assigned tasks between sessions was reviewed and resulted in the existing seven working groups being replaced by ten dealing with the following:

- *Guide* and *Technical Regulations*;
- Hydrometeorological instruments and methods of observation;
- Water levels; measurement and estimation of discharges;
- Machine processing of hydrometeorological data;
- Hydrological forecasting;
- Estimation of maximum floods;
- Terminology;
- Universal Decimal Classification;
- Representative and experimental basins;
- Training in hydrometeorology.

The Commission also adopted recommendations on the exchange of data in real time to permit forecasting for international rivers, on the organization of activities in different countries and on climatological maps for use in operational hydrometeorology.

Third session of the Commission (September 1968)

The Commission held its third session in Geneva in the year that followed the Fifth Congress of WMO (April 1967). Congress had given much valuable guidance to the Commission about involvement in the IHD and had agreed to the valuable contribution of providing staff of the WMO Secretariat to deal with IHD activities and projects for which WMO had assumed responsibility. In addition, the Commission was to a great extent given a free hand to support the IHD in all appropriate ways, including training, forecasting techniques and various measures aimed at developing the economical use of water resources.

A special item of business for the Commission at its third session was to consider representation and documentation at the projected IHD Mid-decade Conference. The purpose of this conference was to review progress, to complete arrangements for the remaining years of the decade and to discuss a long-term plan of action in the field of hydrology. In regard to the long-term plan of action, the Commission was concerned about the uncertainty that still prevailed about the allocation of responsibilities among UN and the Specialized Agencies in the field of hydrology. This question was of long standing and already various UN bodies had indicated that they were looking to WMO to undertake wide responsibilities in hydrology. The Commission, after further discussion, recorded its view that WMO should accept responsibility for organizing international co-operation in collection, transmission and publication of data and for operational aspects of the land phase of the hydrological cycle.

As a further step towards clarification of WMO's role in hydrology, the Commission agreed that the term "hydrometeorology" in its title implied a restricted range of responsibility and was a cause of confusion. It therefore recommended that it should be renamed "Commission for Hydrology" and that its terms of reference should be made specific with respect to the elements of the hydrological cycle and the scope of the resulting activities.

By this time, 1968, it seemed to be clearer than ever that WMO's entry further and further into hydrology was inevitable and could not be reasonably resisted. The Commission fully realized this trend and again expressed its anxiety that national Hydrological Services should be enabled to play a full part in WMO's activities in hydrology. The Commission, therefore, proposed that it would be in the interests of all concerned if arrangements were made to hold an intergovernmental conference of Meteorological and Hydrological Services. This proposal was fulfilled two years later. The Commission also adopted a suggestion of the WMO Executive Council* and, in addition to its normal working group, established an Advisory Working Group to assist the president and advise him on urgent matters that might arise between sessions.

The session marked the completion by Mr Kohler of two terms as president of the Commission. As mentioned earlier, he was succeeded by Professor Evgueni Popov (USSR). Mr Kohler's presidency occurred at a highly creative phase of the Commission's work but, as is found invariably in forward-looking ventures in science and technology, much still remained to be done.

Consolidation and further development

Professor Popov was president of the Commission for two successive four-year terms covering the eight years from 1968 to 1976. During that period the Commission was occupied partly in consolidating the achievements made during the preceding eight years, the first eight years of its existence, and partly in recognizing and coping with new requirements. A lot of the groundwork for international co-operation in hydrology had already been well prepared and the period 1968 to 1976 proved to be years of very rapid development, including a substantial strengthening of WMO activities in many branches of hydrology. It was particularly noticeable that WMO's long experience in promoting international co-operation over the whole vast field of meteorology was of great help in developing a similar organizational framework for international hydrology, notably in relation to operational features. As a result there was ever-increasing acceptance of the soundness of moves to bring meteorology and hydrology closer together for the purposes of international co-operation. The efforts of CHM and its successor, CHy, were proving well justified.

The International Hydrological Decade (IHD), launched in 1965, was the outstanding event at this time as far as the Commission was concerned. All international organizations

* See note on p. 8.

concerned in one way or another with hydrology and the development of water resources were brought together in numerous projects. Many of the activities were entirely within the ambit of CHy whilst others required the Commission to collaborate under the leadership of other organizations. The latter activities included water-balance studies, education and training in hydrology and the organization of symposia and seminars. Altogether the Commission took part in 60 IHD projects which were successfully completed during the decade. WMO issued 21 special *IHD Reports* and in addition some 30 publications on hydrology which had a direct bearing upon the IHD. The following are some examples of the subjects covered by these publications:

- Estimation of probable maximum precipitation;
- Depth-area-duration analysis of storm precipitation;
- Stream gauging;
- Atmospheric vapour flux computations;
- Automatic collection and transmission of data.

During the IHD, WMO and Unesco learned to build up a sound working partnership and jointly sponsored as many as ten international symposia, among which the one on World Water Balance (Reading, 1970) and two held in Canada in 1972 on the Role of Snow and Ice were probably the most important. These and other arrangements were planned in the first place by the joint Unesco/WMO liaison committee for hydrological activities. CHy also played a leading part in the Mid-decade International Conference on the Practical and Scientific Results of the IHD, held in 1969, and there earned wide recognition of WMO's contribution to the success of the Decade. Five years later, in 1974, WMO and Unesco jointly convened in Paris a conference, usually described as the "End of the Decade Conference", to view the results of the IHD and to consider requirements for future international programmes in hydrology. A detailed plan was drawn up for a long-term international programme, to be organized by Unesco, and in the course of its preparation consistency was established and maintained with CHy's existing programme in operational hydrology. The two programmes together constituted the major part of internationally co-ordinated activities in hydrology and water resource development. Other Specialized Agencies associated with these activities included FAO, WHO and UNEP.

A gauging station of the cable-car type over the River Matadero, Amazon Basin

At its third session the Commission for Hydrometeorology had adopted a proposal for the convening of a WMO Technical Conference of Hydrological and Meteorological Services. This proposal was accepted by the WMO Executive Council* and the conference was held in Geneva in September/October 1970 in accordance with arrangements made by the president of the Commission in consultation with the Advisory Working Group. The conference served a very useful purpose in bringing together meteorologists with interests in hydrology, and hydrologists with important requirements in meteorology. A declaration was also adopted with the recommendation that WMO should expand its programme of international co-operation in operational hydrology.

Sixth Congress of WMO (Geneva, 1971)

The decisions regarding hydrology made at the Sixth Congress of WMO heralded the culmination of more than a decade of efforts by hydrologists and meteorologists to persuade the Organization to take its logical place as leader of international co-operation in hydrology. The struggle was long but the wind was favourable in spite of many frustrations. However, a little consideration may show that the apparent delay was not without beneficial results for it ensured effectively that the whole subject of WMO taking hydrology as well as meteorology under its international umbrella was discussed widely and in depth so that strongly opposed views were gradually moderated and, in many cases, won over, whilst the strongly supportive views were encouraged, as seemed desirable, to appreciate the realities and practical consequences if and when WMO agreed to such a decisive step. On the whole it appears that WMO's Sixth Congress in 1971 provided just about the appropriate time and place for these far-reaching decisions in operational hydrology.

Sixth Congress defined the scope and activities to be included under "operational hydrology", finally agreed that the Commission should henceforth be known as the Commission for Hydrology (CHy) and revised its terms of reference to take account of the new outlook. Congress also appreciated that the new responsibilities that it had formally accepted would need to be recorded in the official documents of WMO. Accordingly, it requested the Executive Council* and the Secretary-General to

* See note on p. 8.

study the need for changes in the Convention and to report to Seventh Congress with any proposed draft amendments.

The Commission for Hydrology, under its new name, held its fourth session in Buenos Aires in 1972, a year after Sixth Congress. The Commission understandably had an atmosphere of increased confidence and proceeded to undertake new work in groundwater, soil moisture and water quality. By this time the Commission and its working groups were active in many sectors of hydrology and the related aspects of meteorology. The following selection gives some of the major results of projects on which action was taken at the fourth session.

- Implementation of a project for the comparison between rainfall flood forecasting models used in different countries;
- International codes for the transmission of hydrological basic data and hydrological forecasts;
- International system for identification numbers for hydrological observing stations;
- Publication in the WMO *Technical Regulations* of the chapter entitled "Meteorological services for hydrology";
- Publication of the third edition of the *Guide to hydrological practices*;
- Publication, jointly with Unesco, of an international glossary in four languages listing about 1 300 hydrological terms with their definitions.

The Commission also completed or set in train preparations for participation in the following conferences:

- The United Nations Stockholm Conference on the Human Environment (1972);
- The "End of the Decade Conference" of the IHD (1974);
- The United Nations Water Conference (1977).

Further progress in the status of operational hydrology

In the early 1970s activities in hydrology increased substantially not only within the Commission for Hydrology but also in regard to joint projects with other Specialized Agencies and in projects with national Meteorological and Hydrological Services. It was also apparent that very helpful collaboration existed with other WMO Technical Commissions, notably with the Commission for Instruments and Methods of Observation, the Commission for Basic Systems and the Commission for Atmospheric Sciences. By the time of Seventh Congress (Geneva, 1975), therefore, hydrology was very securely based in WMO although a few formal steps had still to be taken. Action on these was taken by Congress. First, a number of important changes were made to the WMO Convention so as to state explicitly that the promotion of international co-ordination of operational hydrology was one of the objectives of WMO. The Convention has the status of an intergovernmental agreement, ratified and adhered to by all countries—well in excess of 100 at the time—becoming Members of the Organization. Thus the incorporation in the Convention of the acceptance by WMO of responsibilities in regard to operational hydrology gave this important discipline permanent standing as an integral component of one of the UN Specialized Agencies. An additional step was to mention national Hydrological Services in relevant sections of the WMO *Technical Regulations*. The Operational Hydrology and Water Resources Programme, approved by Congress, was also included among the major programmes of WMO.

Collaboration between Meteorological and Hydrological Services had been greatly enhanced by the Advisory Committee in Operational Hydrology (ACOH), which put forward a series of effective proposals that contributed to the smooth functioning of the Commission for Hydrology. A particularly valuable proposal, endorsed by Congress, was the appointment of hydrological advisers to the Permanent Representatives of Member countries with WMO. The Permanent Representative is generally the head of the national Meteorological Service. In countries where the Hydrological and Meteorological Services were separate, it was obviously essential that the Permanent Representative with WMO should have adequate professional advice on all hydrological matters. Such advice would readily be provided by the hydrological adviser who would have contacts with all bodies concerned in the country. It was, therefore, logical to make the arrangement formal and permanent. A very worthwhile by-product was that hydrological advisers, attending Congress with Permanent Representatives, were able to meet as a sub-committee of Congress to consider in detail the Organization's activities in hydrology and water resources.

The amendments to the WMO Convention represented to hydrologists and meteorologists alike a milestone in the history of WMO. In response to many requests which gradually approached a consensus among all interested parties, WMO had constantly expressed its willingness to promote international co-ordination in hydrology, either on its own or with other organizations concerned, and was determined not to appear in any way as a predator. It may be said that in the course of time the problems associated with the entry of hydrology into WMO tended to solve themselves as everyone involved acquired a deeper understanding of the different facets of a highly important question. Above all, however, it must be stated that the work of the Commission for Hydrology, since its formation in 1960, was the principal instrument in bringing meteorologists and hydrologists together and demonstrating that a fruitful partnership could develop in efforts over the whole range of water resources, a subject whose importance cannot be overestimated.

In 1976 Mr R. H. Clark became president of the Commission and, like his predecessors, served two four-year terms ending in 1984. As a specialist in water resources engineering, with particular emphasis on hydroelectric and flood control engineering, Mr Clark brought valuable experience to the leadership of the Commission. He appreciated that much had been gained in the previous sixteen years of the Commission's existence, particularly with the change in the WMO Convention. However, the new president felt that further advances were needed in order to ensure that operational hydrology, which he described as representing the socio-economic value of hydrology, should take its place in full partnership with meteorology in WMO. This objective was greatly helped by some major events which took place during the latter half of the 1970s and in the 1980s.

WMO, through the agency of the Commission for Hydrology, played a prominent part in the United Nations Water Conference, which was held in March 1977 in Mar del Plata, Argentina. The conference was one in a series of highly important meetings arranged by the United Nations and was designed to promote a level of preparedness—nationally, regionally and internationally—that would help avoid a water crisis of global proportions, a possibility that could not be ruled out in view of the increasing demands of the world's population

* See note on p. 8.

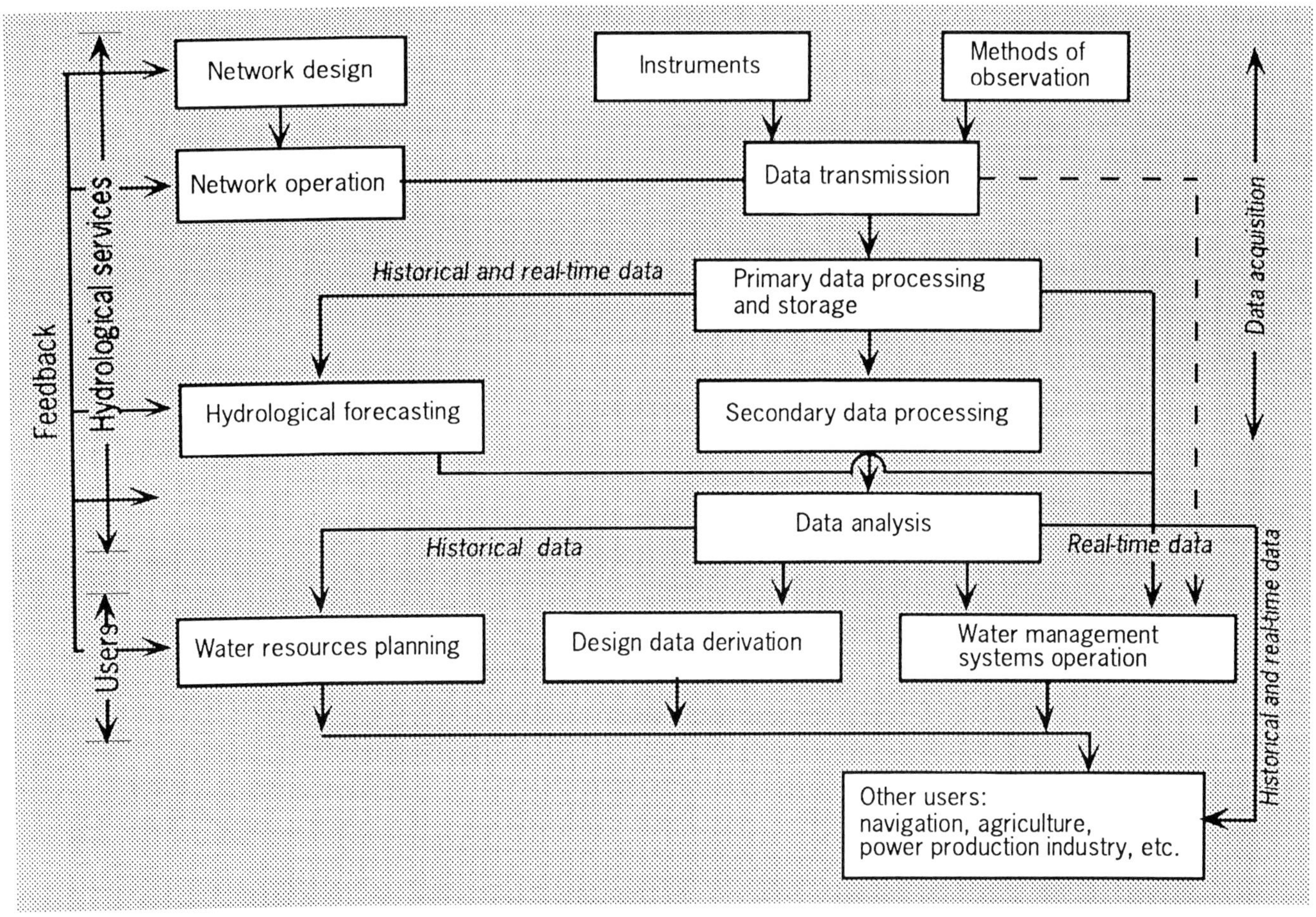

System of operational hydrology

for natural resources of all kinds. The conference recognized the close interrelationship of hydrology and meteorology, particularly in dealing with such natural disasters as floods and droughts. Further discussion of this conference is to be found in a later section (see page 69).

Another very significant event was the World Climate Conference, which took place in the early part of 1978 and led to the establishment of the World Climate Programme (WCP). An important component of the WCP reflects the contribution to be sought from hydrology and, to emphasize the point, is entitled "World Climate Programme—Water". This subject is treated in greater length in a later section (page 71).

In the eight-year period 1976 to 1984 the scientific and technical work of the Commission was pursued with remarkable vigour. The most significant achievement was the development of the operational system known as HOMS; in addition, numerous memoranda and reports were published. The *Guide to hydrological practices* reached its fourth edition and was issued in two volumes. A comprehensive manual on stream gauging was published, also in two volumes.

In 1984 M. Clark was succeeded as president of CHy by Dr O. Starosolszky (Hungary), who had served as vice-president of the Commission during Mr Clark's second four-year term. In an important statement as president, Dr Starosolszky emphasized the dynamic nature of hydrology as a science and a technology and pointed out the importance of assessing the directions in which hydrology should develop. Noting that operational hydrology was the central feature of the WMO Hydrology and Water Resources Programme, he drew up a schematic representation of operational hydrology, reproduced in the block diagram above, and discussed the importance of monitoring the complete system so that weak points maybe identified and remedied. The importance of studying feedbacks was also stressed.

The history of the Commission has been marked by continuous and, from many points of view, remarkable progress in technical developments and their applications. However, the president emphasized that temporal and spatial differences existed in the rate of evolution of operational hydrology and urgent attention needed to be given to those countries where the subject was in an early, rudimentary stage compared with other countries which were already making full use of computers and other advanced facilities.

Dr Starosolszky also included among the objectives of his presidency the need to study with due priority the influence of human activities. He pointed out that in view of possible environmental effects, observations of water quantity and quality and of surface- and groundwater should be examined as a whole in order to detect interrelations and prevent harmful effects.

One of the major scientific and technical events in the term of the new president was the third WMO/Unesco intergovernmental conference held in Geneva in March 1987. It had become an established arrangement for the two organizations to hold a meeting every five or six years in order to review their main programmes over the whole field of hydrology, namely, WMO's Operational Hydrology Programme (OHP) and Unesco's International Hydrological Programme (IHP). Since their inception the two programmes have been co-ordinated at various levels in the course of planning and implementation and in work of this nature, with revised or new procedures being introduced from time to time, the conference was able to carry out a worthwhile task in taking an overall view.

In considering the future activities of the two programmes, OHP and IHP, the conference set out to identify areas in which greater attention would be needed in the years ahead. It was appreciated that the world's freshwater resources would be subject to increasing pressures from a variety of socio-economic factors. For example, projections for the distribution of population pointed strongly to steadily increasing urbanization, a development threatening adverse consequences in the form of drainage and flooding problems and effects on the water balance generally. It was, therefore, becoming more and more important that water requirements should be properly assessed and for this and other reasons the conference stressed the urgent need to strengthen hydrological facilities in vulnerable areas, notably with regard to observing networks, forecasting systems and the organization of the management of water resources.

Another event, partly social and partly scientific and technical, had taken place a year earlier than the WMO/Unesco conference. The year 1986 heralded the twenty-fifth year since WMO created a Hydrological Commission. This silver jubilee provided an adequate reason for celebration and there were some other closely associated anniversaries as well.

The Commission celebrates

For the hydrological community the year 1986 produced a concentration of anniversaries. The Commission itself had been established by WMO twenty-five years earlier and IMO in its final years had set up a Commission for Hydrology in 1946 so the scene was set for a ruby anniversary as well as a silver jubilee. There were thus two excellent reasons for celebration and arrangements went ahead with a degree of enthusiasm for which hydrologists are well known. The scene of the celebration was the Research Centre for Water Resources in Budapest and it is appropriate to acknowledge the courtesy of the government of Hungary in offering to hold the ceremony in the home country of the Commission's president, Dr Starosolszky. Remarkably, a third anniversary was much in evidence—the centenary of the establishment of the Hydrological Service of Hungary. Continuing the close link with IAHS, the jubilee celebrations took place immediately after the Association's Budapest Scientific Assembly.

The two-day meeting opened on 11 July 1988 with congratulatory addresses by senior officials of the Hungarian Government, by the Secretary-General of WMO, Professor G. O. P Obasi, and by representatives of Unesco, UNEP and IASH. The four presidents of the Commission in turn gave accounts of the activities and highlights during their terms of office. There then followed workshops on long-term planning of WMO activities in hydrology and water resources, on the state of the art in operational hydrology, and on such topics as instruments and methods of observation, transmission and processing of data, and hydrological forecasting. The final session took the form of a discussion of water and climate. The meeting was attended by more than 100 delegates from numerous countries and from governmental and non-governmental international organizations.

In their speeches the presidents gave some interesting statistics illustrating the progress of the Commission for Hydrology over the years. The support of countries, Members of WMO, is shown in the following list:

Session of CHy	Year	Countries represented	Number of delegates
I	1961	27	59
II	1964	35	65
III	1968	39	73
IV	1972	32	74
V	1976	44	89
VI	1980	60	131
VII	1984	51	108
VIII	1988	71	146

The silver jubilee celebration was most successful and well worth while. It highlighted what is probably a unique event: how members of a scientific and technical discipline, recognizing its growing practical importance and the need for international co-ordination, co-operated with an existing international organization, WMO, in collaborative efforts which brought meteorology and hydrology ever closer together so that, eventually, activities in operational hydrology became officially a component of WMO's responsibilities, acknowledged officially in the WMO Convention. The precedent thus created may suggest to other branches of geophysics that their international interests could be well catered for by association with an existing international organization of very wide scope, with interrelationships with most other branches of geophysics and with a world-wide membership.

Using time as a co-ordinate, the silver jubilee of 1986 came very close to the end of this historical account which has largely taken the form of a chronicle of events and meetings—Congress, Executive Council*, the Commission and its sessions, relations with other organizations and so on—on the route to decisions which resulted in changes to the WMO Convention. These events and meetings were in major or minor ways occupied with what might be described as high policy or strategy in its widest sense affecting the practice and international potentialities of hydrology. However, all the time CHy's working groups were striving towards the development and transfer of technology and techniques, standardization, exchanges of high-quality data and technical training, to mention but a few of the aspects in which progress was sought. Each working group had a substantial commitment and each member of a group acted as a rapporteur for an assigned topic. Thus the Commission, helped by an enthusiastic team in the WMO Secretariat, accomplished a great deal and its reports and results became well known not only among WMO Members but also among all other international organizations with interests in hydrology. It would be true to say that the Commission, by its wide outlook, its work and its enthusiasm, did much to show that operational hydrology would find a secure and sympathetic place within WMO. It is now fitting, therefore, to select some examples of the Commission's activities and to explain the objectives and the potential value of achieving good, scientific progress.

Comparison of hydrological forecasting models

Hydrological forecasts are one of the most important practical aspects of applied hydrology. Such forecasts are an essential requirement in connection with flood warnings, the control or regulation of runoff, hydroelectric power generation, navigation in rivers and canals, irrigation and the management of water supplies and water quality. In addition to these long established requirements, problems of pollution in rivers resulted in new demands for forecasting services to be extended to predict low river flows. During periods of low water it is important, notably from the health point of view, to know whether the available volume of water is sufficient to dilute the polluting agents to a safe concentration. For the management of water resources, low-flow forecasts at one end of the scale have an importance equal to that of flood forecasts at the other end.

In the past, up to about the later 1950s, hydrological forecasting techniques were fairly rudimentary because of shortages of data and also because systems for rapid processing of data had still to be developed in Hydrological Services. When these Services brought high-speed electronic computers into use,

* See note on p. 8.

A water-level recorder and stream gauging on the Maule River in Chile

research began in many countries into the possibilities of producing more complex forecasting models aimed at making full use of all available data and at offering improvements in reliability, timeliness and accuracy.

As experiments proceeded in various Hydrological Services and institutions and as information became available about the forecasting techniques under development, the Commission proposed that an intercomparison of conceptual forecasting models should be undertaken under arrangements to be co-ordinated by WMO. The proposal was widely agreed and the comparison took place in the period 1968–1976. The objectives were:

(*a*) To compare and evaluate computer-based hydrological forecasting models which provide short-term estimates of streamflow;

(*b*) To provide information and guidance on the applicability of such models in relation to different specified conditions and accuracy criteria.

The comparison was investigative rather than competitive. It was assumed that each forecasting model provided for the trial had already reached operational status; otherwise it would have been discarded. It was intended to assess the accuracy of each model in a variety of conditions, with particular reference to climate and geography, and leave it to other potential users to decide whether any model, or a refinement of it, would meet their purposes.

Forecasting models were entered by six countries and a six-year calibration period (1968 to 1974) was followed by a verification period of two years (1974 to 1976). The whole comparison project was managed by WMO. For each data set the Secretariat evaluated the simulated discharges produced by the tested models for both calibration and verification periods and comparisons were made by means of agreed graphical and numerical criteria.

WMO published a comprehensive report describing the techniques that were involved in the comparison and commenting on the results. Copies of the report were widely distributed and the intention to stimulate interest in the development and use of hydrological forecasting models was fulfilled.

United Nations Water Conference

WMO played a prominent part in the UN Water Conference, which was held in Mar del Plata in 1977. The conference was one of the most important of the series of meetings organized by UN for the purpose of studying problems concerned with the social and economic aspects of development, population and the environment. Water is a vital resource for all nations but there are many deficiencies, some almost intractable, among both developed and developing countries, the latter usually having the most pressing requirements. The plan of action adopted by the conference was of the nature of a 20-year plan although, in effect, the specified period was indicative of the urgency of a host of needed developments and thus an attempt to impose a time limit for significant progress to be attained.

The targets set by the conference were fully considered by the Commission for Hydrology and as a result WMO's Operational Hydrology Programme was adapted as necessary in order to achieve consistency with the conference resolutions and plan of action. In addition, WMO in conjunction with Unesco prepared the way for international efforts on the assessment of water resources by submitting a document to the UN Committee on Natural Resources. The document contained proposals for international projects to provide countries with technical and practical support in determining the steps required to improve national organizations for dealing with water resources. These steps were concerned with efficiency and the introduction of new technology in routine operations.

HOMS

The title of the Hydrological Operational Multipurpose Subprogramme is admirably descriptive but is somewhat lacking in elegance. Now usage and familiarity have dictated that the acronym HOMS has virtually superseded the cumbersome title. The concept of HOMS derived from directives given by Seventh Congress (1975) to the Commission for Hydrology. The first directive was that CHy should adopt as a prime objective the provision of maximum assistance to national Hydrological Services, especially to those of developing countries. Secondly, Congress requested the Commission to study the hydrological implications of certain resolutions and recommendations approved at the UN Stockholm Conference (1972). CHy's attention was drawn in particular to questions of water distribution, water quality and water pollution to which the UN conference had attached major importance.

In response, at its fifth session (Ottawa, July 1976), CHy reviewed its programmes and projects and concluded that the most promising course would be to develop a long-term system of technology transfer that would incorporate all the basic elements of technology necessary for dealing with problems related to water resources. In this way the Commission would be in the best position to help national Hydrological Services and also to give proper attention to the requirements highlighted by the Stockholm Conference. The Commission decided that the system in its agreed form would have to pursue the following aims:

- To provide an efficient means of technology transfer;
- To improve the quantity and quality of hydrological data;
- To provide an international, systematic framework for the integration of techniques and procedures in the collection, dissemination and processing of hydrological data;
- To assist in the selection and application of appropriate technology and in the associated training;
- To advise and assist in the field projects of Member countries;
- To promote special attention to projects in national and international river basins.

The success of HOMS was immediate and has been well sustained since its approval by Eighth Congress in 1979. The whole concept was considered by Congress to be remarkable for its ambitious overall objective and HOMS may well be regarded as one of the most fertile and constructive sub-programmes designed within WMO.

Organization of HOMS

The long-term future of HOMS is assured and it is therefore appropriate to describe it in the present tense. HOMS is organized as a co-operative effort of the Members of WMO. Any country wishing to participate in HOMS does so by designating a HOMS National Reference Centre (HNRC), usually in the national Hydrological Service. The international activities associated with HOMS are supervised and co-ordinated by a steering committee appointed by the Commission for Hydrology.

The *HOMS Reference Manual* (HRM) contains detailed descriptions of the structure and functioning of HOMS. The available technology is provided in the form of separate components, typical examples of which include a set of computer programs for maintaining a bank of hydrological data and a manual on methods of estimating probable maximum precipitation. By 1988 the number of HOMS components was approximately 350, each one presented in a two-page summary description written in a standard format. The HRM provides guidance in the use of components and also covers user requirements for such projects as the design of a reservoir, flood forecasting and flood-plain zoning.

The comprehensive nature of HOMS as an operational multi-purpose sub-programme is well illustrated by the various sections concerned in the organization, structure and operation. These sections are:

A Policy, planning and organization;

B Network design;

C Instruments and equipment;

D Remote sensing;

E Methods of observations;

F Data transmission;

G Data storage, retrieval and dissemination;

H Primary data processing;

I Secondary data processing;

J Hydrological models for forecasting and design;

K Analysis of data for planning, design and operation of water-resource systems;

X Mathematical and statistical computations.

The field of interest for HOMS is very wide. HOMS is aimed at serving national, regional and international agencies engaged in operational hydrology and in hydrological research.

HOMS is very simple in concept, well tried and extremely sound in its applications. It aims to provide hydrological technology to those in need, particularly the Hydrological Services of developing countries, in the form of techniques, methods and software supplied mainly by the Hydrological Services of developed countries. The *HOMS Reference Manual* contains the descriptions of the components available for transfer and each of the 100 or more HOMS National Reference Centres in Member countries has a copy of the *Manual* at its disposal. About 400 components are available for transfer, a component being a detailed description of an instrument, such as a sediment meter, including drawings enabling it to be constructed, or a listing of a computer program, e.g. one for the statistical analysis of extreme values. The hydrologist

Flooding in Haiti

needing the component looks up the supplier and asks him to send it so that it can be applied in the service where it is required. By mid-1989 some 1 800 components had been transferred since HOMS was established in 1982.

World Climate Programme—Water

The World Climate Programme is described in a separate chapter. However, in this chapter on hydrology and water resources it is fitting that the hydrological aspects should be mentioned since their potential importance in the event of changes in the global climate is very great. For many years, centuries perhaps, it was widely believed that the average climate, regionally as well as globally, was unchanging except for small annual variations which for the most part were regarded as random oscillations and not pointers to climate change. Climatologists were, of course, on the watch for any variations which, if long maintained, might prove to be important. For example, it was noted that changes in sea temperature in a small area had caused a fish population to migrate to a neighbouring, more hospitable region and to remain there apparently indefinitely. Such symptoms of possible changes in climate have always been monitored as soon as they were detected but there were few, if any, calls for action by governments or the United Nations.

The picture began to change around the middle of the present century when scientists showed increasing concern about the possibility of major, irreversible changes in the global climate resulting from either natural causes, e.g. changes in the intensity of the sun's radiation, or from the effects of man's industrial activity, e.g. the continued and increasing burning of coal and oil. The evidence produced was sufficiently strong for WMO, in conjunction with other international organizations, including Unesco, FAO, WHO, and UNEP, to convene an international conference, the World Climate Conference, in 1979 for a wide discussion of climate and possible changes and variations and to consider the implications for governments and populations.

One of the principal results of the conference was an agreement that WMO, with the co-operation of other international organizations, should organize a World Climate Programme which would be concerned with data collection and processing, basic research into global climate and a study of the practical consequences of climate change. This programme is essentially long-term and its results must be expected to have far-reaching consequences for mankind. If significant and permanent changes in the global climate are to take place, the sooner they are foreseen and their implications recognized, the better it will be.

Indeed, water is one of the key components of the climate system and it has many roles in the environment in addition to the vital part it plays in the atmospheric heat engine. It is essential to modern civilization as water to drink, for power production, agriculture and a host of other purposes. However, knowledge of the global water cycle remains rather poor; it is still very difficult to measure precipitation and evaporation over the land surface of the globe and impossible over the seas.

The role of hydrology and water resources is obviously crucial in any fundamental changes in the global climate. Temperature and rainfall regimes would be redistributed on a global and regional scale and it seems sensible to visualize the prospect of large, densely populated areas with abundance of water gradually or perhaps rapidly becoming regions of inadequate rainfall and dependent upon the transport of water from more fortunate places. The Commission for Hydrology is deeply involved in the World Climate Programme—Water and will be concerned in the economic and social decisions that may have to be made relating to such matters as irrigation and hydro power, drought relief and energy planning.

Prospect

Since the establishment of the Commission in 1961 and the development of the Hydrology and Water Resources Programme some time later, this area of endeavour has received an increasing measure of support from Members, from the different organs of the Organization and from the Secretary-General and other parts of the Secretariat.

The current Hydrology and Water Resources Programme addresses a wide range of pressing issues through the activities of the Commission for Hydrology, the various meetings it promotes and the wide range of scientific and technical publications it has produced. Foremost amongst the latter is the *Guide to hydrological practices* which, in its 4th edition, is contained in two volumes. HOMS continues to play a vital part in the essential technology transfer process, while two recent activities recognize the importance of ready access to global data. In the first of these, the INFOHYDRO contains, on PC and in a published manual, details of the world's Hydrological Services (more than 400 of them), the instrument networks they operate, the data they collect and how they store these data. In the second, at the Global Runoff Data Centre at the Federal Institute of Hydrology, Coblenz (FRG), are held the flow records of the world's rivers—an archive which is growing rapidly.

Hydrology faces many problems that are different from those encountered by meteorology. The science is essentially river-basin-based rather than globally based, but that is not to say it is less international in character. Indeed, there are more than 200 international river basins which demand co-operation between Hydrological Services, but in a manner that may be somewhat different from the co-operation between Meteorological Services. Meteorologists are not faced, for example, by the reluctance of some upstream countries to share their data with those downstream. Thus, despite the promotion of institutions for co-operation in international river basins, there are problems of this type which remain unsolved. Others, which are by their nature science- or engineering-based, make different demands on the ingenuity of hydrologists for solutions.

Of course, the promotion of co-operation between Hydrological and Meteorological Services has been a central theme in WMO's activities over the years and there is a good record of such collaboration. Use of the WWW for hydrological purposes, provision of quantitative precipitation forecasts, co-operation under the Tropical Cyclone Programme and some examples from the field of remote sensing provide the necessary evidence. It may be, however, that hydrologists have been reticent in detailing their needs to the meteorologists, particularly for areas which are currently developing. For example, while the requirements for the resolutions and precisions of remotely sensed hydrological variables have been set out alongside those for meteorology, the hydrologist may not have established the needs for archiving these data. Nevertheless, remote sensing from satellites and by ground-based radar is about to make a large impact on hydrology and assist in the development of the so-called mesoscale hydrology which is the vehicle for linking global circulation models and the usual river basin models. Remote sensing will help to control the rising costs of data collection, which is a burden being faced by Hydrological Services. Use of data-collection platforms will increasingly provide a flexible alternative to the conventional means of communication and offer more real-time possibilities. This will be very useful,

The Salto Grande barrage spanning the Rio Uruguay between Argentina and Uruguay

especially in the forecasting of water quality, where there will be increasing demands during the years to come.

Within these areas and others hitherto less well defined, the WMO Hydrology and Water Resources Programme will find growth at the interfaces with other international programmes. It will take on increasing significance for the non-governmental science fostered through ICSU and its tributary bodies. The International Decade for Natural Disaster Reduction is to be an important initiative during the 1990s where hydrological forecasting and the efforts that lead up to it will play a significant role. Then there is the follow-up to the WHO-based International Drinking Water Supply and Sanitation Decade, which will depend even more on greater precision in the water-resource assessment activities that are central to the Hydrology and Water Resources Programme. Indeed, in many parts of the world the availability of water is the limiting factor for water supplies and sanitation plans. Of course close links will be continued with the International Hydrology Programme of Unesco, as its fourth phase unfolds and proposals are made to continue the programme into the next century. The hydrological activities of WMO and Unesco are essentially complementary, attempting to satisfy rather different sets of aims.

These pointers to the future and others which will undoubtedly materialize will ensure that the Hydrology and Water Resources Programme, with its base in operational hydrology, must have an expanding horizon and secure scientific prospects.

CHAPTER X

WMO RESEARCH PROGRAMME

From its very modest beginning in 1952, the WMO Research Programme has developed spectacularly and it is now among the most important of the Organization's activities. The present chapter is concerned with the manner in which technological developments, such as electronic computers and artificial Earth satellites, have provided opportunities for major advances in understanding the atmosphere and its ways, and how WMO has responded to the challenges thus presented. The resulting growth in meteorological research throughout the world was accompanied by changes in the role of WMO, including that of the Secretariat.

The attitudes of WMO Members to the responsibilities of WMO in meteorological research at the time of First Congress were summarized in Chapter II. The outcome was that in its early days, the Organization did little more than *encourage* research, for example, by adopting resolutions indicating those areas in which it was felt that further research was urgently required. The research itself was left entirely in the hands of national Meteorological Services, universities and research institutes. The research activities of the WMO Secretariat were practically limited to providing support to the working groups of the Technical Commissions. Of these, the Commission for Aerology (CAe) had the principal responsibilities for meteorological research although it must not be forgotten that the other Commissions had a general brief to "keep abreast of and promote meteorological developments in both the scientific and practical fields".

As no funds were earmarked by First Congress to finance meetings of working groups, their tasks had to be fulfilled mainly, if not entirely, by correspondence. The undoubted success of many of the groups was due to the unstinted devotion of the experts serving on the groups; special tribute must be paid to the chairmen without whom little would have been achieved. In many instances, especially within CAe, a working group was asked to prepare a report on the present state of knowledge in a particular specialized subject and to highlight those areas in which further research should be concentrated. The report had to be submitted for consideration by the parent Commission, which would then decide what further action was desirable.

At the first session of CAe, in Toronto in 1953, several scientific discussions were organized outside the formal business meetings. These provided an opportunity for individual members to present scientific papers on various topics of interest to the Commission. It was felt that some of these papers merited wider distribution and ought therefore to be published in some suitable form by WMO. This led to a decision by the Executive Council* in 1954 to start a new series of WMO publications, namely the *Technical Notes*. This series also provided a good medium for publishing appropriate reports of working groups and studies carried out in the new Technical Division of the Secretariat. The first Technical Note, entitled "Artificial inducement of precipitation", was in fact prepared in the Secretariat on the basis of information provided by experts in different parts of the world on the results of rain-making experiments. We shall return to this subject later.

* See note on p. 8.

Collaboration with ICSU

Perhaps the most far-reaching decision of the first session of CAe related to the meteorological programme for the International Geophysical Year. Its significance will be better appreciated after a brief description of the relationship between WMO and the other international organizations (non-governmental) with special interest in meteorology. In the days of the former IMO, there had been close collaboration with the International Association of Meteorology, established in Brussels in 1919 as one of the four major associations within the International Union of Geodesy and Geophysics (IUGG). The objects of this Association, whose name was changed to International Association of Meteorology and Atmospheric Physics (IAMAP) in 1957, are:

(*a*) To promote the study of all problems on the physics of the atmosphere;

(*b*) To initiate, facilitate and co-ordinate in this field those researches which require international co-operation;

(*c*) To stimulate discussions and provide for the publication of results of the researches.

The traditional means by which IAMAP strives to achieve these aims are by convening international meetings and by providing limited financial assistance for particular investigations. An example of IMO/IAMAP collaboration was the grant made in 1936 by IAMAP for the purchase of a number of radiosondes to enable some national Meteorological Services to take part in an investigation of the upper atmosphere. The grant was made on the condition that the countries concerned would carry out the observational programme as laid down by IMO.

The relations between WMO and IUGG were formalized by the signing of a Working Agreement in 1953. IUGG is one of the unions belonging to the International Council of Scientific Unions (ICSU) and a similar Agreement between WMO and ICSU was signed in 1960. The substance of these Agreements is that ICSU (and its subsidiary IUGG) is recognized by WMO as the international forum for the advancement of meteorology while WMO is recognized by ICSU as having the primary responsibility for the international organization of meteorology. In 1960, Dr R. C. Sutcliffe, who at the time was both president of CAe and Secretary of IAMAP, paraphrased the Agreement as "assigning to WMO all the hard work of international organization and to IAMAP the arranging of scientific meetings". He felt that there was scope for improving the relations between the two organizations. For example, he considered that WMO did not "draw as freely as it might upon the scientific strength of the universities" while, in its turn, IAMAP paid scant attention to "the scientific problems of professional applied meteorology". Events were about to happen which would dramatically change this situation.

During the International Geophysical Year an increased programme of weather observations was carried out all over the world. Left: *launch of a rocket to carry out an upper-air observation;* above: *taking a solar observation on an ice shelf in Antarctica*

International Geophysical Year

At the time when WMO was coming into being, the most important development within IUGG was the follow-up to a proposal made in 1950 that the 75th anniversary of the First International Polar Year (1882/83) should be commemorated by organizing a similar venture in 1957. This proposal was approved by IUGG in 1951 on which occasion the International Association for Meteorology insisted that WMO be invited to participate. At its second session, in 1951, the WMO Executive Council* accepted this invitation and at the same time suggested that the project should not be limited to the polar regions; it should be extended to cover relevant geophysical phenomena all over the world. These ideas found support and the name International Geophysical Year (IGY) was adopted by ICSU in 1952. WMO was then invited to co-operate in the preparation of the meteorological aspects of the programme.

A preliminary draft programme was drawn up at a meeting of the ICSU Special Committee for the IGY (SCIGY) at Brussels in July 1953. Some meteorologists felt at the same time that the scope of the proposed meteorological programme was so comprehensive that it could not possibly be completed in the 18 months of the IGY. As a result, within WMO, the task of assigning priorities was given to CAe and, as already mentioned, the matter was discussed at the first session of the Commission later that year. The session adopted a series of recommendations based on the principle that, during the IGY, priority be given to problems which could be solved only by effective world-wide collaboration and, in particular, to problems relating to the large-scale dynamical and thermodynamical processes. The CAe recommendations found favour with the Working Group for the IGY which had been set up by the WMO Executive Council* in 1952 and the programme proposed by this group was approved by the Executive Council and, with a few additions, by IUGG.

In the best traditions of IMO, the Second Congress of WMO decided that the primary responsibility for implementing the IGY meteorological programme should be left to individual Member countries. There was however a significant departure from tradition when Congress asked the Executive Council* to develop a plan whereby the Secretariat could act as an international centre for the essential IGY meteorological observational data. As Congress did not allocate any funds for operating this centre, the Executive Council decided that it would have to operate on a self-financing basis, in other words, the costs would have to be covered by the income resulting from the sale of publications containing the IGY meteorological data. The initial capital for establishing and running the centre was obtained by deferring the repayment to Members of the US $140 000 surplus from the first financial period. In this way, the IGY Meteorological Data Centre came into being in October 1956 as part of the Technical Division of the Secretariat. For the first time, the Secretariat was to be given an important supporting role in a research project.

The primary function of the centre was of course to collect, collate and publish the millions of observations to be made at meteorological stations during the IGY. Prior to its establishment, agreement had been reached on the layout of the standard forms for recording the surface and upper-air observations. It had also been decided that the main observations would be published on microcards, a careful study having shown that, at that time, this was the most economical and convenient method. The first task of the Centre was therefore to develop a system for cataloguing the observations and to plan the layout of the microcards, each of which could reproduce up to 96 of the

* See note on p. 8.

standard forms. The full story of how the Meteorological Services of the world responded and of how the Centre successfully completed its job has been told elsewhere. Here, let it suffice to say that the main IGY surface and upper-air observations were published on some 16 500 microcards, which were sold at US $5 990 for a complete set. The sums realized by the sales slightly exceeded the overall cost of running the Centre.

So successful was the whole operation that when ICSU decided in effect to extend most of the IGY programme for a further year (1959), under what was called the International Geophysical Co-operation (IGC), the WMO Third Congress agreed that the Centre should also collect and publish the IGC meteorological data. This time the necessary initial capital was provided by a loan from the Munitalp Foundation, which was duly repaid in full from the income from the sale of the IGC microcards.

In the discussions on this item at Third Congress in 1959, some delegates expressed the view that the WMO Secretariat should continue to discharge the functions of the IGY Centre on a permanent basis. The prevailing view was however that the responsibility for the regular publishing of meteorological observations, and thus for making them readily available for research workers, should be in the hands of national Meteorological Services. Congress decided that the most useful function for the Secretariat would be to maintain a catalogue showing how meteorological data may be obtained for research purposes from different countries. The IGY Centre was accordingly closed at the end of 1961 and the Catalogue of Meteorological Data for Research was published progressively in three separate volumes in the course of the next few years.

The publication of the IGY meteorological data was not of course an end in itself. The ultimate success of the IGY depended on the use made of these data for meteorological research. It is difficult to assess precisely the value of all the research carried out by meteorological institutes and others, but, judging by the quality and quantity of the scientific papers published in the ensuing few years, based on the IGY meteorological data, it can confidently be asserted that thanks to these data the science of meteorology made very substantial progress.

Tropical meteorology

As already mentioned, the IGY differed from the earlier Polar Years in that the observational programme was not limited to the polar regions. Special efforts were in fact made to improve the observational networks elsewhere, and especially in the tropics, in order to obtain the data needed to study some of the outstanding problems relating to the behaviour of the atmosphere as a whole. The results provided a stimulus to the steadily growing interest in tropical meteorology, which at that time was much less well understood than the meteorology of the temperate latitudes. This was reflected within WMO by the development of a research programme in tropical meteorology. Even during the IGC in 1959, WMO had taken its first significant step in this direction by convening a Symposium on Tropical Meteorology; this was held in Nairobi and was largely financed by the Munitalp Foundation. On this occasion, scientists from many parts of the world presented the results of their most recent researches in tropical meteorology and at the end the participants drew up a list of topics on which they considered that further research was urgently required.

Delegates from tropical countries at Third Congress in 1959 pressed for increased efforts by WMO in tropical meteorology with the result that it was decided that the Organization "should do everything possible to initiate, sponsor and encourage the establishment and operation of one or more research institutes for tropical meteorology". Doubtless some of those present recalled the earlier discussion at First Congress on the subject of international meteorological research institutes (see Chapter II). As a first step, the Secretariat prepared a report on the activities of the four existing institutes for tropical meteorology and on the plans for establishing two new institutes. Valuable assistance was later provided to these two institutes, at Karachi and Pune, under the WMO Technical Co-operation Programme (see Chapter XIV).

Research on tropical meteorology is of course conducted in many other places in addition to the research institutes. In 1962, it was decided that the Secretariat should prepare and distribute annually a report, based on information to be provided by Member countries, giving a list of research projects in tropical meteorology under way with a brief description of each project. These reports have proved to be of great value, especially to those engaged in research or about to undertake a research project and wishing to find out what is already being done elsewhere.

Following the undoubted success of the Nairobi symposium, Congress also decided that an important part of the continuing effort of WMO to encourage research in tropical meteorology should be the convening of further gatherings of this kind. Thus it was that a second symposium was held at Rotorua (New Zealand) in November 1963. Again there was a high standard of scientific papers and discussions, greatly appreciated by the participants from 23 different countries. From that time on, WMO has organized such symposia at fairly regular intervals, sometimes covering the whole field of tropical meteorology and at others limited to some specialized aspect, such as the Symposium on Meteorological Aspects of Tropical Droughts, held in New Delhi in December 1981.

The above activities were all within the traditional role of WMO of encouraging research, leaving the planning and execution to national bodies. In the 1960s, however, there was a significant departure from tradition when WMO and ICSU jointly assumed the responsibility for planning and managing a major international research project, the Global Atmospheric Research Programme (GARP). The GARP Atlantic Tropical Experiment (GATE), the field phase of which was carried out in 1974, was by far the most important contribution made by WMO to tropical meteorological research up to that time. More will be said about this in the section on GARP. For the moment, the main point is that GATE produced the most complete set of observational data for a large part of the tropics ever to be published and that the availability of these data provided a great stimulus to research workers in this field. At the International Conference on Scientific Results of GATE, held in Kiev in September 1980, it was reported that about 1 000 scientific papers had already been published thanks to the GATE data and the magnitude of the interest in the problems concerned.

The upsurge of interest in tropical meteorology associated with GATE was reflected in the relevant activities of WMO. In 1975, the Working Group on Tropical Meteorology of the Commission for Atmospheric Sciences (the new name of the Commission for Aerology adopted by Fifth Congress in 1967) assessed the state of knowledge in several specific areas of tropical meteorology and outlined a possible long-term WMO research programme directed primarily towards topics of greatest economic significance. It was realized that many aspects of the programme were already being implemented, partly as a result of GATE, but the group felt that there would be fuller participation by Meteorological Services if the programme could be seen as an essential part of an overall WMO strategy taking account of the economic needs of tropical countries as well as of the wider scientific aims. These ideas were supported by WMO Congress, and led ultimately to the adoption of the WMO Programme on

Research in Tropical Meteorology, which by 1983 had six main components, namely tropical cyclones, monsoons, semi-arid-zone droughts, tropical rain-producing systems, interactions between tropical and mid-latitude weather systems, and tropical limited-area weather prediction and modelling in the tropics. Progress in all these components is monitored by the CAS working group. To illustrate the kind of activity undertaken within this programme, let us consider briefly three of the components.

The *tropical cyclone* component is closely related to the WMO Tropical Cyclone Programme described in Chapter V, but is more limited in scope in that it deals only with the research aspects. The principal aim is to improve our understanding of tropical cyclones (known in some parts of the world as typhoons or hurricanes) with a view to obtaining more accurate forecasts of the development and movement of individual storms. An important workshop was held in Bangkok in December 1985 with participation of 80 specialists from all parts of the world. They had in-depth exchanges of experience on research trends and considered that the meeting had helped to promote collaboration between researchers and operational forecasters.

The potential economic benefits of the *monsoon* component are enormous, dealing as it does with research to improve the accuracy of forecasting the onset, duration and intensity of the monsoons which bring vital rains to vast areas of Asia and elsewhere. Separate studies are being carried out for the summer monsoon and for the winter monsoon. Valuable data for both these aspects were provided by the Monsoon Experiment (MONEX) executed in 1978/79 as part of GARP, but the project needs observational data over a longer period of time. Such data are being collected at the activity centres in New Delhi and Kuala Lumpur, where research on numerical modelling of the monsoons is being intensified. Plans are also in hand for studies of the East African monsoon.

In recent years, serious food shortages have arisen from rainfall deficiencies in the semi-arid regions of Africa, Asia and elsewhere. Under its Technical Co-operation Programme, WMO has been involved since 1973 in the planning of studies of the recurrent droughts of the Sahel (see Chapter XIV); this project is aimed primarily at obtaining the best possible applications of existing knowledge to the planning and execution of agricultural work in the area. The *semi-arid-zone drought* component of the WMO long-term research programme in tropical meteorology on the other hand is directed towards improving drought-forecasting techniques by increasing our understanding of the causes of rainfall deficiencies. In the long run, this component will doubtless benefit from the work being done under the World Climate Research Programme, described in Chapter XII. In the meantime, much can be achieved by enabling researchers to meet in order to exchange information about their current investigations and ideas for further research. Such was the main purpose of the first and second Symposia on Meteorological Aspects of Tropical Droughts, held in New Delhi (1981) and Fortaleza (1984). Participants called for numerical experiments with improved models of physical processes in the tropics, continuation of synoptic/statistical studies and better international data-exchange facilities; work on all these items is under way.

Weather modification research

Reference has already been made to the first publication in the WMO *Technical Note* series, entitled "Artificial inducement of precipitation". At the time when this was issued, in 1954, the results of attempts to stimulate rainfall artificially, generally by seeding rain-producing clouds with crystals of silver iodide, had been very inconclusive. Most of the work up to that time had been operational, with commercial firms trying to meet the needs of farmers and others for more rain. In the above *Technical Note* it was recommended that further carefully controlled scientific—rather than operational—experiments should be undertaken. The subject was kept under review by CAe and its successor CAS, and further *Technical Notes* were published in 1956, 1969 and 1977. Official statements on the subject were also issued by WMO from time to time in response to a need expressed by directors of Meteorological Services for an authoritative opinion which they could use when faced with claims of success by commercial rain-makers.

The statement approved in 1970 confirmed that it was still not possible to give a final assessment of the results of experiments on precipitation stimulation, warm fog dissipation, hail suppression and hurricane modification and accordingly recommended that operational efforts should be undertaken only after the most careful study of the particular situation by experts and with the understanding that the desired results may not always be achieved. In a revised version of this statement, approved in June 1974, reference was made to the apparently contradictory results obtained, even from carefully controlled experiments on stimulation of precipitation. More positive results were reported regarding the dissipation of warm fog. Although hail-suppression experiments were very difficult to assess, there appeared to be "promising prospects of success in the near future". Although in some cases the seeding of hurricanes had been followed by reduced maximum wind velocities, confirmation that this was due to the seeding was still required from further experiments. In the latest revision of the statement, approved in 1985, it is concluded that "detection of the consequences of human intervention in tropical cyclones would not be simple" as the central pressure and maximum winds during their life cycle had been found to be more variable than had earlier been thought.

In 1975, the Seventh WMO Congress adopted plans for the Precipitation Enhancement Project (PEP), the aim of which was to conduct under WMO management an international experiment, each stage being carefully and objectively planned, executed and evaluated so that the results, no matter the outcome, would be accepted by the scientific community. After lengthy analysis of information from several locations, a site for PEP was selected in north-western Spain. Field measurements were made at the site in 1979, 1980 and 1981 with the object of determining whether clouds suitable for seeding occurred with sufficient frequency in the selected area so that scientifically acceptable results could be achieved with a seeding experiment no longer than five years in duration. The observational data were carefully analysed in a variety of ways, but the results were not sufficiently positive to justify proceeding with the actual rain-making experiments and the project was therefore discontinued by decision of Ninth Congress in 1983. The data from the field activities proved to be a rich source for research workers and formed the basis for several papers presented at subsequent scientific gatherings, such as the Fourth WMO Scientific Conference on Weather Modification, held in Honolulu in August 1985. Furthermore, the experience gained in planning and carrying out the field phase of PEP, as reported in *Synopsis of the Precipitation Enhancement Project—1985*, should be of great value for anybody contemplating further scientific efforts in weather modification.

In developing the scientific basis of weather modification, WMO has encouraged research in various aspects of cloud physics, such as theoretical simulations of cloud behaviour. An International Cloud Modelling Workshop was held in the Federal Republic of Germany in 1985 and a second was held in France in 1988. Participants at these workshops represented all the centres conducting research on cloud-scale numerical simulation. Recognizing the importance of cloud physics to many other areas than weather modification, Tenth Congress in 1987 established the WMO Cloud Physics and Weather Modification Research Programme. The fifth in the series of weather modification conferences (to be held in China in 1989) will also be expanded to include the application of cloud physics to additional topics, such as transport and removal of toxic materials, remote sensing, etc.

International Years of the Quiet Sun

As already described earlier in this chapter, WMO played a major role in what basically were IUGG/ICSU projects, namely the IGY in 1957–1958 and the IGC in 1959. WMO also collaborated with these organizations in the International Years of the Quiet Sun (IQSY) 1964–1965. For this, the meteorological observing programme was much less ambitious than for the IGY, being limited almost entirely to upper-air data, including rocketsondes and ozone measurements. By this time, most countries were publishing their own upper-air observations (radiosonde and radiowind) and a permanent international data centre for ozone data had been set up under WMO auspices by the Canadian Meteorological Service. There was thus no longer the same problem about making the data readily available for research workers as there had been for the IGY. Under the circumstances, it was decided that it would not be necessary to set up a data centre in the WMO Secretariat for the IQSY. A standard form for the rocketsonde data was introduced by WMO in consultation with the countries concerned and the president of CAe, and countries making this kind of sounding were requested to enter the results on these forms for sending to one or other of the IQSY data centres in the USA or the USSR.

Global Atmospheric Research Programme

At the time of the IQSY, planning had already started for the Global Atmospheric Research Programme, which was to involve a radically different form of collaboration between WMO and the ICSU family of organizations. In view of the historic importance of this unprecedented type of joint effort between a governmental organization (WMO) and a non-governmental organization (ICSU), it is worth describing in some detail the steps which led to the signing of the relevant WMO/ICSU Agreement.

Since the pioneering work of V. Bjerknes and L. F. Richardson in the early 1900s, meteorologists had dreamed that some day it might be possible to use a more scientific approach, based on the well-established dynamic and thermodynamic equations which govern the behaviour of the atmosphere. Two things were lacking; adequate world-wide weather observations, especially of conditions in the upper air, and a means for making the calculations sufficiently rapidly for operational purposes. Things began to change in the 1940s with the expansion of networks of radiosonde and radiowind stations. At the same time, the first electronic computers were being developed. A milestone was reached in 1950 when J. von Neumann and his collaborators produced the first reasonably successful numerical weather forecast. This stimulated further research at an accelerating pace and within a few years numerical methods were being used operationally. The experience thus gained led many to the conviction that the full potential of these methods was not being realized, partly because of gaps in the global observing networks. It was also accepted that further research into mathematical models of the atmosphere was required and that computers of even greater power than those currently available would be required both for research and for operational forecasting.

The next breakthrough came during the IGY with the successful launching of the first artificial Earth satellite, SPUTNIK-1. Here at last was a tool which, with development, opened up the real possibility of obtaining truly world-wide weather observations at a reasonable cost. On the political side, the growth of space technology was seen as an opportunity for international collaboration, especially between the Great Powers, and hence for improved international relations. Scientists and politicians had thus a common objective: to take full advantage of these exciting new developments. This realization led to the adoption by the General Assembly of the United Nations of a series of resolutions on the peaceful uses of outer space. From the WMO point of view, the most important of these was Resolution 1721 (XVI), which gave a leading role to WMO in international efforts to advance the state of atmospheric science so as to provide greater knowledge of climate and to develop existing weather forecasting capabilities.

In response to this resolution, the Secretary-General of WMO arranged with the two leading countries in space research, the USA and the USSR, to send top-level experts to Geneva to assist in the preparation of the report. Among them were Academician V. A. Bugaev and Dr H. Wexler, whose names will always be remembered in connection with the World Weather Watch (see Chapter III). The WMO report included the following preliminary list of research problems as worthy of efforts at the international level:

- Problems in connection with solar and other external influences on the Earth's atmosphere and the interaction between upper and lower atmosphere;
- Problems on the general circulation and heat balance;
- Problems on numerical weather prediction;
- Medium and long-range forecasting.

This report was duly considered at the next UN General Assembly, which adopted Resolution 1802 (XVII), Part III of which states the following:

1. *Notes with appreciation* the prompt initial response of the World Meteorological Organization to the request of the General Assembly, as embodied in Resolution 1721 (XVI), that it report on a programme to advance atmospheric science research and to develop improved weather forecasting capabilities in the light of developments in outer space;
2. *Calls upon* Member States to strengthen weather forecasting services and to encourage their scientific communities to cooperate in the expansion of atmospheric science research;
3. *Recommends* that the World Meteorological Organization, in consultation with other United Nations agencies and governmental and non-governmental organizations, should develop in greater detail its plan for an expanded programme to strengthen meteorological services and research, placing particular emphasis on the use of meteorological satellites and on the expansion of training and educational opportunities in these fields;
4. *Invites* the International Council of Scientific Unions through its member unions and national academies to develop an expanded programme of atmospheric science research which will complement the programmes fostered by the World Meteorological Organization.

The WMO response to the request to develop its plan for strengthening meteorological services was to intensify the planning of the World Weather Watch, with the results described in Chapter III. On the research side, the WMO Executive Council* established in 1962 a Working Group on Research Aspects of Meteorological Satellites which drew up a list of problems

* See note on p. 8.

requiring urgent attention. The proposals of the group were considered at Fourth Congress in 1963 when it was decided to set up a high-level Advisory Committee to advise the Executive Council* on "principal research problems in the atmospheric sciences, especially in connection with the application of meteorological satellites, and ways of promoting this research ...". On the basis of the list of problems prepared by the previous working group, the Advisory Committee selected a number of items which should receive the highest priority. The items were grouped under the following headings: atmospheric composition and structure; solar and other influences on the Earth's atmosphere; interaction between upper and lower atmospheric layers; interaction of the Earth's surface and atmosphere; general circulation and budgets of energy, momentum and water vapour; cloud and precipitation physics; atmospheric pollution; weather prediction; modification of weather and climate and research in sensors and measuring techniques. This list was approved at the following session of the Executive Council for Members to take into consideration when planning their national research programmes.

In the meantime, ICSU was considering the best way to respond to the UN request to develop an expanded programme of atmospheric science research. There was much debate about whether to assign responsibility for this to an existing ICSU body, such as IAMAP, or to set up a new body. Finally, in late 1964, IUGG set up by correspondence a co-ordinating committee which was later transformed into the ICSU/IUGG Committee on Atmospheric Sciences. The deliberations of this committee were greatly influenced by a proposal, emanating originally from the American scientist Professor Jule Charney, that the nations of the world should undertake a global observational experiment to measure the large-scale circulation of the atmosphere for a limited period of time. At its first meeting in Geneva, in February 1966, the committee fully supported this proposal and suggested that specific details for the experiment should be developed in consultation with individual scientists and with international and national organizations, both governmental and non-governmental. Collaboration between the ICSU/IUGG Committee and the WMO Advisory Committee was facilitated by convening their meetings in the same place with a few days of overlap. In this way, the proposal for a global experiment quickly came to the attention of WMO, where it was received enthusiastically.

The concept of the global experiment was refined by the ICSU/IUGG Committee and at its second session (April, 1966) it was decided to propose that 1972 be designated as a 12-month period for an intensive, international study and analysis of the global circulation in the troposphere and lower stratosphere (below 30 kilometres). It was on this occasion that the name Global Atmospheric Research Programme (GARP) was suggested.

Progress thereafter was rapid. At a meeting in November 1966, under the chairmanship of Professor G. O. P. Obasi (later to become the fourth Secretary-General of WMO), the CAe Working Group on Tropical Meteorology formulated plans for a data-gathering experiment in the tropics. This idea was taken up by both the WMO and ICSU committees and was accordingly included in the GARP proposals submitted to the WMO Fifth Congress in April 1967. Both committees had agreed by this time that the best way to complete the planning for GARP would be to establish a joint WMO/ICSU committee specifically for this purpose. This suggestion was duly approved by Fifth Congress and by ICSU and was then incorporated in the formal Agreement on GARP signed on 10 October 1967.

In this historic document, it was laid down that GARP would be sponsored jointly by WMO and ICSU. The Joint GARP Organizing Committee (JOC) was given the following tasks:

(*a*) To consider, to endorse and recommend jointly to ICSU and WMO scientific goals and plans for GARP, including sub-programmes, that are considered essential prerequisites in defining the scientific requirements of GARP (this will include defining detailed experimental objectives and operational requirements for their implementation);

(*b*) To recommend to WMO those techniques and procedures developed in GARP programmes that may be applied in the operation of WWW;

(*c*) To recommend to WMO the manner in which the scientific requirements of GARP can best be supported by the operation of WWW.

To facilitate its work, JOC was given authority to select and convene working groups and special study groups and to select consultants within its field of responsibility. The twelve scientists to serve on JOC were selected by mutual agreement between WMO and ICSU. Perhaps most importantly, JOC was requested to guide the activities of a small full-time Joint Planning Staff, which was located in the WMO Secretariat with personnel recommended by JOC and approved by WMO and ICSU. The Secretary-General of WMO was made responsible for the administration of the finances and the provision of services for JOC and JPS.

Even prior to the signing of the Agreement, considerable progress had of course been made on defining the objectives of GARP and on ways of achieving these objectives. Of special importance was the GARP Study Conference held at Skepparholmen, near Stockholm, in June/July 1967; it was convened by the ICSU/IUGG Committee with co-sponsorship by WMO. The 53 meteorologists from 13 countries who attended this conference made valuable suggestions regarding the research that should be undertaken before carrying out a global experiment, the observational requirements for the experiment, and the technical feasibility of its implementation.

The conference report was carefully examined at the first meeting of the JOC, held in April 1968 under the chairmanship of Professor B. Bolin. On the basis of the JOC proposals, the Executive Committees* adopted a statement in which GARP was defined as a programme for studying those physical processes in the troposphere and stratosphere that are essential for an understanding of:

(*a*) The transient behaviour of the atmosphere as manifested in the large-scale fluctuations which control changes of the weather; this would lead to increasing the accuracy of forecasting over periods from one day to several weeks;

(*b*) The factors that determine the statistical properties of the general circulation of the atmosphere which would lead to better understanding of the physical basis of climate.

The programme was envisaged as consisting of two distinct but closely related parts:

- The design and testing by computational methods of a series of theoretical models of relevant aspects of the atmosphere's behaviour to permit an increasingly precise description of the significant physical processes and their interactions;

- Observational and experimental studies of the atmosphere to provide the data required for the design of such theoretical models and the testing of their validity.

Over the next few years a dialogue on the following lines developed between the scientists doing research on atmospheric

* See note on p. 8.

models and instrument technologists who understood the practical limitations of the most advanced observing systems, including those based on artificial satellites. The experts in atmospheric models stated what elements should be observed and with what accuracy, frequency and spacing between the observations. The instrumental experts, who served on a working group set up by COSPAR (the ICSU Committee for Space Research), then indicated the extent to which these requirements could be met by existing techniques or by improved systems expected to become operational in time for the GARP experiment. Using this information, the modellers, working under the auspices of the JOC Working Group on Numerical Experimentation, carried out further numerical experiments simulating and thereby testing the adequacy of the proposed observational system. At each stage, all concerned had to bear in mind that in the end the governments must decide whether the experiment should go ahead; the cost therefore had to be kept as low as possible without compromising the scientific integrity of the experiment.

GARP Atlantic Tropical Experiment

From an early stage in the planning of GARP, it had been foreseen that one or more preliminary experiments on a regional scale might be necessary to provide data which would be needed for formulating models of particular atmospheric processes. Such experiments could also serve as a test of some of the observing systems planned for the global experiment. To follow up the suggestion originally made by CAe, the JOC organized a Study Group on Tropical Disturbances; the report of this group was considered at the second session of JOC in 1969 when it was decided to submit a formal proposal for a GARP tropical experiment to the Executive Committees* of WMO and ICSU. It was agreed that this proposal provided an adequate basis for proceeding with the planning of the experiment and that the time was ripe to convene a planning conference on GARP to enable nations willing to take a significant part to exchange information about their possible contributions.

This conference was held in Brussels in March 1970 with about 100 participants from 25 countries. Strong support was expressed for an experiment over the Atlantic between 20°N and 10°S approximately, where it seemed likely that better facilities would be available than in the area over the Pacific originally envisaged by JOC. The scientific objectives, as summarized later by JOC, were to provide a basis upon which to develop appropriate schemes for estimating the effect of the smaller tropical weather systems on the larger-scale circulations and to provide data against which the validity of numerical prediction methods could be tested in the tropics.

The stimulus thus provided by the Brussels Conference for the further planning of the GARP Atlantic Tropical Experiment (GATE) led speedily to the establishment of the necessary mechanisms to complete the arrangements and to manage the whole operation. All interested countries were invited to serve on the Tropical Experiment Council while those making major contributions were represented on the Tropical Experiment Board. A full-time management group was also set up under Dr J. P. Kuettner. After three years' hard work by all these bodies and by many other groups and individuals, the GATE operational phase was successfully implemented from June to September 1974.

It is not possible here to describe in detail the many different aspects of this exciting example of international co-operation in a scientific enterprise. Some 70 countries participated and there was an imposing array of ships and aircraft to supplement the observing facilities of the World Weather Watch; geostationary and polar-orbiting satellites also played an important role. The operation headquarters at Dakar in Senegal were manned by an international team of scientists from the main participating nations. Many difficulties were encountered, such as unexpected gaps in the WWW network and the unreliability of a new and relatively untested system for measuring upper winds. These were largely overcome and the final result was that, as mentioned earlier, unique data sets were obtained for the area of the tropical Atlantic. Many lessons were learned which proved to be useful in the planning and implementation of other GARP experiments. In summarizing some of these, Dr Kuettner added that "the human factor is more important than technology in international field research of such magnitude and complexity. If human relations between the scientists of many nations are as good as they were in GATE, success is almost assured".

* See note on p. 8.

Joint planning staff at WMO Headquarters discussing FGGE operations

Artist's impression of observing systems used for the First GARP Global Experiment

Global Weather Experiment

The excellent spirit of international co-operation manifested in GATE gave great encouragement to those engaged in planning other GARP experiments, especially the First GARP Global Experiment, later given the simpler name Global Weather Experiment but frequently called by its acronym FGGE. It had originally been thought that the FGGE could take place in 1972 but this proved to be unduly optimistic and in the end the field phase did not commence until December 1978. As for GATE, the surface-based observations for the FGGE consisted primarily of the WWW network. This included about 9 200 stations making surface observations and about 850 stations making upper-air observations. Over the oceans, there were five ocean weather stations and about 50 commercial ships making both surface and upper-air observations while some 7 400 merchant ships made regular surface observations during the course of their normal operations. Special steps were taken to bring the above WWW system up to maximum efficiency for the FGGE. For example, several special upper-air stations were installed on remote islands and arrangements were made to collect ship observations which had not been received via coastal radio stations. The artificial satellites forming part of the WWW also played a key role and here again a major effort was made to ensure that they met the enhanced requirements of the FGGE.

Of outstanding importance for the FGGE was the system of five geostationary satellites which provided complete coverage of the world between latitudes 30°N and 30°S and also partial coverage for a further 20 latitude degrees to the north and south respectively. Two of these were provided by the USA, one by Japan and one by the European Space Agency. The fifth was a joint effort with the USA providing the satellite and processing the data and the European Space Agency providing facilities for system control and data acquisition. The earliest weather satellites provided mainly cloud pictures, but in the 20-year period from SPUTNIK-1 up to the FGGE there had been remarkable technological developments. In particular, it had become possible to derive the vertical temperature profile from infra-red observations by polar-orbiting satellites while the cloud images from the geostationary satellites could be used for determining the wind field at cloud height. For the FGGE, a special communication facility was provided on the USA polar-orbiting satellites, the French ARGOS system, which made it possible to determine accurately the position of moving platforms, such as instrumented balloons or buoys drifting in the oceans, and also to collect the observations from these platforms.

At the outset of the planning for the FGGE it had been envisaged that a major role would be played by super-pressure balloons carrying an electronic thermometer and a small radio transmitter. These balloons rise to a height of 15 km or so and then drift along with the wind at this level. With a system such as ARGOS, it is possible to locate these balloons accurately and hence to determine the wind speed at the height at which they are floating. It was in fact the inventor of this balloon system, Vincent Lally, who convinced Jule Charney that a global weather experiment was feasible. In the event, the constant-level balloons were used to a much lesser extent than Lally had proposed; they nevertheless helped to improve the definition of the flow pattern in the upper air over the tropics. In all, some 300 such balloons were launched from Ascension Island and Canton Island.

An important contribution to the FGGE was also made by aircraft. In the first place, special arrangements were made to collect the observations made by commercial aircraft. The regular system whereby these observations are handed in at the end of the flight was supplemented by a system in which the data were automatically recorded on cassette tapes and another in which they were transmitted automatically via geostationary satellite to ground receiving stations. In this way, several thousand additional aircraft reports were yielded daily.

The second contribution by aircraft was the provision by the USA of about ten aircraft equipped with dropwindsondes. In this system, a radiosonde is dropped from an aircraft in flight and descends by parachute, transmitting back to the aircraft pressure, temperature and humidity data. The upper winds are determined by accurately locating the sondes with the OMEGA navigation system.

A unique feature of the FGGE was the deployment of more than 300 instrumented floating buoys over the oceans of the southern hemisphere south of latitude 20°. The need for this arose from the large gaps in the WWW surface stations in this part of the world, which is largely covered by the ocean. All the buoys measured the atmospheric pressure and the sea temperature, while some also measured the air temperature and wind speed. The data were transmitted by radio to the TIROS-N satellite carrying the ARGOS data-collection and location-finding system. A wide variety of means were used to deploy the buoys. Some were launched from Antarctic supply ships, commercial vessels, research ships and naval vessels; others were released from islands; about 20 were dropped from aircraft. In such a novel system, it was inevitable that there should be some failures. On the whole, however, the buoys worked remarkably well and there is no doubt that this was one of the great successes of the FGGE observing programme.

To meet the scientific objectives of the FGGE, it was essential to have a dense network of stations measuring the detailed vertical wind profile in the equatorial tropics. The WWW network was inadequate and the wind observations from satellites, aircraft and floating balloons could not provide the required vertical resolution. A fleet of tropical wind observing ships was therefore organized for this purpose. It consisted of a mixture of research vessels already carrying the necessary equipment and various other vessels for which a special system, the NAVAID wind-finding system, was developed under WMO auspices. In all there were more than 40 ships provided by 16 countries, with 25 NAVAID equipments supplied by WMO from a special fund.

The research meteorologists for whom the FGGE had been planned would of course have liked to have had data from all the above observing systems for the whole duration of the experiment. Unfortunately, the resources available were insufficient for this and it was therefore decided to concentrate the special observing systems during two special observing periods. Furthermore, as some time was needed to bring some of the systems up to full potential, it proved desirable to have a build-up period. The final schedule was: build-up year 1 December 1977–30 November 1978; operational year 1 December 1978–30 November 1979; special observing periods 5 January 1979–5 March 1979 and 1 May 1979–30 June 1979.

It can well be appreciated that to collect and analyse all the FGGE data was a task well beyond the capabilities of the telecommunication and data-processing facilities of the WWW. In any case, whereas the WWW operates essentially in real time to meet operational needs for day-to-day weather forecasting, many of the FGGE data from the special observing systems only became available after an interval of days or weeks. It was therefore necessary to design a special data-management plan for the FGGE. The main features of this highly complex scheme may be summarized as follows. Four existing meteorological centres in Japan, the UK, the USA and the USSR collected from a specified area all the data from the surface-based WWW stations, including any which were not transmitted in real time and had therefore to

A radiosonde about to be carried aloft by balloon to obtain upper-air data

A NOAA ship with a tethered balloon for taking upper-air meteorological observations

be gathered by mail or other means. All these data were transcribed in prescribed formats on magnetic tape and a complete set of tapes was assembled in the USSR centre. The data from the space-based and special observing systems were collected by about 15 centres and the resulting data sets were forwarded to a centre in Sweden where they were entered onto magnetic tape. The centres in Sweden and the USSR exchanged all their data and arrangements were made for complete data sets to be archived at the World Data Centres in Washington and Moscow, from where they could be obtained by research groups.

From what has been said earlier, it will be clear that much of the FGGE research has to be carried out by computer. For testing numerical models of the atmosphere it is desirable that the data should refer to observations made at fixed times from a fixed network of stations. Many of the FGGE observations were however taken between the regular stations of the WWW and at times other than the internationally agreed synoptic hours. To facilitate this kind of research, it was therefore necessary to estimate from the real, irregularly distributed observations the values of the meteorological variables at a regular network of points over the globe at evenly spaced times throughout the year of FGGE operations. This mammoth task, calling for highly sophisticated computational techniques, was carried out at the Geophysical Fluid Dynamics Laboratory in Princeton, USA, and at the European Centre for Medium-range Weather Forecasting at Shinfield Park, England.

At first sight, it might appear that such a highly complex undertaking as the FGGE would require a large central management body, but in fact this was not necessary. Each of the observing systems had its own arrangements for operational management and there was thus little scope or indeed any real need for a strong central control. It was nevertheless foreseen that contingencies could arise for which limited but useful remedial action might be possible and a modest Operations Centre was accordingly set up in the WMO Secretariat. The staff of the Centre performed their tasks on the basis of status reports on all the components of the observational system. In this way, the Centre was able to follow especially the operation of those components which were complementary; for example, it proved possible to avoid a certain amount of duplication between the tropical ships scheme and the dropwindsonde programme.

The principal aim of the FGGE was of course to tackle the problems on a global scale, but the opportunity was taken in the course of the operational phase to attack some weather problems of a regional nature. For example, reference has already been made to MONEX, an experiment concerned with both the north-east monsoon of the countries of south-east Asia and the south-west monsoon of India and its neighbours. During the FGGE, there was also an experiment on the south-west monsoon of West Africa, WAMEX. Quite apart from the scientific value of these regional experiments—there was also one for the polar regions, POLEX—they had the advantage of giving additional incentives to the countries immediately affected to make their maximum contribution to the FGGE as a whole.

Mention has already been made of the research ships used to provide observations over the vast expanses of the tropical oceans. These ships are normally employed for oceanographic research and those responsible for their programmes were naturally keen that during the FGGE they should also carry out oceanographic studies in addition to making the meteorological observations. Furthermore, the oceans are of great importance in meteorology as they act as a heat reservoir and thus have a significant moderating effect on the weather and climate. In view of these and other considerations, close collaboration was maintained through all stages of GARP between the oceanographic and meteorological communities. For example, two of the twelve members of JOC were oceanographers and WMO paid close attention to the deliberations of the Scientific Committee on Ocean Research (SCOR) of ICSU. The oceanographic programme established under the auspices of SCOR was carried out concurrently with the operational phase of the FGGE.

When the field phase of FGGE ended on 30 November 1979, the most immediate remaining task was to complete the collection and processing of the mass of observational data. The raw data from the satellites, consisting of telemetry signals, had to be converted into meteorological parameters which then had to be combined with the more conventional direct observations into complete data sets. Most of this work was accomplished through the WWW system but there remained all the observations which for one reason or another were not received sufficiently quickly for operational weather forecasting. The most difficult part was to produce from all these data internally consistent data sets in grid-point form.

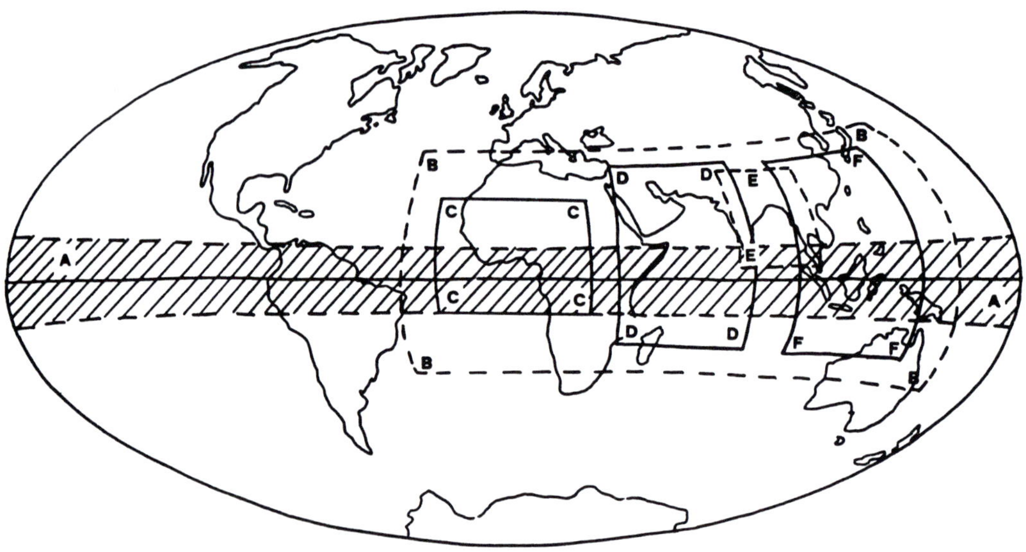

A FGGE tropical wind observation belt
B Monsoon subprogramme region
C West African monsoon (WAMEX) region—May to August 1979
D Arabian Sea studies (MONEX)—May and June 1979
E Bay of Bengal studies (MONEX)—July and August 1979
F Winter monsoon studies (MONEX)—December 1978 to February 1979

Monsoon sub-programme, regional experiment areas and the FGGE tropical wind observing belt

As might be expected, handling the WWW data proved to be relatively straightforward and complete sets in grid-point form were available at the World Data Centres within a few months. Unforeseen problems were encountered in trying to produce the final data sets, which is not surprising in retrospect when one considers that nothing on this scale had ever been attempted before. Some of the data had to be reprocessed to remove some inconsistencies and the work was not completed until 1984. In the meantime, many research projects had been started and some successfully completed. At an international conference on the early results of the FGGE and the large-scale aspects of its monsoon experiments, held in Tallahassee, Florida, in January 1981, the overall conclusion was that the FGGE data were of very good quality and that all the observing systems had by and large performed according to expectations. Even earlier, in December 1979, at a symposium sponsored by the Australian Academy of Science and Bureau of Meteorology, it had been concluded that the extra data available in real time during the FGGE, and especially the observations from the buoys, had significantly improved weather forecasting in the southern hemisphere.

By 1984, it was already possible to start evaluating the results of research based on the FGGE data and to consider their practical implications for the WWW. For this purpose, four seminars were convened to review the progress in specific fields. These were: data assimilation and observing systems experiments (ECMWF, Shinfield Park, September 1984); tropical meteorology, numerical forecasting and data assimilation in the tropics (Tallahassee, Florida, October 1984); global diagnostic studies (Helsinki, August 1984); and numerical modelling and predictability (Sigtuna, Sweden, October 1984). In all these areas, striking advances were reported.

The above seminars were arranged in conjunction with a major conference on the results of the FGGE and their implications for the WWW, which was held in Geneva in May 1985. Here there was a comprehensive review of progress made in all aspects of meteorological research relevant to the FGGE, of the scientific and technical results and of the derived benefits to meteorology. The conference approved a statement on the overall accomplishments of the experiment, the gist of which was well expressed in the following quotation from the *Annual Report* of WMO for 1985:

> ...the Global Weather Experiment was the most ambitious and complex international scientific undertaking ever realized in the field of meteorology. Six years after the field phase, it can now be judged a real success, fully achieving its objectives of observing and measuring, more intensively than ever before, the development of weather systems over the entire globe and of amassing an unprecedented data set, leading to improved understanding of atmospheric motions and weather forecasts of greater accuracy and range. The Global Weather Experiment produced the resources necessary to establish a global integrated system of geostationary and polar-orbiting satellites that can now be used by all nations in support of their weather and climatic services and will continue to form the backbone of the World Weather Watch to the year 2000. It also catalysed the development and demonstrated the capabilities of several powerful new observing techniques which are likely to make cost-effective additions to the future global observing system.

Mention has already been made of several of the GARP sub-programmes, such as GATE, MONEX, and WAMEX, which were particularly relevant to the FGGE. There were also two other sub-programmes of more limited regional extent. The first of these was the Air Mass Transformation Experiment (AMTEX), designed to clarify the transfer processes by which energy and momentum are supplied from the sea surface to the air and transported to the free atmosphere through the planetary boundary layer. Two field experiments were made in February 1974 and February 1975 in the vicinity of the south-west islands of Japan, where the Kuroshio current flows towards the north-east. The work was carried out mainly by Japanese scientists with help from Australia, Canada and the USA. The resulting research threw much new light on the problems in question. The last of the sub-programmes was the Alpine Experiment (ALPEX), the general objective of which was to determine the airflow and mass field over and around mountain complexes under various synoptic conditions. Some 20 countries, mainly European, participated in the project, which included a special observing period in March/April 1982. The results were reviewed at a Scientific Conference in Vienna in October 1985. In brief, it was concluded that greatly improved knowledge had been obtained of local mountain winds in the Alps and that much progress had been made in representing the effects of orography in atmospheric models, a subject of considerable importance for operational numerical weather prediction.

Physical basis of climate

The Global Weather Experiment of 1978/79 and its related experiments (GATE, MONEX) were designed to meet the first objective of the GARP, namely the improvement of weather forecasts. It was recognized from the outset that experiments of such short duration could make only a very limited contribution towards the second objective, the understanding of the physical basis of climate. For this it would clearly be necessary to conduct research over a period of years, if not decades. Even before the completion of the field phase of the FGGE, JOC and other bodies had started to turn their attention to the problem of how best to tackle the longer-term problems associated with climate. Apart from the scientific interest in climatic research, the whole subject of climate, and in particular the practical effects on mankind of possible changes in climate, started to receive international attention in the early 1970s, partly as a result of the serious droughts in the Sahel and elsewhere. For example, at the United Nations Conference on the Human Environment at Stockholm in 1972, it was recommended that WMO, in co-operation with ICSU, should undertake activities aimed at improving our understanding of the causes of climatic change, whether natural or man-made.

WMO had been involved for many years in studies of climatic fluctuations, within both CCl and CAS. The growing interest in the research aspects was reflected in 1971 by the transfer from CCl to CAS of the Working Group on Climatic Fluctuations. In 1974, the WMO Executive Council* raised the status of the topic by establishing a Panel of Experts on Climatic Change which drew up proposals for coherent international action in this field. Following a proposal by JOC and doubtless remembering the great success of the GARP Study Conference of 1967, WMO and ICSU jointly convened an International Study Conference on the Physical Basis of Climate and Climate Modelling at Wijk near Stockholm in July/August 1974. The conference was attended by a cross-section of leading climatologists, oceanographers and atmospheric modellers who together proposed a series of observational programmes to obtain the data needed for climatic studies on various time scales. These proposals were further developed by JOC in a recommendation for a GARP Climate Dynamics sub-programme.

All the above activities, and in particular the plan prepared by the EC Panel on Climatic Change, led to the convening in 1979 of the World Climate Conference, which in turn gave rise to the World Climate Programme with its important research component. More will be said about this in Chapter XII.

---

* See note on p. 8.

WMO Research and Development Programme

It is evident that WMO research activities have expanded over the years, especially in relation to the GARP and its successor, the World Climate Research Programme, both of which are carried out jointly with ICSU. In parallel with these two major joint undertakings, WMO has also developed its own research programme in such a way that instead of being a miscellaneous collection of individual projects the latter has now become a systematic comprehensive programme in which priorities are assigned at each session of Congress. In between these sessions, a close watch on progress is maintained by the Executive Council while CAS is responsible for giving guidance on the execution of each of the priority items. The overall WMO Research and Development Programme is at present divided into the following main groups: weather prediction research; tropical meteorology research; weather modification research; and environmental pollution monitoring and research. The work in tropical meteorology and weather modification has already been described and it remains to deal with the last two groups, namely weather prediction and environmental pollution.

Six study projects are at present under way in activity centres supported by 14 countries. They cover the following items: application of high-resolution quantitative satellite data; limited-area weather prediction modelling; phenomenological studies; objective interpretation methods; very short-range forecasting; and mediterranean cyclones. For each of the projects, a steering group drawn from activity centres is responsible for its implementation, including organizing symposia, workshops and scientific meetings, preparing and publishing technical reports, and organizing catalogues and data sets. The CAS Working Group on Short- and Medium-range Weather Prediction Research oversees all this work and produces a highly appreciated annual report on progress in numerical weather prediction, based on information provided by Member countries.

Another CAS body, the Working Group on Long-range Forecasting, carries out similar activities related to weather prediction for longer periods, such as months or seasons. The latest WMO Conference on Long-range Forecasting was held in Toulouse, France, in June 1987. The topics discussed on this occasion included the effects of sea-atmosphere interaction,the use of analogue and statistical methods, the application of dynamic models, and methods of verification and testing. The proceedings of such conferences and other relevant papers are published in the *Long-range Forecasting Report* series.

Environmental pollution research

Much of the WMO work on environmental pollution relates to the network of stations for monitoring background air pollution which is described in Chapter XI. Here we are concerned more with the research aspects of the subject. There are several important facets of this, such as developing mathematical models for use in studies of the transport and dispersion of atmospheric pollutants, the role of contaminants in modifying physical, chemical and biological processes in the troposphere and of course the possible climatic effects of atmospheric pollutants as discussed in Chapter XI. To illustrate the role of WMO in such research work, let us consider what is being done with regard to atmospheric ozone.

The pioneering work by G. M. B. Dobson in the early 1930s led gradually to the establishment of a world-wide network of stations to measure atmospheric ozone. To the extent that international support was necessary, this was provided by the International Ozone Commission of the International Association of Meteorology. In the spirit of the WMO/IUGG Agreement of 1953 mentioned earlier (see page 73), the responsibility for the operational aspects of ozone measurement was transferred to WMO in 1959; one result of this is that the World Ozone Data Centre in Toronto was opened in 1960 and has since been operated by the Canadian Meteorological Service under WMO auspices. For many years, this work was considered to be mainly of scientific interest with little practical significance. It was of course realized that by absorbing harmful solar ultra-violet radiation, the ozone played a vital role in human well-being, but there was no reason to suppose that this protective shield might someday be destroyed. The improved observations of the vertical distribution of ozone during and after the IGY showed that the processes whereby the balance of ozone is maintained in the stratosphere could not be entirely explained by contemporary theory and this led to the discovery of the important role of trace gases, such as the oxides of nitrogen. It seemed therefore that ozone could be affected by man-made influences, such as the release of nitrogen oxides by supersonic aircraft. In the early 1970s, a further threat to the ozone layer was postulated, namely the release into the atmosphere of chlorofluoromethanes used in refrigerators and aerosols. This gave rise to increasing public concern, as regards not only the effects on health of a reduction of ozone, but also the possible climatic changes.

In the light of these developments, WMO gave higher priority to the relevant activities. For example, in 1973, CAS established a new working group to study stratospheric questions and called for increased monitoring of ozone. At a session of the group in September, a statement was prepared on "Modification of the ozone layer due to man's activities and some possible geophysical consequences". Following the release of this as an official WMO statement, the WMO Executive Council* approved in 1976 a plan for a Global Ozone Research and Monitoring Project aimed at enabling WMO to provide advice to Members and to the United Nations concerning:

(*a*) The extent to which man-made pollutants might be responsible for reducing the quantity of ozone in the stratosphere, with particular emphasis on the role played by chlorofluoromethanes and nitrogen oxides;

(*b*) The possible impact of changes in the stratospheric content of ozone on climatic trends and on solar ultra-violet radiation at the Earth's surface;

(*c*) The establishment of the basis and identification of the needs for strengthening the long-term monitoring programme of the ozone system for determination of trends and of future threats to the ozone shield.

Support for this project was provided by UNEP and scientific guidance was given by the CAS Working Group on Atmospheric Ozone, set up especially for this purpose. At the request of UNEP, WMO prepared a comprehensive review paper for a meeting in Washington, DC, in March 1977, on the ozone layer. In the World Plan of Action on the Ozone Layer adopted by this meeting, WMO was designated as the leading agency for the relevant aspects of the plan.

The subsequent WMO activities in this field have included: assisting countries in the development and maintenance of stations for measuring ozone; designating the NOAA

* See note on p. 8.

laboratories in Boulder, Colorado, as the World Dobson Spectrophotometer Central Laboratory; organizing international comparisons of Dobson and other ozone-measuring instruments; and convening international scientific meetings to discuss various specialized aspects of the ozone problem. The results of all these efforts are being published in the WMO *Ozone Project Report* series, of which more than 20 have now been issued.

Concluding remarks

It will be clear from the preceding paragraphs that WMO has, from the outset, given great attention to research activities and has steadily expanded its efforts in this field. It will be equally clear that these activities have been highly successful and that great progress has thereby been achieved in expanding man's knowledge and understanding of the atmospheric processes on a truly global basis. The scale of these activities has been such that they could not have been undertaken without full support at the governmental level, support which it would have been difficult to envisage if WMO had not been a governmental body enjoying the full confidence of its Members.

It is important to note, however, that from the outset WMO has co-operated fully with the appropriate non-governmental international scientific bodies, notably the International Council of Scientific Unions and its constituent Unions. From this point of view, the successful formulation of the programmes and their implementation have been essentially joint endeavours. Indeed, this happy combination of scientific efforts was doubtless an essential element in winning the confidence and support of the Members of WMO. The scientific input has thus been of the highest possible standard and the execution of the various programmes has been commensurate with their importance.

The many practical benefits already derived from these research activities will be evident from the information given in the chapters of this book which deal with the operational and other practical applications of meteorological knowledge. Much further practical assistance in such fields will doubtless become available in the years to come as the data acquired in completed research programmes are studied further and as still more information becomes available from research programmes already in progress and those now being planned.

CHAPTER XI

PROTECTION OF THE ENVIRONMENT

Introduction

This chapter is designed to serve a twofold purpose—first, to describe the work of WMO aimed directly at the problems and care of the environment and, second, to show how the efforts of the UN family, including WMO as a Specialized Agency, and of other governmental and non-governmental organizations to prevent further degradation of the environment and to make improvements wherever possible are combined into a major world-wide programme. In recent decades there has been a strong upsurge of interest in the human environment with many countries giving great attention to environmental questions. However, the subject is very complex and has numerous aspects. In many respects, therefore, it can be treated in a satisfactory manner only by an international approach.

The London smog of November 1952 brought to a standstill a vast area of the city and its suburbs and was blamed for a loss of life regarded as totally unacceptable in an advanced country. Within two years the Government introduced the Clean Air Act and the problem of smoke in the atmosphere of industrial areas was virtually solved. Similar, and in some cases wider, legislation was enacted in Europe and North America when atmospheric pollution reached an intolerable level for a fairly lengthy period. These examples were associated with particular, well-known weather conditions. In addition, industrialized countries were gradually compelled to take measures against other forms of pollution affecting rivers, coastal waters and the high seas, and most recently the upper atmosphere.

In the present century the number of industrial or near-industrial countries has continued to increase and thus more and more countries have become aware of environmental pollution problems. Although some examples of pollution were seen as widespread, e.g. in the case of a contaminated river flowing through a number of countries, it was generally considered that the effects from a single source of pollution were not measurable beyond a radius of one or two hundred kilometres. Eventually, however, the view was expressed in a number of countries, mostly among workers dealing with various aspects of environmental pollution, that the phenomenon was of wide scope and could possibly affect not merely individual countries but larger regions of the world. Atmospheric scientists in particular had long noted that atmospheric pollution was capable of modifying the climate of a locality and had also demonstrated that, when measures such as the British Clean Air Act came into force, significant improvements in local climates generally occurred.

The next step for atmospheric scientists and associated workers in the field of pollution was to consider whether man's industrial activities could produce more permanent changes in the atmospheric composition and long-range transport of pollutants and affect regional and even the global climate. The questions inevitably came to the forefront because, after World War II, the growth of industry and the increase in energy consumption were on an accelerating scale, promising beneficial results but also giving rise to concern about the likelihood of dire effects upon the world community, such as acid rain and climate warming. Great attention was paid to the rate of consumption of fossil fuels, principally coal and oil, and to the amounts of carbon dioxide released into the atmosphere year by year. The problem of atmospheric warming caused by the so-called greenhouse effect became a subject for discussion by politicians and administrators as well as by technologists. Man's impact upon climate became a live topic for discussion among governments and within the United Nations. It was soon realized that the subject was not a simple one concerned only with climate but that it contained many ramifications of vital importance to the welfare of mankind. The subject was clearly one for all countries to consider within the framework of the United Nations; it was necessary to plan international courses

Industrial activities can jeopardize the health of populations

of action. In the late 1950s WMO began to show concern about changes in the composition of the atmosphere and the possibility of adverse consequences and it began co-ordination of a world ozone network and of a background air-pollution network. In the late 1960s additional problems of a serious nature concerning the state of the environment came to the forefront and were widely discussed in the United Nations as a whole.

Strict observance of effluent laws is becoming increasingly necessary

WMO and the environment

The responsibilities of WMO are concerned not only with meteorology, the science of the atmosphere, but also with operational hydrology, the study and management of water resources. WMO is also involved in the practical applications of meteorology and hydrology affecting most sectors of the industrial, economic and social life of a country. All WMO's activities are concerned with the environment, whether they take the form of fundamental research or applications in specific projects, and it is therefore not unusual to see WMO described as an environmental organization. WMO's specific responsibility is to provide authoritative scientific information and advice on the condition and behaviour of the global atmosphere and climate and the factors that affect them.

When IMO, the predecessor of WMO, was created in 1873, the purpose was to develop networks of meteorological observing stations so that progress could be made in the science of weather forecasting, the primary objective being to contribute to the safety and efficient operation of shipping. Over the years the range of meteorological services expanded and the requirements became more exacting, most notably with the advent and rapid growth of aviation. This work of weather forecasting together with the increasing use being made of climatology was mainly concerned not with the protection of the environment but with assisting the efficiency and safety of operations taking place in the environment. At the same time a significant amount of information about the atmosphere, such as its composition, was being accumulated although applications for user interest were not clearly foreseen. For example, long before the ozone layer became a topic for public discussion, a small number of meteorologists in various parts of the world were using Dobson ozone spectrophotometers to measure the characteristics of the ozone layer and to study their variations in order to construct theories of atmospheric ozone as well as for stratospheric circulation studies. Thus when meteorologists began to discuss whether changes in atmospheric constituents resulting from industrial processes might cause changes in climate extending beyond local areas into large regions or even the Earth as a whole, the observations required and the means of obtaining them were already largely available.

For several years in the latter part of the 1950s the WMO Commission for Aerology (now called the Commission for Atmospheric Sciences) discussed the need for a comprehensive programme of research into the composition of the atmosphere and the changes revealed by monitoring. The importance of atmospheric chemistry as a branch of meteorology was clearly recognized and in 1961 the Commission adopted a series of recommendations for submission to the Executive Council* of WMO regarding observations of the chemical composition of the atmosphere and the need to collect and publish the data so obtained. The Executive Council approved the proposals and since that time has consistently accepted the Commission's view that studies in atmospheric chemistry and in the meteorological aspects of air pollution are so interrelated that they should be considered as a single subject. In 1982 the Executive Council stated that the study of tropospheric chemistry, including the life cycles of chemical substances, and of anthropogenic factors contributing to acid rain, as well as of the transport of pollutants in the atmosphere, should be given priority by atmospheric chemists in co-operation with meteorologists. Furthermore, in 1989 the Executive Council reiterated that atmospheric chemical observations should be carried out with the same attention as the measurements of other meteorological parameters and urged national Services to ensure that the chemical composition observations become an integral part of atmospheric observations in general during the 1990s.

WMO's concern about effects on the environment, including the possibility of inadvertent changes in climate resulting from variations in the composition of the atmosphere, was reflected in proposals put forward by the Commission for Atmospheric Sciences (CAS) for the development of a Background Air Pollution Monitoring Network, generally known by the acronym BAPMoN. The programme received wide support from the Member countries of WMO and the number of observing stations increased rapidly after its establishment in the late 1960s.

WMO was also well ahead of world concern in the measurement of the concentration of ozone in the atmosphere. In preparation for the International Geophysical Year (1957-1958) it became the focal point for collecting and publishing ozone data, a responsibility taken over on its behalf by Canada in 1960. From about 1957 WMO had assumed responsibility for guidance and practical help for establishing standard international procedures and thereby ensuring uniformity and high quality of ozone observations. In the early 1970s widespread concern arose over the possibility that supersonic transport flights might have destructive effects on the ozone layer. WMO studied this question and issued a statement saying that the main threat to atmospheric

* See note on p. 8.

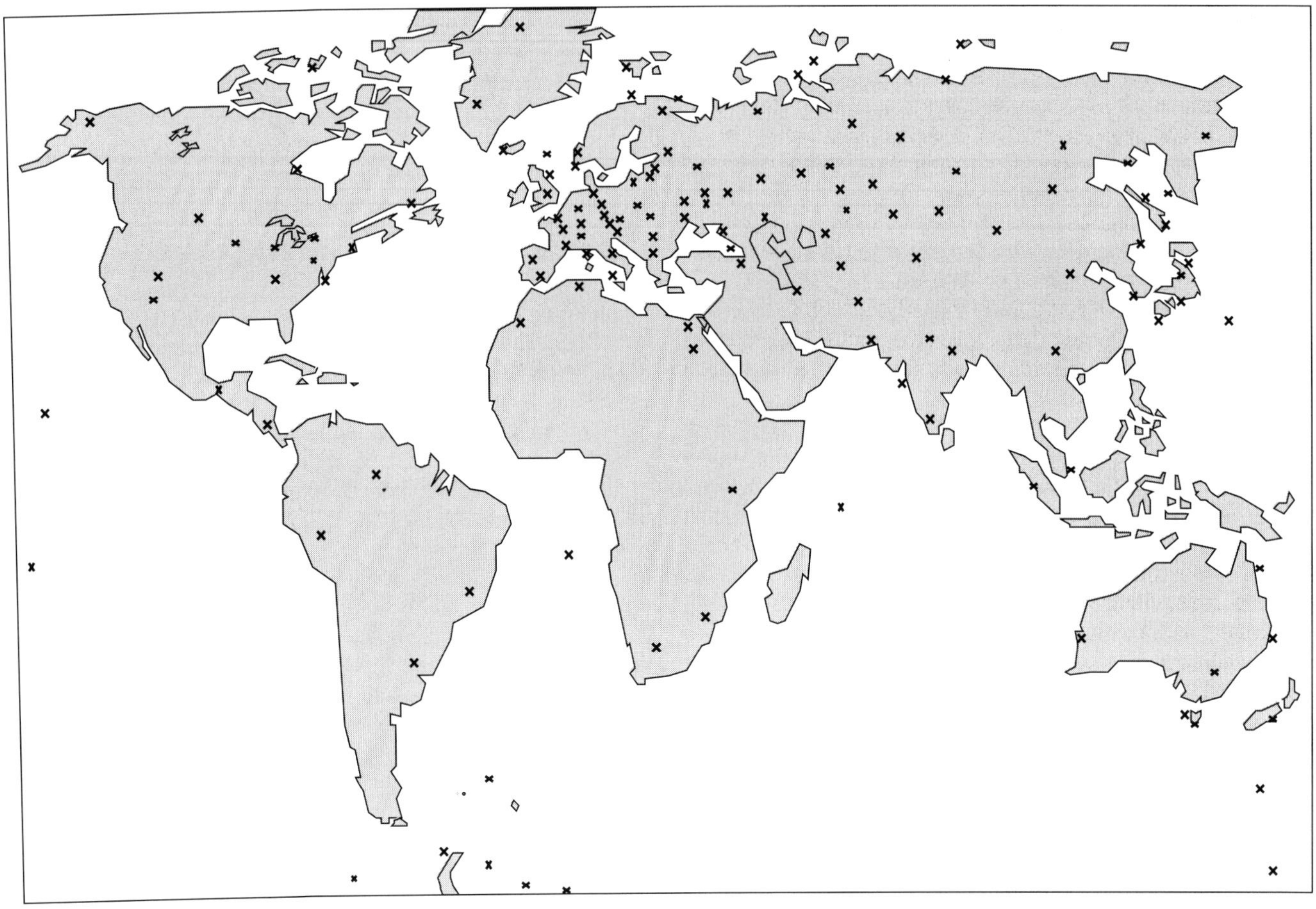

The Global Ozone Observing System (GO_3OS) stations

ozone was not aviation but the growing release of chlorofluorocarbons. In 1976 the Executive Council* of WMO established a Global Ozone Research and Monitoring Project which has from time to time issued valuable assessments and benchmark reports. At the present time more than 140 ground-based ozone stations, operated by nearly 60 Member countries of WMO, constitute the foundation of the Global Ozone Observing System (GO_3OS).

There has been continuing public concern about the possibility of a reduction in the amount of ozone in the stratosphere as a result of human activities, which, it is recognized, could have serious consequences for mankind. An authoritative statement on the subject was first issued by WMO in 1975. This statement was brought up to date and re-issued in 1978, followed by another in 1982. In addition, comprehensive assessment reports on the state of the stratospheric ozone were published in 1981 and 1985. In 1981 and 1982 the WMO Ozone Project issued two benchmark reports on the potential climatic effects of ozone and other minor trace gases, thus starting to focus scientific attention on this important subject of potential ozone change. At the same time WMO was at the forefront of providing the necessary scientific information to the intergovernmental negotiations which in 1985 culminated in the signing of the Vienna Convention for the Protection of the Ozone Layer. In 1988 a comprehensive report was drawn up by an international Panel on Ozone Trends established by NASA/WMO, in collaboration with scientists from a large number of countries and also from UNEP and other international bodies.

A WMO Project on Research and Monitoring of Atmospheric Carbon Dioxide was approved in 1977. Incorporated in the description of the project were elements of a first WMO statement on the possible effects on climate of increasing concentrations of CO_2. A formal statement was published in 1979. During the 1980s the carbon dioxide problem has been in 1979. During the 1980s the carbon dioxide problem has been tackled in connection with environmental research and monitoring and under the World Climate Research Programme. BAPMoN provides the only continuous measurements of carbon dioxide which have shown that it increased dramatically from ~315 ppmv in 1957 to just over 350 ppmv in 1988. WMO arranged for continuous provision of standard calibration gases by supporting a Central CO_2 Laboratory hosted by the Skrips Institute of Oceanology. It also supported the first international assessment of the pre-industrial levels of CO_2 in 1983 and a number of conferences on the observation and analysis of CO_2.

WMO has also been active, in collaboration with IOC and UNEP, in marine pollution monitoring, particularly in the studies relating to ocean-atmosphere interchange of pollutants. The meteorological component was accorded a prominent place in the GESAMP activity and a detailed study of the role of the atmosphere in the pollution of the oceans was published by WMO in 1989 as GESAMP Report No. 14.

From the late 1960s WMO and its Commission for Atmospheric Sciences encouraged studies of the long-range transport of pollutants and the Organization pioneered in promoting the relevant programme as an essential prerequisite to understanding the world balance and distribution of pollutants. Its efforts were rewarded when in 1977 the Economic Commission for Europe (ECE), WMO and UNEP launched the Co-operative Programme for the Monitoring and Evaluation of the Long-range Transmission of Air Pollutants in Europe, known as EMEP. Under this programme WMO established two meteorological data assessment centres (in Moscow and Oslo), which calculate the long-range trans-boundary transport of primary pollutants in Europe. The EMEP became a project under the Convention on

* See note on p. 8.

Long-range Trans-boundary Air Pollution adopted by the intergovernmental conference organized by ECE in 1983. WMO continued to collaborate on the meteorological component of the project and has carried out periodic assessments, the most recent of which was conducted in 1989.

The projects initiated by CAS are carried out by working groups consisting of scientists from Member countries of WMO supported by scientists in the Research and Development Department of WMO. Reports prepared by these groups and published after approval by the Executive Council earn considerable influence in the scientific community and elsewhere. (Some of the topics mentioned above are further discussed in Chapter XII—The World Climate Programme.)

More recently, in order to respond to the challenging tasks in the field of the environment, the Executive Council at its forty-first session (1989) considered that, in coming years, atmospheric environmental research and monitoring activities should be given very high priority and established the Global Atmospheric Watch (GAW) as an umbrella system integrating many monitoring and research activities such as those of GO_3OS and BAPMoN. The GAW will strengthen and co-ordinate activities aimed at the study of further changes in atmospheric concentrations of greenhouse gases, in the ozone layer and in the long-range transport of pollutants, including the acidity and toxicity of rain and the atmospheric aerosol load. The GAW will further provide a framework within which it will be possible to plan, design, advise on, co-ordinate and scientifically evaluate global atmospheric-composition monitoring activities that will affect decisions regarding the environment in the 21st century.

Action by the United Nations

While WMO was principally concerned with the prospect that mankind might inadvertently cause adverse effects upon global and regional climates by the burning of coal and oil, the United Nations was beginning to show wider concern about obvious signs of the continuing degradation of the environment. Some of these indications were clearly man-made, others appeared to be due to natural causes. The encroachment of deserts into semi-arid areas, the destruction of tropical rain forests (themselves a factor in the climate balance) and evidence that in various areas ecological systems had already been eroded and were under continuing threat—all these factors caused anxiety among governments and led to considerable discussion within the United Nations. It became obvious that the widest possible international action was required involving all governments and all interests in the field of science, technology, industry and the socio-economic sphere.

The development of major international activities concerned with the protection of the environment may be said to have begun in July 1968 in the form of a proposal to the Economic and Social Council of UN (ECOSOC) that UN should organize a world conference on internationally significant problems of the human environment. Later the same year Unesco, in co-operation with FAO and WHO, took steps to arrange a world conference of scientists to discuss problems of the biosphere. The objective was to identify problems calling for united international and scientific efforts into the degradation of the biosphere as a result of various human activities. At first the organizers of the conference adopted a narrow definition of the biosphere to such an extent that the atmosphere was not accepted as playing a significant role. This restricted outlook was readily amended when it was pointed out that any English dictionary would describe the biosphere as "the part of the Earth and its atmosphere in which living things are found".

This temporary omission of WMO was however turned to advantage because it led to complete acceptance of the significance of the atmosphere in relation to the environment in general and to the biosphere in particular. Other Specialized Agencies of UN, some of which had not hitherto been involved closely in weather and climate, gained an appreciation that the atmosphere must be regarded as one of mankind's most important natural resources and that its degradation by man's industrial and other processes constituted a fundamental environmental problem. It was also recognized that climate had to be included among the principal ecological elements liable to be influenced by human activities and thus contributing to the degradation of many ecosystems. As a result WMO, both at the national level through Meteorological and Hydrological Services and as an international organization with the status of a Specialized Agency of UN, became more and more involved in the growing international efforts to provide improved protection of the human environment and more enlightened use of the world's natural resources.

Continuing discussions within the UN family and among other international bodies and also within individual countries led to a decision by the United Nations to hold a Conference on the Human Environment. It was accordingly arranged that the conference should be held in Stockholm in 1972 on the invitation of the Government of Sweden.

Before proceeding to the conference itself it seems necessary to mention an aspect of the environment that might appear to have been neglected. The main impetus for a conference on the human environment arose from concern over mankind's activities that might have adverse effects upon the conditions in which people, plants, animals and institutions exist. It was recognized that the global climate and certain regional climates were important factors. However, climate is not by any means the only branch of meteorology with an important role in consideration of the environment. Natural disasters—tropical cyclones, tornadoes, droughts, floods, earthquakes, volcanoes, etc.—all have serious impacts upon the environment as well as causing heavy losses of human life.

Of the disasters mentioned above, tropical cyclones (also known as hurricanes in the Western Atlantic and as typhoons in the Pacific and China Seas) and tornadoes fall into the realm of synoptic meteorology, which is the source of weather forecasts and warnings of adverse phenomena. Droughts and floods are of atmospheric origin but involve hydrologists to a very great extent in addition to meteorologists. Earthquakes and volcanoes are, of course, dealt with in other branches of geophysics.

Thus the environment is a system of extraordinary complexity involving not only climate but all branches of geophysics and also the particular activities which have aroused deep concern.

UN Conference on the Human Environment (1972)

Once the decision to hold the conference had been made, the United Nations set up an organizing committee consisting of representatives from 27 Member States forming an appropriate geographical distribution. The Specialized Agencies, including WMO, and certain other international bodies were invited to attend the meetings of the organizing committee as observers, a non-policy-making role.

The representative of WMO described the Organization's range of activities that would fall within the scope of the human environment and, in particular, mentioned air pollution, climatic changes due to man's impact on the environment, urban climates and building climatology, and the national disasters of meteorological and hydrological concern. On behalf of WMO he offered, and this was agreed, to contribute to the world conference major papers on the following topics:

1. The quality of air as a natural resource to support life;
2. (*a*) Transport of pollutants through the atmosphere and resulting contamination, control of atmospheric pollution;
   (*b*) Implications of global weather and climate modifications for the human environment;
   (*c*) Effects of air pollution on materials;
   (*d*) Urban climates and building climatology;
3. Climatic effects of air pollution;
4. Natural disasters, e.g. tropical cyclones, tornadoes, droughts.

WMO also contributed a number of shorter papers related to specific items on the conference agenda. The whole WMO contribution for the conference was the aggregate of papers and advice from national Meteorological and Hydrological Services, from experts specially commissioned and within the WMO Secretariat, which was also responsible for co-ordination.

It is of interest to note that in 1970 while all the arrangements for the conference were in progress, involving so many nations and sectors of social and economic activity, a tropical cyclone occurred in the Bay of Bengal. In Bangladesh the high winds, heavy rainfall and floods caused the loss of more than 250 000 lives as well as extensive damage. UN immediately expressed grave concern and WMO, in co-operation with the Economic and Social Commission for Asia and the Pacific (ESCAP), set up a Tropical Cyclone Project to prepare recommendations for prevention and mitigation of such disasters. Work under this project was subsequently extended to all areas of the world vulnerable to attack by tropical cyclones, hurricanes or typhoons.

As an integral part of the planning for the conference, the organizers established in 1970 a Functional Group on the Human Environment with the task of preparing a document describing how the UN system, with special reference to the Specialized Agencies, might effectively deal with the wide range of problems requiring action. The Secretary-General of WMO, Dr D. A. (later Sir Arthur) Davies, was elected chairman and convener of the group, which contributed to the conference a comprehensive document drawing attention to the collaborative projects, involving several Agencies, that needed to be mounted and also listing the efforts required from each Agency within its own individual responsibilities.

In a separate address to the conference, the Secretary-General of WMO pointed out that, although delegates were primarily concerned with the evils of pollution, however caused, it should be borne in mind that there existed another, more positive side to the environment. From ancient time man had studied the natural environmental processes in order to fulfil his needs more and more efficiently. WMO and its predecessor IMO had been active since 1873 in the acquisition and practical application of knowledge of atmospheric processes, especially of those phenomena called weather and climate. He stressed that the weather forecasts and other information supplied to sectors of industry and commerce by national Meteorological Services were concerned with operational activities in the environment, e.g. aviation, shipping and agriculture. Dr Davies described the development of the World Weather Watch with its three-pronged structure—the Global Observing System, the Global Telecommunication System and the Global Data-processing System—which reaches every country in the world and collects and distributes raw and processed data.

Dr Davies mentioned the continuous co-ordination of ozone measurements and emphasized the recent establishment of a global network of special stations for monitoring the background pollution of the atmosphere. In 1972, at the time of the conference, 59 stations in 25 countries had come into operation and the number was steadily increasing. Dr Davies gave two main reasons for extending the World Weather Watch to include measurements of atmospheric pollution and remarked that both reasons were of direct importance to the conference. In the first place, possible changes in the Earth's climate must be monitored by studying over long periods any variations in the quantities of pollutants, particulate as well as gaseous, which might modify atmospheric processes. The second reason, also of major importance, concerned the action of the atmosphere in transporting pollutants from their source and depositing them in other places without regard to man-made boundaries or frontiers. The Secretary-General also explained that the atmosphere, as research had shown, was a highly important source of marine pollution, in some cases the primary source. This evidence underlined the need for co-ordinated monitoring systems for atmospheric and oceanic pollution. He also drew attention to the importance of hydrological monitoring of river basins and inland water, a task which could be pressed forward only by co-ordinated meteorological and hydrological observing networks.

The conference was undoubtedly a great success and it acquired a lasting importance and authority. The Conference Director stated that the first steps had been taken to bring hope to mankind for a clean environment. He added: "The task has been to take the political decisions that will enable the community of nations to act together in a manner consistent with the Earth's physical interdependence."

The conference approved the establishment of the United Nations Environment Programme (UNEP) which, from its headquarters in Nairobi, Kenya, has been working with strong leadership and vigour to achieve results that are so dependent on enthusiastic support from national and international efforts.

The conference adopted 109 recommendations and not surprisingly the importance of meteorology was reflected in no fewer than 35 of them. These recommendations and other matters arising from the conference were considered by the General Assembly of the United Nations in December 1972. A Governing Council for UNEP was established and at its first meeting (Geneva, June 1973) it laid down a series of programme areas on which action was to be initiated without delay. WMO had already begun discussions with UNEP and projects were soon in course of development on background monitoring of air pollution, the physical basis of climate, marine pollution, the forecasting of drought, the development of water resources and desertification. Many ideas emerged from the conference and were pursued in a series of UN World Conferences during the ensuing few years covering such subjects as: Food (Rome, 1974), Population (Bucharest, 1974), Women (Mexico City, 1975), Human Settlements [Habitat] (Vancouver, 1976), Water (Mar del Plata, Argentina, 1977) and Desertification (Nairobi, 1977).

Another conference should also be mentioned on the subject of climate. At the Stockholm Conference on the Human Environment, much attention was given to the possibility of man's industrial activities leading to changes in the global climate. The importance of this aspect of climate change cannot be emphasized too strongly, but soon after the Stockholm gathering meteorologists and other members of the scientific community were expressing concern about indications that the global climate might change as a result of natural causes. Much public interest was aroused and it became clear that questions of climate change and climate variability had to be considered in the widest possible context. Accordingly WMO, with the support of other concerned international organizations, convened a World Climate Conference, which was held in Geneva in February 1979 (see Chapter XII).

UN Conference on World Food

The World Food Conference was held in Rome in November 1974 and, although arranged independently, came to be regarded as a sequel—one of several—to the Conference on the Human Environment. The subject, world food and its production, vividly illustrates the point made by the Secretary-General of WMO in his address to the Stockholm meeting—namely, that the environment should be used for beneficial purposes and not merely to fill the atmosphere with the harmful output from factory chimneys.

WMO played a full part in the preparations for the conference. Major factors of concern were the effects of variability in weather and climate in causing fluctuations in world food production. It was also recognized that efforts to increase food production throughout the world must take into account the weather and climate characteristics of each region.

The proceedings of the conference were conducted in two parts: the short-term problem of dealing with current food shortages; and long-term requirements to increase food production and to institute a world food security system that would keep track of the world-wide situation and, whenever necessary, give early warning of any shortages likely to occur.

The resolutions adopted by the conference were almost all within the scope of WMO's responsibilities and, with the approval of Seventh Congress (Geneva, 1975), were absorbed into the programmes and projects of the Technical Commission for Agricultural Meteorology. World food, with its enormous number of problems of great variety and complexity, has since 1974 been the dominant subject in the work of the Commission. A notable feature has been that many countries have made requests for WMO to provide consultants to advise on setting up agrometeorological departments in their national Meteorological Services.

UN Conference on Human Settlements (HABITAT)

Another sequel to the Stockholm Conference on the Environment took place in Vancouver, Canada, from 31 May to 11 June 1976 in the form of a World Conference on Human Settlements, generally referred to as the HABITAT Conference. Among the main reasons for the conference were the need to prevent further degradation of the environment and the urgency to provide dwellings to millions living under deplorable conditions in consequence of catastrophic population imbalances. The preparatory work had concentrated on such pressing issues as population growth, migration from rural to urban areas, inadequate control over land use, social injustices and the need for planned urban development. The HABITAT meeting was one of the largest conferences ever convened by the United Nations. It was attended by over 1 000 delegates, representatives and observers from 131 countries, 29 UN and other world and regional agencies, and 142 non-governmental organizations.

WMO submitted a background paper to the conference under the title "Weather, climate and settlements", one of eleven papers prepared by UN and its Specialized Agencies. At the conference itself the Tropical Cyclone Project, operated by WMO with the support of a number of other members of the UN family, was given a great deal of attention. The representative of the Office of the UN Disaster Relief Co-ordinator (UNDRO) said that in countries liable to natural disasters, settlement plans should be based on vulnerability analyses and pre-disaster preparedness studies in all of which WMO's participation with other concerned organizations would be indispensable. He referred in particular to such meteorological and hydrological phenomena as severe local storms, tropical cyclones, floods and droughts. Representatives of the Philippines, Bangladesh and Mauritius spoke about the death toll and damage resulting from tropical cyclones; the Sudano-Sahelian countries and Somalia referred to the effects of droughts which had occurred in the preceding few years.

The recommendations of the conference clearly pointed the way for much involvement on the part of WMO. Since the conference, and even before, WMO has been rendering expert assistance on planning for mitigation of weather disasters, on planning for water supplies and use and also on the application of climatological information to urban and housing design, reduction of air-pollution hazards and the utilization of solar and wind energy.

UNEP and WMO

In view of the fact that WMO's responsibilities are almost wholly concerned with what is happening in the environment and with making efficient use of the environment as a natural resource, the Executive Council* of WMO sought an opportunity to invite the Executive Director of the United Nations Environment Programme (UNEP), Dr M. K. Tolba, to visit the WMO Headquarters in Geneva. The visit took place in June 1976, some three and a half years after the creation of UNEP, and Dr Tolba addressed the twenty-eighth session of the WMO Executive Council. He stressed that UNEP was not so much an operational agency for environmental issues but was responsible, as a catalyst where desirable, to co-ordinate and to give common objectives to the many environmental activities under way within the UN system and elsewhere. UNEP would add new, additional directions to environmental programmes already in existence and would endeavour to promote new ones whenever any gaps or deficiencies were found. The whole environment programme was to be regarded not as the exclusive concern of UNEP but as the shared responsibility of the entire UN system and the international community at large. For this reason UNEP had initiated, and would maintain indefinitely, a comprehensive exercise of joint programming with other UN organizations.

The Governing Council of UNEP had identified a number of priority subject areas because with so many problems of the human environment it was necessary to be selective in order to achieve maximum results with limited resources. Dr Tolba referred to subjects in which UNEP regarded co-operation with WMO to be of great importance:

(*a*) *Human settlements*
WMO contributed valuable technical reports on urban and building climatology to the Vancouver Conference on HABITAT. These reports were published for wider distribution and UNEP hoped for further co-operation in this field.

(*b*) *Terrestrial systems—management and control*
UNEP wanted major projects aimed at improving the productivity of arid and semi-arid ecosystems, the management of tropical woodlands and forests, water quality and water management in rural areas, soil

* See note on p. 8.

condition and measures for the conservation of genetic resources and biospheric sites. In all these objectives WMO was providing valuable assistance by means of agrometeorological and hydrological studies.

(*c*) *Oceans*
WMO had a particularly important contribution to make in the scientific research and monitoring components of the plans for the protection of regional seas in the development, with the Intergovernmental Oceanographic Commission (IOC) of the Integrated Global Ocean Station System* (IGOSS).

(*d*) *Energy*
The Executive Council** of WMO had approved a programme on the meteorological aspects of energy. There was much work to be done on the utilization of less-polluting and renewable energy resources such as solar energy, wind power and biogas.

(*e*) *Natural disasters*
UNEP's priorities in this areas were said to be improved warning systems, prevention and mitigation. WMO was praised for its Tropical Cyclone Project which was set up with terms of reference that ranged widely from meteorology into all aspects of disaster prevention, preparedness and mitigation.

(*f*) *Functional tasks*
UNEP placed great importance on a number of functional activities in which WMO was closely involved. These included Earthwatch with its Global Environmental Monitoring System (GEMS) and its projected International Referral System. The "Outer Limits" component of Earthwatch incorporated three subjects for which WMO carried the basic responsibility—*climate change, weather modification* and risks to the *ozone layer*.

United Nations Conference on Desertification

The UN Conference on Desertification was held in Nairobi, Kenya, from 29 August to 9 September 1977. WMO was fully involved in the preparations for the conference, which was attended by representatives of 95 countries and over 60 international organizations, both governmental and non-governmental. The conference was one of a number of offshoots of the UN Conference on the Human Environment (Stockholm, 1972).

Desertification is defined as the increase in an area or region of conditions by which biological productivity is reduced with a consequent decrease in plant biomass and of the land's capacity to support livestock and crops. During the early 1970s world-wide concern was aroused by estimates of the rate at which desertification was proceeding, roughly equivalent to an area the size of Belgium and the Netherlands each year. The cause was at first generally attributed to climate change and as a result it was some time before governments accepted that urgent action was required for countermeasures to be applied.

In advance of the conference a number of studies were carried out with the purpose of providing guidelines as to the decisions and recommendations that might be reached by the conference. The main studies consisted of:

- Case histories of areas where desertification was a problem;
- Background documents on climate and desertification, ecological change and desertification, population, society and desertification, technology and desertification;
- A world map of desertification prepared by FAO with the collaboration of Unesco and WMO.

It was very clear from the results of the studies that, whereas in most cases marginal climatic conditions were conducive to desertification, climate change as such was not the root cause of the problem. A primary cause was shown to be over-exploitation of the land, whether agricultural or pastoral, in circumstances where the ecological balance was highly sensitive. Another cause was shown to be lack of overall planning regarding land use. The studies identified 24 countries where all or nearly all the land not already desert was at risk and 23 countries where much of the land was at risk. It was clear that large-scale action was required by governments in accordance with plans to be instituted by UN.

An encouraging feature, emphasized during the conference discussions, was that the case studies had shown that desertification could be halted and that degraded land could be reclaimed by systematic plans, taking account of weather and climate, and applying such measures as afforestation, revegetation, control of pastures and livestock numbers, and so on.

The conference adopted a plan of action with a target date of the year 2000 for implementation. Many of WMO's existing programmes were referred to as providing valuable support and there were also many of the recommendations in technical matters which had a close bearing on WMO's field of interest and responsibility. Subjects covered included:

- The impact of recurring droughts and climate fluctuation on land management;
- The need for adequate networks of meteorological, climatological and hydrological stations in areas of concern;
- Monitoring of atmospheric changes, dust transport and changes in irrigated lands;
- Assessment of water needs and the reduction in water losses by evaporation;
- Flood control and warning systems;
- Building climatology and associated information for human settlements;
- Forecasting of various extremes, e.g. frost, floods, snow, hurricanes, sand and dust storms.

The Global Environment Monitoring System

Co-ordinated international arrangements to collect data on the state of the atmosphere were launched in 1873 when the International Meteorological Organization, the predecessor of WMO, was created. From that date the world-wide network of meteorological observing stations grew rapidly and, as technology advanced, the state of the atmosphere was measured with greater depth and increasing precision. Later, towards the end of the nineteenth century, many countries became aware of the damage caused to health and buildings by the burning of coal, and national (rather than international) networks were introduced for the measurement of smoke and sulphur gases in the atmosphere. A proportion of the recording stations were positioned close to weather-observing stations so that air-pollution measurements could be interpreted in the light of

* In 1983 IGOSS was renamed Integrated Global Ocean Services System.

** See note on p. 8.

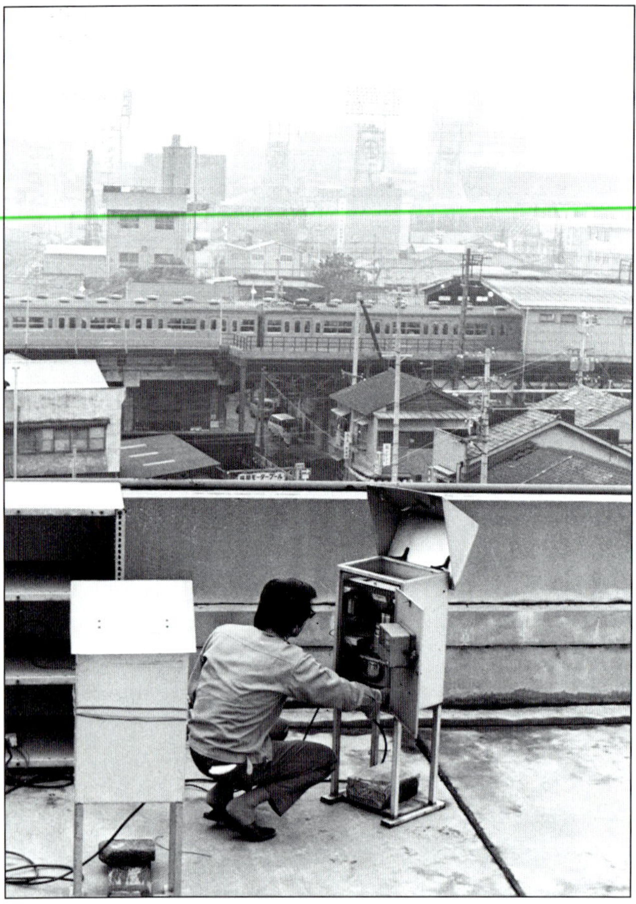

Measuring the pollution level in Japan

meteorological conditions. International arrangements for the measurement of certain features of atmospheric pollution were initiated by WMO in the 1950s when it began to develop a world-wide network of stations for monitoring background levels of pollution. At about the same time WMO took steps to organize an international network for the routine measurement of atmospheric ozone concentrations.

At the UN Conference on the Human Environment it was acknowledged that an internationally co-ordinated effort was required to collect, analyse and evaluate comprehensive data on those variables which determine the state of the environment and the changes it undergoes in space and time. It became the responsibility of UNEP to co-ordinate international monitoring activities at the regional and global levels. In this way the Global Environmental Monitoring System (GEMS) was created. There was no intention of duplicating any atmospheric monitoring already in operation. In fact, it was recognized and welcomed that several UN agencies such as FAO, WHO, Unesco and WMO were all engaged in global monitoring activities for their own purposes. The task was therefore to integrate individual systems and to fill any gaps in order to bring into force a comprehensive, world-wide monitoring system that would meet all the requirements specified at the Stockholm meeting.

From the nature of their direct interests it was clear that the UN agencies would have a major role in the GEMS. For example, WMO's Background Air Pollution Monitoring Network (BAPMoN), the World Weather Watch, the Global Ozone Observing System (GO_3OS) and the Integrated Global Ocean Station* System (IGOSS) under development jointly by WMO and the International Oceanographic Commission were regarded as indispensable to GEMS, the role of which was to develop and supervise additional monitoring activities, supported by the expertise of the UN Agencies and their effective channels of communication. Furthermore, the participants in GEMS had to ensure comparability of the total volume of data that became available.

The monitoring arrangements developed and partially funded by UNEP fell into five main categories:

- Atmospheric composition changes in climate-related monitoring;
- Monitoring the long-range transport and deposition of air pollutants;
- Ocean monitoring;
- Health-related monitoring;
- Monitoring renewable natural resources.

Since 1974 various sub-projects of this programme have been developed by WMO, with support from UNEP, all within the GEMS arrangements. UNEP also contributed substantially to the GARP Global Weather Experiment because of its potential for a deeper understanding of the global climate and for identifying the data requirements for climate research. Furthermore, from 1978 to 1982 UNEP partially supported the WMO Global Ozone Research and Monitoring Project, the WMO CO_2 Research Project and the study of the possible environmental impact of precipitation enhancement (within the framework of the WMO Precipitation Enhancement Project—PEP). WMO guides and implements the meteorological aspects of the MED POL programme also with partial support from UNEP.

GEMS has been in existence for less than two decades but has made encouraging progress in co-ordinating a number of data-acquisition projects, which often involve long lead times before reaching maturity. It is pleasing to place on record that there has always been a fruitful partnership in the environmental field between UNEP and UN Specialized Agencies such as FAO, Unesco, WHO and WMO.

UN Symposium on Interrelationships

In August 1979 WMO was represented at a symposium in Stockholm convened by the Government of Sweden on behalf of the United Nations for discussion of the interrelations among resources, environment, population and development. The need for such a symposium became clear in debates on world problems in the UN General Assembly. However, to a great extent the meeting could also be seen as a logical and indeed inevitable consequence of the succession of international conferences which took place during the 1970s as a follow-up to the Conference on the Human Environment and were concerned with matters of the greatest importance to the future of mankind. These conferences, some of which have been discussed above mainly within a WMO context, and the resulting action programmes gave rise to strong convictions that development, whether on a global, regional or national scale, must be viewed in the broadest possible manner with full regard to the multitude of factors arising not only from economic and social aspects but also from questions of population, the environment and the use of natural resources.

During the discussions at the symposium it was clear that the World Climate Conference (Geneva, February 1979) was regarded as a most valuable and opportune addition to the series of United Nations conferences. The proceedings of the World Climate Conference, which was convened by WMO,

* See note on p. 93.

provided useful information by means of the comprehensive assessment that was made of the implications of climate and climate change for many problems concerning the natural world and human society. The importance of the WMO Tropical Cyclone Project, with its objective of saving lives and minimizing destruction to property, was also acknowledged.

One of the primary tasks of the United Nations has been that of establishing conditions throughout the world that would promote economic prosperity and general well-being in the developing countries. The task, essentially a long-term one, revealed numerous problems which underlined the necessity of multi-disciplinary efforts for the formulation of development policies and the associated research and operational programmes. The widespread interest aroused by the World Climate Conference among governments and many sectors of national communities offered an assurance that the United Nations would invite WMO participation in studies of interrelations in which hydrology, weather and climate would be involved in any way. It appeared at the symposium that the numerous and varied programmes of WMO, which are environmental and aimed at quite specific objectives approved by the WMO Congress, would be applicable directly or, in come cases, with slight adaptation to a wide range of studies into the interrelationships considered by UN to be of major importance.

WMO progress report

Two resolutions of the United Nations General Assembly in 1987 requested members of the UN family to report on "progress made in achieving environmentally sound and sustainable development" and "on progress made in their organizations towards sustainable development". A report on behalf of WMO was submitted to UN in September 1988 by the Secretary-General; this is summarized in the three paragraphs which follow.

The fundamental "renewable" natural resources are climate and water. The Earth's atmosphere can also be considered as a natural resource, one on which man's activities are having a significant impact. WMO has therefore a vital interest in sustainable development and has the capacity, in scientific strength and world-wide organizations with 160 Member countries, to make important contributions to planning for sustainable and environmentally sound uses of natural resources. The specific climate and environmental issues ensure a major role for WMO and, in addition, its more traditional programmes in weather, climate and water resources provide an essential scientific basis for achieving sustainable development of all renewable resources. Actual and potential benefits available from the applications of climatic and hydrological data and of weather forecasts are well illustrated in the sectors of agriculture, the management of water resources and in energy for conservation, for the operation of production and distribution systems and for the assessment of renewable (e.g. wind, solar and hydroelectric) energy sources.

The organization and programmes of WMO are all related to sustainable development, as a few examples can illustrate. The *World Weather Watch* system consists of observing networks and facilities operated by the 160 Member countries and integrated and co-ordinated in three global components—the Global Observing System of surface stations and weather satellites, the Global Telecommunication System for the collection of weather data and the broadcast of weather analyses and forecasts generated by the Global Data-processing System. The *World Climate Programme*, co-ordinated by WMO and with other agencies involved, operates in four divisions: data, research, applications of climate information in all economic sectors and also climate impacts, conducted by UNEP, for the assessment of the social and economic impacts of climate variability and climate change. *Hydrology and water resources* activities are directed to improving technical procedures and raising observing standards in all countries. Hydrological information serves a wide variety of interests. *Technical co-operation* projects are currently under way in 41 countries with support and advice from WMO. *The Global Atmosphere Watch* environmental pollution monitoring and research activities include the operation and analysis of data from the Global Ozone Observing System, background air-pollution monitoring for world-wide assessment of greenhouse gases and toxic and acid substances in the air and precipitation, and assessing the transfer of pollutants from the atmosphere to oceans, lakes, forests and fields.

WMO attaches major importance to its environmental work since it considers that man has gone beyond the stage where the atmosphere may be used as a refuse dump of infinite capacity without the danger of serious consequences. The dramatic ozone depletion during the Antarctic springs of the 1980s is one of the first demonstrations that the uncontrolled release of man-made chemicals in the atmosphere could cause substantial damage to the natural state of the environment.

Looking to the future

WMO is very conscious of the vital contribution it can make to investigating and assessing the possible outcome of mankind's impacts upon the natural environment. The Global Atmosphere Watch as a programme for environmental pollution monitoring and research is to be maintained with very high priority, as stated by the Executive Council at its forty-first session.

The main long-term objectives of the WMO environmental pollution monitoring and research activities incorporated in the GAW system are:

- To provide authoritative scientific information and advice on the composition and behaviour of the global atmosphere and the factors that affect them and thus contribute to the relevant WMO responsibility;
- To establish and co-ordinate an operational system to determine global and regional levels and long-term trends of natural and man-made atmospheric constituents (including those with a possible impact on climate) in order to forecast future states of, and stresses on, the environment and to enable governments to take prompt action to reduce pollution;
- To further the understanding of the chemistry and physics of the environment and climate-related atmospheric constituents and properties, and to apply this knowledge in the fields of meteorology and climatology, especially through the application of atmospheric models;
- To promote studies of the interaction of the atmosphere with the marine and terrestrial biosphere;
- To meet the responsibilities of WMO to provide leadership and guidance in international efforts directed towards the protection and management of the atmospheric environment.

The programme depends largely upon the support of the national Meteorological and Hydrological Services of the Member countries of WMO. These Services have their own responsibilities in regard to the environment and the WMO Secretariat, in addition to its co-ordinating role, will provide

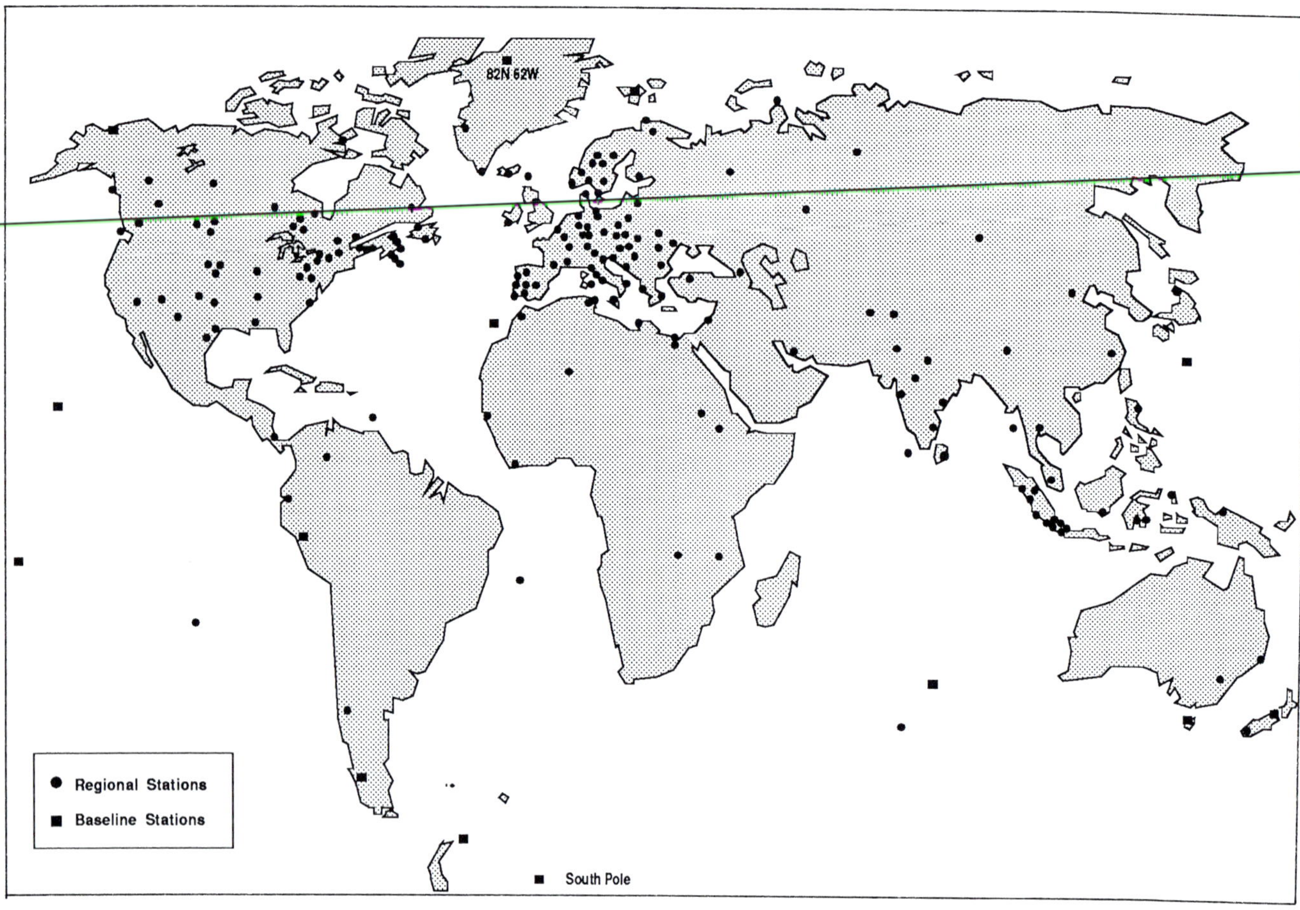

The Background Air Pollution Monitoring Network (BAPMoN)

advice and other assistance so that objectives will be reached at the highest levels of technical efficiency.

An item of special importance is the implementation of the Global Atmosphere Watch, which should serve as an early warning system to detect further changes in atmospheric composition and will enable the prediction of the future state of the atmosphere and of the related Earth systems. The GAW will be based on an improved BAPMoN and on the Global Ozone Observing System (GO_3OS).

With regard to ozone some recent events, listed below, have demonstrated the growing concern of countries regarding changes in the ozone layer and indicate a strengthening of support for the efforts of WMO and its collaborators:

(*a*) The adoption by a large number of governments of the Vienna Convention for the Protection of the Ozone Layer, which came into force in September 1988;

(*b*) The Montreal Protocol on Substances that Deplete the Ozone Layer, which came into force on 1 January 1989;

(*c*) An International Conference on Saving the Ozone Layer, organized by the Government of the United Kingdom and held in London in March 1989. More than 120 governments participated in the conference, no fewer than 80 of them being represented by ministers, and many international bodies concerned by the problem also took part.

Major items in research include better understanding of the bio-geochemical cycles of the most important components of the atmosphere's changing chemistry and their potential effect on weather and climate. In parallel will follow the development of techniques for modelling the transport and dispersal of atmospheric pollutants and to make comprehensive studies of all available data in order to improve knowledge of the exchange of pollutants between environmental compartments, e.g. atmospheric and ocean. Precise knowledge and interpretations of major alterations of background concentrations of atmospheric constituents are other aspects of intensive environmental research.

The whole programme is highly challenging in both its scientific and practical aspects. It is very long-term and in the period of the Second Long-term Plan, 1988–1997, there is every prospect that significant progress will be made.

CHAPTER XII

WORLD CLIMATE PROGRAMME

Introduction

The World Climate Programme (WCP) was a sequel to the World Climate Conference which was organized by WMO and held in Geneva from 12 to 23 February 1979. The Conference itself was a response to concern being expressed with increasing emphasis by climatologists and certain other scientists regarding indications from observations and from theoretical studies about the possibility of changes in the global climate. Such changes might occur as a result of natural causes or as a result of man's industrial activities or through a combination of the two. Furthermore, if it appeared likely that changes in the global climate would take place in the next few decades, or even in a hundred years' time, it would be essential for governments, the United Nations and other concerned bodies to be informed so that political, economic and social consequences could be assessed and action planned in order to mitigate any adverse effects.

The interest arising in the scientific community regarding climate change stemmed less from the possibility of naturally occurring variations than from the likelihood that the burning of fossil fuels and other industrial processes would eventually, and perhaps inevitably, result in the inadvertent modification by mankind of the world climate. Some writers have suggested that the industrial revolution, which began in the first half of the nineteenth century, was indirectly and unwittingly the start of a massive experiment in testing the capacity of the Earth and its atmosphere to disperse and remove pollution. The experiment has continued on an ever-increasing scale and there is little evidence that mankind has the will or the ability to arrest it or to bring about a reversal.

It had long been recognized that the question of climate change was not merely a highly complex scientific problem. An aspect of major importance was whether it could be brought home to peoples and their governments that the world's climate could not be regarded as being in a steady state or unchanging and that, out of concern for future generations, the possibility of permanent changes had to be taken seriously. The general public, of course, has always been familiar with the variability of local, national or regional climates. Experience has shown that the weather can depart noticeably from the average during months or seasons, producing wet or dry winters, hot or cool summers. However, the public appreciates that, over a long period, averages of temperature, rainfall and other elements are restored to their normal values.

In applied climatology, it has been customary to treat the climate, whether on a local or larger scale, as constant but containing an internal variability caused by effects which may be periodic, quasi-periodic or simply non-periodic. Climatology has applications in most sections of a country's economic and social activity. Some examples are agriculture, all forms of transport, water conservation and drainage, health and disease, forestry and fisheries and offshore developments. To the ordinary person, climate may be described as the expectation of weather. The individual's plans are made on the assumption that weather through the year will proceed along familiar lines. However, in scientific terms, climate has usually been defined as the average condition of the atmosphere over a long period of time, in other words, the long-term summation of the atmospheric elements and their variations that, over short time periods, constitute weather.

For applications of climatology and in some areas of research, climatologists, after discussion within IMO, the predecessor of WMO, adopted the practice of averaging the meteorological elements over thirty-year reference periods and making fresh calculations every ten years, e.g. 1931–1960, 1941–1970 and so on. It was found that small differences occurred between the processed data of successive reference periods, but it was recognized that the variable character of the climate's behaviour made such differences inevitable. Their existence does not indicate a real climate change, but such variability is a fundamental characteristic of climate and has an important economic impact. The statistics provide not merely averages but additional information of the greatest value in applied climatology where many highly complex problems arise. To quote a simple example, in the building of a dam for flood control, the engineering designer has to assess the strength of structure needed for the dam to last for 50, 70 or 100 years. The statistics on climate readily provide estimates of the worst conditions of wind, temperature and rainfall to be expected within these periods, or, if preferred, the return periods of severe conditions as specified by the designers.

Although climatological data, analysed over periods of years and taking into account the variability of climate rather than the question of climate change, have numerous, very valuable applications in advancing national prosperity, it must be emphasized that meteorologists do not regard climate as a static phenomenon. Indeed their concept of climate is a dynamic one and they are ever on the alert to identify and monitor any evidence that might indicate that a climate change is taking place. Such a change might be suspected if a series of fluctuations of one or more meteorological elements were to move in one direction sufficiently long and effectively to cause modification in other environmental parameters. In these circumstances, it is advisable to suggest a possibility rather than to announce a change since there are many examples of long-term trends which have gone into reverse for comparable lengths of time. Sometimes a striking but isolated variation may occur, e.g. a severe summer drought which climate statistics suggest has a probability of occurrence of once every two or three centuries. This event may indeed be part of a real change towards a drier climate, but it is more likely to be part of the variability of the existing climate. Even so, the late 1960s and early 1970s were notable for a decade of single and seemingly unconnected climate events which were included among the many factors which prompted WMO to take a concentrated and deep interest in the possibility of major changes in the global climate.

These events occurred all over the world and were disruptive of human society. In the ten-year period, the southern border regions of the Sahara desert, known as the Sahel, experienced a five-year drought with famine and death on a vast, continental scale. In the year 1972, there occurred a world-wide series of costly climatic events, including drought in the Soviet Union and

The result of drought in Venezuela

the incidence of *El Niño* off Peru. In 1974, poor monsoons severely reduced food production in India; in 1975 cold spells in Brazil badly damaged coffee crops. In the United States, a series of cold winters forced many industries to close with very serious social consequences. More recently, in 1984/85, drought in the Horn of Africa caused severe famine and mass emigration.

If governments and their citizens are to be given adequate warning of likely changes in the world's climate, it is essential that such possibilities should be identified well in advance and that reliable predictions should be available as to the manner in which events would unfold. The climatic events referred to above were not foreseen and action to mitigate their effects had to be taken with the greatest urgency. In each case, the occurrence could be explained within the statistics of the natural variability of climate, but so many disastrous episodes being experienced within a relatively short period gave rise to speculation that developments were taking place that might have causes outside natural variability. WMO, in common with other concerned UN agencies, was involved in the international action that followed those disasters, in particular the prolonged drought in the Sahel, and was in a position to appreciate the widespread concern about climatic changes.

In addition to natural events such as drought, monsoons with little rainfall and prolonged cold spells, there were other factors which raised questions of climate change and claimed the attention of WMO and of certain other international bodies. Governments and scientists generally were also beginning to take a close interest. These other factors had their origins in industrial activities and focused attention upon the possibility that mankind could be responsible for changes in the global climate. It had long been appreciated that mankind was changing the radiative processes of the atmosphere and perhaps its general circulation by the emission of the waste products of industry but the potential consequences were not widely discussed until about the 1960s or somewhat later. Then the whole subject of climate variability and climate change came under close consideration by WMO.

Action by WMO towards a World Climate Conference

It need hardly be said that the study of climate has been one of the major interests of WMO as it was of its predecessor the International Meteorological Organization, founded more than a century ago. Information about climate and, as knowledge steadily expanded, the implications for services to human activities were constantly under review in each WMO Congress, held at four-year intervals, and in other constituent bodies such as the Executive Council, the Regional Associations and Technical Commissions. Among the Technical Commissions those for climatology, agriculture, hydrology, marine meteorology and the atmospheric sciences embodied wide interests in climate on all scales—local, regional and global—in their fields of responsibility.

In the late 1960s and early 1970s, concern increased markedly within WMO about the world's climate from three main directions. In the first place, there was the question of the growing amounts of carbon dioxide and other so-called greenhouse gases in the atmosphere. As shown in the diagram below, the concentration of carbon dioxide in the atmosphere in 1959 was about 313 parts per million, and, in a period of 20 years, i.e. up to 1978, the concentration had increased to about 330 parts per million. In the figure annual averages or twelve-month running means have been plotted and it seems evident that the long-term trend is well established.

Secondly, in addition to the carbon dioxide problem, recent research had shown that gases from sources other than fossil fuels were capable of reinforcing the effects of carbon dioxide. The extensive use throughout the world of nitrogen fertilizers to maintain and expand agricultural production involved the release into the atmosphere of nitrogen compounds with the potential for altering the photochemical balance of the ozone layer in the stratosphere. This layer absorbs ultra-violet radiation from the Sun. If changes in the ozone layer reduced this absorbing capacity, the transmission of short-wave radiation direct from the Sun to the Earth would constitute a serious health hazard in the form of skin cancer. Additional threats to the ozone layer could, it was

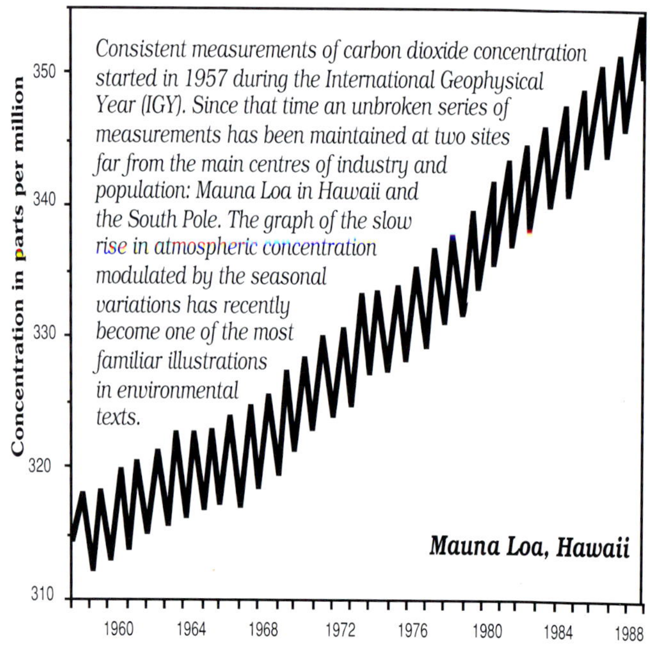

Consistent measurements of carbon dioxide concentration started in 1957 during the International Geophysical Year (IGY). Since that time an unbroken series of measurements has been maintained at two sites far from the main centres of industry and population: Mauna Loa in Hawaii and the South Pole. The graph of the slow rise in atmospheric concentration modulated by the seasonal variations has recently become one of the most familiar illustrations in environmental texts.

suggested, result from oxides of nitrogen released in very high-altitude supersonic flight or passing into the atmosphere through the agency of chlorofluoromethanes used in refrigeration and propellants.

In the third place, expert projections became available as to world requirements in the future and emphasized in striking terms the importance of climate, whether with regard to food, energy or the state of the world environment. For example, it was anticipated on a very conservative basis that the population of the world would increase from 3.5 billion in 1970 to about 6.5 billion by the turn of the century. Estimates made by the Food and Agriculture Organization (FAO), a Specialized Agency of the United Nations, indicated that world aggregate food demand would double by the year 2000 and that it was already essential to face the challenge of a corresponding increase in food production, a requirement involving detailed climatic considerations. It appeared that a most critical question would be to achieve maximum world food production by exploiting the characteristic climate of each portion of agricultural land.

The position with regard to energy seemed to be no less ominous. Increasing energy needs would far outstrip oil production by the end of the century and result in a growing dependence on coal throughout the world. The addition of carbon dioxide to the atmosphere in increasing quantities must be expected to create a very serious threat to the global climate.

Leaving aside extremes that might arise from the natural variability of climate, the main reasons for possible climate change, as they appeared to WMO, stemmed from man's industrial activities, even if these continued at a constant rate as from about 1970, and from the expansion of activity that would inevitably result if the projections into the next century for population, food demand and energy consumption were found to be soundly based.

Another consideration of great concern to WMO, as the world authority on meteorological matters and as a Specialized Agency of the United Nations, was that in the early 1970s a number of statements had been made in quite catastrophic terms that the climate of the world was undergoing important changes which would affect peoples of all countries. These statements received wide and continuing publicity. They were not invariably well considered, but nevertheless Meteorological Services in every part of the world received inquiries as to whether the statements had a sound scientific basis or carried the authority required for acceptance. Inevitably there were suggestions and requests that WMO should issue a statement on climatic change, setting out the existing state of knowledge on the subject and discussing ways in which changes of a short- or long-term nature might occur.

WMO was not going to be rushed into the issue of a statement as a result of publicity, however widespread, which was of a speculative nature. For one good reason, the subject of climatic change had long been recognized by meteorologists as a problem of extraordinary complexity and possibly not soluble to an extent that would be of practical value. Moreover, the climate is not simply controlled by the atmosphere alone. The climate system has five physical components: the atmosphere, oceans (hydrosphere), ice-masses (cryosphere), land surfaces (lithosphere) and biosphere. These components have quite different physical characteristics and are linked to one another and to conditions outside the system by a variety of physical processes as shown in the following figure. The subject is therefore a multi-disciplinary one calling for wide consultation and collaborative effort.

Climate change was widely discussed at WMO's Seventh Congress (Geneva, 1975) and among the ideas exchanged was the advisability of holding a World Climate Conference. There was also considerable interest in a forthcoming United Nations Conference on Desertification, prompted by the disastrous drought in the Sahel region of Africa, and a review of the urgent measures set in motion. At this conference, WMO would clearly have an important role and it was assumed that questions of climate would be prominent. Congress adopted a resolution giving guidance to the Executive Council* and to the Secretary-General on the action to be taken in the context of climate variability and climate change.

Meeting immediately after Congress for its twenty-seventh session, the Executive Council* initiated the action required. It re-established its Panel of Experts on Climatic Change and asked the Secretary-General to arrange for additional experts to be nominated to serve on the panel by the United Nations Environment Programme, the United Nations Educational, Scientific and Cultural Organization, the United Nations Food and Agriculture Organization and the International Council of Scientific Unions.

* See note on p. 8.

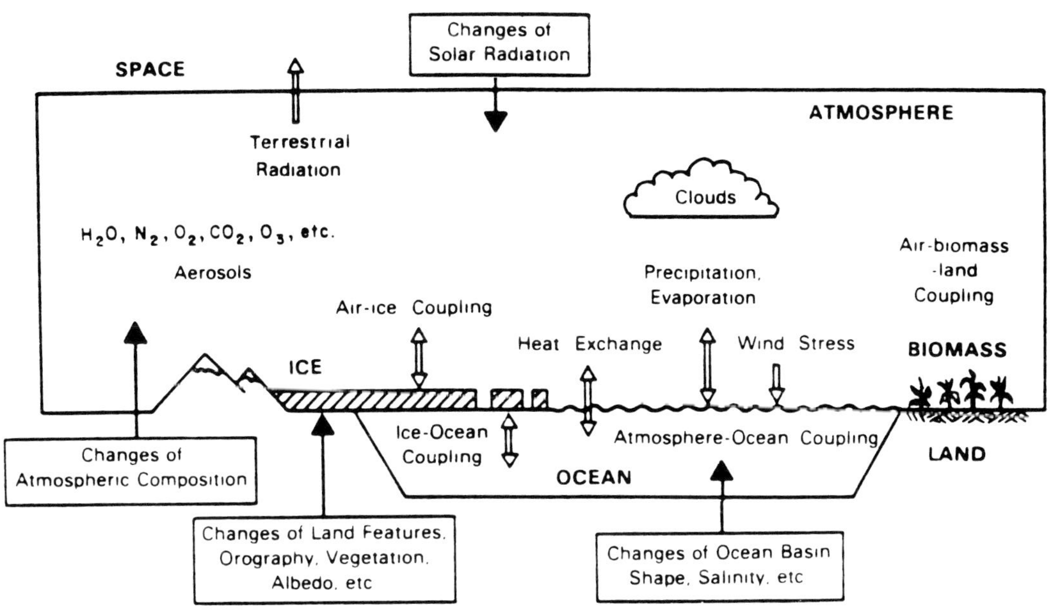

Atmosphere-ocean-ice-land-biomass climatic system showing physical processes responsible for climate and climatic change

The tasks given to the EC panel, acting in collaboration with the president of the Commission for Atmospheric Sciences, included the following:

- To review activities concerned with climatic change at national and international levels in order to identify topics requiring further investigation;
- To develop plans for an integrated international effort to study climatic changes and their implications for man's natural environment and for world food production;
- To review data requirements;
- To prepare statements on climatic changes due to (*a*) natural causes and (*b*) effects of man's activities.

A year later, at its twenty-eighth session (Geneva, 1976), the Executive Council* had before it the statement submitted to it by the Panel of Experts on Climatic Change. After some discussion, in which there was some divergence of opinion, it agreed that an authoritative statement issued officially by WMO would not only provide much-needed information and opinion but would also help pave the way for a major international programme on global climate. Accordingly, it approved for publication the statement on climatic change drawn up by the panel.

The Executive Council* also gave further consideration to the question of convening a World Climate Conference. It was urged that such a conference, organized on a comprehensive basis, would help to promote wide-ranging discussion between government administrators and economists on the one hand and those engaged in climatological research and in the applications of climatology on the other. It was accordingly agreed in principle that WMO should convene a Conference on Climatic Change in order to review the whole subject—its research aspects and its impacts for the world at large. The Secretary-General was requested to hold consultations with other concerned international organizations to enlist their co-operation and to discuss the general and detailed arrangements.

In its report to the twenty-eighth session of the Executive Council*, the Panel of Experts on Climatic Change stressed the importance and even the dangers of the continuously increasing amounts of carbon dioxide in the atmosphere as a consequence of the unrestricted use of fossil fuels for industrial and domestic purposes. The Council therefore asked the Secretary-General to make urgent arrangements to call a scientific meeting for a comprehensive examination of the carbon dioxide problem.

This particular session of the Executive Council* may be regarded as marking a milestone in the history of WMO and in the prominence which the Organization was giving to the question of climate. Notable features of the session were clearly the decision in principle to hold a World Climate Conference and the adoption for world-wide publication of a statement on climatic change. A summary of the latter, which was issued in June 1976, is given in the following section.

Summary of WMO statement on climatic change

In spite of man's advances in technology, his economic and social welfare are still highly dependent upon climate. It is not only the demand for food which illustrates this dependence; floods, droughts and extremes of temperature disrupt urban communities and hamper economic and social development.

Since the climate has been so continuously variable in the past, it must be assumed that it will continue to vary in the future. However, long-term trends are masked by shorter-term fluctuations and by regional changes. Recent climatic events have led to speculation that a major climatic change may be occurring on a global scale. Such a global change could occur from natural causes, but it would be gradual and almost imperceptible because shorter-term variations are likely to be large enough to obscure the long-term trends. The shorter-term climate changes, which may be due to natural or man-made causes, require urgent attention and further studies.

Natural shorter-term variability has been highlighted in the droughts and weather extremes which have caused so much human suffering. Governments could respond to these changes if sufficient warning could be given. Of equal concern are possible changes of climate resulting from man's industrial activities, but with the present state of knowledge of the atmosphere, an accurate assessment of the magnitude of such changes cannot be made. Meteorologists and other scientists are seeking to improve the availability of data relating to past behaviour of the atmosphere, the oceans and other relevant climatic factors. They are endeavouring to improve assessments of the impacts of climate changes and to intensify research aimed at a more complete understanding of climatic processes.

Planning for economic and social development should make full use of knowledge of the shorter-term variability of climate. Governments may expect to face difficult decisions if future research should reveal that man's activities could produce changes in climate of wide scope and serious consequence.

WMO's views were expressed concisely in the concluding paragraphs of the statement on climatic change:

(*a*) Although, in the long term, a major natural change to a different climatic regime must be expected, it is unlikely that any trend towards such a change would be perceptible in the short term as it would be obscured by the large shorter-term climatic variability;

(*b*) The shorter-term natural or possible man-made changes in climate are of immediate concern because of their important impact on human welfare and economic development;

(*c*) An improved ability is needed to predict short-term natural changes in climate to enable governments to consider appropriate action;

(*d*) An improved ability to predict the impact of man's activities on the global climate is needed in view of their possible consequences;

(*e*) Existing knowledge of natural short-term climatic variability, although limited, should be used more effectively in planning economic and social development.

Organization of the World Climate Conference

At its twenty-ninth session (Geneva, May–June 1977), the Executive Council* of WMO decided that the Organization should convene a scientific and technical World Climate Conference to be held in February 1979. This decision confirmed the agreement in principle reached at its preceding session and resulted from a favourable report by the Secretary-General, who had been asked to consult appropriate international organizations and to invite their co-operation. The main purposes of the Conference were to be:

(*a*) To review knowledge of climatic change and variability, due both to natural and to anthropogenic causes;

(*b*) To assess possible future climatic changes and variability and their implications for human activities.

* See note on p. 8.

The Executive Council* also decided that a major international effort would be required to investigate problems of the global climate and therefore set in motion the preliminary arrangements for a long-term World Climate Programme. It was confidently anticipated that the Conference would recommend such a programme and that it would be approved by the eighth WMO Congress, to be held in Geneva in May–June 1979, a few months after the Conference.

An organizing committee for the Conference was immediately formed comprising representatives from a number of international bodies and also scientists serving as individual experts and not as national or institutional delegates. Dr Robert M. White was elected chairman of both the committee and the Conference. The arrangements proceeded on the basis that, in order to meet its objectives, the Conference should consist of two phases each lasting one week. During the first phase (12–16 February 1979), specially invited speakers would present overview papers covering current knowledge of climate and the interactions between climate variability and change and human society. These papers would be discussed at length by all participants.

In the second week (18–23 February 1979) groups of designated experts representing many disciplines and regions of the globe would draw upon the information, including the discussions, presented in the first week together with their own knowledge in order to assess the understanding to date of climate and its interactions with mankind and to formulate general recommendations for international action. The experts would be divided into working groups each dealing with a component of the proposed World Climate Programme which would be submitted to the Eighth Congress of WMO to be held in Geneva some three months after the end of the Conference.

The arrangements for the Conference were carried out with the utmost efficiency, thus demonstrating the inspiring leadership of Dr White and the massive efforts of the WMO Secretariat.

The World Climate Conference

The Conference, organized and convened by WMO with the wholehearted co-operation of other international organizations as well as the Member countries of WMO, took place in Geneva from 12 to 23 February 1979 and was an outstanding success.

During the first week of the Conference, twenty-six overview papers on a variety of climate-related topics were presented and discussed by the world's leading specialists. The first week attracted over 350 experts from more than 50 countries. During the second week more than 100 experts from all parts of the world remained to engage in detailed discussions which resulted in a decision to issue a World Climate Conference Declaration in the form of an appeal to nations:

(*a*) To take full advantage of man's knowledge of climate;

(*b*) To take steps to improve that knowledge significantly;

(*c*) To foresee and to prevent potential man-made changes in climate that might be adverse to the well-being of humanity.

The Declaration also contained a preamble, a discussion of the problem of climate and the future, and finally an appeal to all nations to support the World Climate Programme to be developed by WMO.

In his keynote address, entitled "Climate at the millenium", Dr White drew attention to an important new concept arising from the Conference documentation. He pointed out that climate should be regarded as a resource even though it did not conform to the accepted definition of a resource. Furthermore, while access to climatic resources was restricted by national boundaries and property rights, climate also had some of the characteristics of a common property resource in that it could be modified by remote action. While the consequences of a global warming must be a matter of speculation at the time (1979), it was clear that such a change would have vastly different kinds of impact in various regions of the world. There would be winners and losers. A climate change could be the cause of a major redistribution of wealth and, from the point of view of humanity, an arbitrary one.

Taking an overall view of the Conference, there can be no question that it provided up to that time by far the most comprehensive assessment on the status of our knowledge of climate and its relationship to various aspects of the natural world and human society. The overview papers, published with the keynote address and also the Declaration with its supporting papers (WMO Publication No. 537, Geneva, 1979), formed an outstanding collection of authoritative texts on many subjects and should have a lasting value to all who are interested in climatic problems.

The participants in the Conference were fully representative of the world scientific community and were able to reach a consensus on the possible courses by which the global climate might evolve. It is obviously remarkable that a diverse group of specialists from many disciplines related to climate were able to arrive at an agreed view on this complex subject. Of particular interest was the general accord concerning the impact upon the global climate of an increase in the amount of carbon dioxide in the atmosphere.

The Conference was of the firm opinion that there was already a great deal of valuable information available about the natural variability of climate. It was urged that this knowledge could be put to immediate and continuing use in order to advance economic progress of all nations and, especially, to establish climatic services in the developing countries.

There was considerable discussion of a proposal that the Conference should communicate to all governments and to the United Nations a strong recommendation for the convening of a global ministerial conference on matters related to climate. It was at length agreed, however, that such action should be deferred until research could reduce the existing uncertainties as to the future course of the global climate and provide more specific guidance on the social and economic impacts of anticipated changes in climate and its variability. Nevertheless, in view of the importance of exploiting to the full all available knowledge of climate variability, it was considered appropriate to suggest that regional conferences at ministerial level would serve many a valuable purpose. It was agreed that they could promote vigorous national and international action to stimulate the application of climatological data to various sectors of national economies, notably in developing areas.

An encouraging and most welcome feature of the World Climate Conference was the interest shown by the world press, radio and television. Reporters from well over 100 newspapers, magazines and other media were present throughout the Conference and people in many countries received frequent reports on the deliberations and conclusions.

Some Conference highlights

The World Climate Conference was unique and was unquestionably one of the most important projects ever undertaken by WMO with the encouragement and support of other international bodies. The keynote address of Dr White, the overview papers, the discussions thereon, the reports prepared by the expert working groups and the Conference Declaration all

* See note on p. 8.

helped to provide an inspiring contribution to the World Climate Programme that was to be submitted for approval to the Eighth Congress of WMO. It seems appropriate therefore to place on record in this historical account a selection of topics which were given prominent treatment in the various papers and discussions. Some at least of these items may demonstrate the need for scientists ready and willing to work at the interface between the physical and socio-economic sciences.

(a) Physical basis of climate—climate modelling

Considerable interest was shown in the problems and uncertainties of giving an adequate description of the global climatic system with its five physical components—the atmosphere, ocean, ice masses, land and biosphere. Apart from the atmosphere (which shows the greatest variability on most space and time scales), it was the role of the oceans which predominated in the discussions. Concern was expressed at the limited amount of oceanographic data available. For their studies of climate, meteorologists and oceanographers need data for various depths in the ocean, not merely at the surface. It was appreciated that the establishment of special observing networks for ocean data would be highly expensive and present extremely difficult practical problems.

Good progress was reported in the development of general circulation models for simulating major features of the global atmosphere and of the average state of the present world climate. However, it still remained to produce models providing predictions of change a year or a decade ahead or from one winter to the next. On the other hand, there had been some success with models testing the response of the climate system to specified natural changes, e.g. in the Sun's radiation, and to possible man-made changes in the land surface or in the carbon dioxide content of the atmosphere.

There was a lively debate about the reliability of the models in revealing the "sensitivity" of the climate system to induced changes. It was accepted that the models were not yet able to cope with such important feedback mechanisms as cloudiness or changes in the circulation of the oceans. A consensus seemed to emerge, however, that in spite of their extensive simplifications, the models were approximating the true sensitivity fairly well.

Participants emphasized the need for models that would be able to show on a regional scale the response of the climate to changes of one type or another. Expected changes of both temperature and rainfall distribution must be known on this scale if the results were to be of use to planners at the national level.

Reference was made to the "robustness" of the climate system, in other words, the extent to which stability could be preserved against forced changes by strong feedbacks. The concept of a robust climate system was challenged in the light of palaeoclimatic evidence of large and occasionally sudden changes of climate, the causes of which were unclear. It was also stated on theoretical grounds that a complex non-linear system, such as the global climate, could have several quasi-stable states and could shift from one to another quite suddenly. It was pointed out, however, that if the climate system did possess any actual instabilities, they were extremely difficult to detect and, in fact, climate models did exhibit a degree of stability or robustness when boundary conditions were varied.

Consideration of the increasing concentrations of carbon dioxide in the atmosphere made clear the overriding importance of climate modelling. The history of past climates became less valuable as a prediction technique if mankind's alteration of the environment introduced previously unknown influences on climate. If that were to occur, the modelling approach would probably be the only method which could be used to predict the future course of climate.

(b) Some aspects of climatic variability

It was accepted that climatic variability was a key influence on the human economy, the main impacts being caused by short-period anomalies of precipitation or temperature. These parameters showed large variability, but not abnormally so, in the 1970s; the consequences were nevertheless highly significant because cereal-growing regions and pastoral belts were affected.

Several of the overview papers referred to the possibility of making useful extended-range weather forecasts giving trends and departures from climatic norms for up to a year ahead. One method of trying to meet this requirement would involve the identification of periodicities in the weather and the assumption that they would continue. Reservations were expressed, and were well supported, on the grounds that the periodicities that might be used for seasonal forecasts would have amplitudes too small for practical use.

(c) Climatic scenarios

A study of warm and cold episodes in climate history shows the different effects likely to arise from natural causes. A global cooling could result from a sequence of heavy volcanic eruptions, which would infuse dust into the stratosphere and disturb the radiation balance. On the other hand, a global warming could be caused by a lull in volcanic activity. Historical studies are valuable for the construction of climatic scenarios, a technique that can also be used with climate models, which has been adopted by workers in the field of palaeoclimatology. It was reported that geochemical techniques, based mainly on isotopes, had rapidly expanded the volume of information about past climates, yielding absolute age datings and palaeotemperatures.

It was clear that the scenario method had helped to draw attention to the probability of a gradual global warming around the year 2000. The Late Tertiary (between about 12 million and 2.5 million years before present) provided an interesting pointer with a fully glaciated Antarctic and virtually ice-free Arctic Ocean having a marked influence on the atmospheric and oceanic circulations. It was felt that if the concentration of atmospheric carbon dioxide were to rise above a certain level, there would be an increase in the probability of global warming to an extent that could cause the thin Arctic pack-ice to disappear.

(d) Climate monitoring and data collection

There was considerable emphasis upon the vital necessity of setting up adequate services for climate monitoring and data collection. The spectrum of data collection is vast, comprising meteorological, oceanographic, hydrological and geophysical data in one category and biological/ecological and economic/sociological data in others. The existence of well-established and efficient data bases, by means of WMO's World Weather Watch and in special climatological networks set up by national Meteorological and allied Services, was recognized but it was clear that all the requirements of an adequate climate monitoring programme called for an expansion of existing networks and the creation of new ones. Particular concern was expressed from two main standpoints with regard to the lack of data from ocean areas. First, the quality of ocean data needed to be improved since the standard error of sea-surface temperatures was unacceptably high. Second, there was a great need for regular data from those sea areas where time and space variability of temperature were known to be high.

(e) Climate and the biosphere

The biosphere reacts with the other four components of the climatic system on time scales characteristic of life cycles of the Earth's vegetative cover. Although the Earth's ecosystem is controlled to a considerable degree by the climate, many activities of mankind have significantly altered the Earth's vegetative cover and thus interfered with the natural biospheric component of climate. It was even suggested that agriculture and land clearance over thousands of years might have had a greater influence upon the global climate than the more recent growth of industrial activity. Man is constantly modifying important bio-geochemical cycles, e.g. the main elements present in organic matter, and it was urged that monitoring should be carried out so that any changes could be explained.

(f) Climate and the composition of the atmosphere

Within the context of climate change and variability, human activities which change the concentrations of certain gases and particles in the atmosphere appeared to be regarded by many participants as the most important. The cutting down of forests, which serve as a sink for carbon dioxide, was listed as an important factor and stressed the need for reliable data about rates of deforestation, particularly in the tropics.

The increasing concentration of the greenhouse gases such as carbon dioxide, of aerosols and of gases which might affect the ozone layer led to additional emphasis being placed upon monitoring the atmospheric composition in order to keep a close watch on changes actually taking place. It was also noted that since the build-up of concentrations would be slow, climate models should give timely indications of probable changes, which could subsequently be checked by monitoring—and assist in assessing the reliability of the models.

Since the effects of increasing carbon dioxide concentrations would be on a global scale, it was evident that any plan of action to restrict the use of fossil fuels and to reverse deforestation would have to be organized internationally. World-wide agreement and compliance would be essential.

(g) Climate and energy

The discussion on climate and energy was introduced by a projection to the year 2030 by when, it was estimated, the population of the world would have doubled from its size in 1975, the food requirement would have trebled and the consumption of energy would have increased fourfold or more. All available energy sources were discussed including nuclear power and solar and ocean thermal electric conversion systems. It was agreed that fossil fuels presented the main risks to the global climate whereas alternative energy sources could have effects of only local significance. However, the whole problem was considered to be so complex that much work needed to be done in both theoretical and applied research in order that energy policies could be given a better foundation in relation to climate-energy interactions. Apart from energy production, attention was also drawn to the manner in which climate affects practically all energy-associated activities.

(h) Climate and agriculture

The importance attached to the likely effects of climatic change upon agriculture was shown by the allocation to this subject of two sessions and part of a third. Agriculture was regarded as an excellent field for exploiting climatological data tailored to meet specific requirements. Agricultural scientists appeared to be confident that sufficient food could be produced to feed the Earth's growing population provided that all countries used modern techniques and kept up to date as new knowledge was acquired.

The discussion on climate and agriculture was obviously very wide and it was emphasized that agricultural systems, which include cultivation, harvesting, storage, marketing and other aspects, had reached a remarkable level of resilience and efficiency. It must be mentioned that agroclimatologists as a group seemed to be much more concerned with extended-range weather forecasts and with climatic variability than with climate change.

(i) Climate and water resources

It was assumed, partly out of necessity, that water engineers engaged in project design would use climatological statistics for the past 30, 50, 100 years to meet their requirements for similar periods in the future. However, hydrologists showed their concern to acquire adequate information about water resources as a global problem. It was therefore suggested that the successful development of climate models would be given a special welcome by hydrologists.

(j) Climate and health

The overview paper and the discussion on health aspects brought out in a striking way the problems that whole communities would experience in the event of significant changes in their climates. If the climate in a region became more tropical, a variety of new diseases would occur to which the local population had developed no immunity or other defences. The strong influence of climate on insect-borne diseases was considered to be a factor of special importance in tropical regions where malnutrition reflects another of the serious impacts of climate on human health. Prediction of climatic anomalies or even reliable medium-range weather forecasts could help in preventing outbreaks of disease among human beings and animals.

The conclusion was that whereas mankind might be able to lessen the impact of small climatic changes, any changes which were severe or abrupt could have serious consequences. The aim must therefore be to foresee such climatic changes in good time.

(k) Economic and social aspects

Efforts to describe and quantify the impacts of climatic change upon economic and social activities gave rise to a discussion of outstanding interest. It was noted that the cost or the benefit of climate variability fell inequitably upon different social or economic sectors of society and that it would be difficult to restore equity. A practical approach might be to try to identify, region by region, those who had gained and those who had lost as a result of a particular change in climate. Decisions would then be required as to how far those who had gained should assist in compensating the losers.

Earlier theoretical work in social and economic development had taken no account of climatic change. In future, such a possibility had to be accepted as a factor in socio-economic studies and it appeared that governments would be presented with a number of alternative courses, of which few would be free from potential regional or class conflict. The obvious course would be to arrange for these studies to be undertaken by international groups.

Economists urged strongly that climate variability and climate change should not only be studied in a general global framework. They referred to potentially disastrous phenomena

Exploiting natural resources.
Above: *solar energy*
Left: *wind energy*

such as droughts, floods, tropical cyclones and tornadoes, all of which are features of the climate and should be studied in connection with national and regional planning.

WMO's Eighth Congress launches the World Climate Programme

After the success of the World Climate Conference, it was a fairly straightforward process a few months later for the Eighth Congress of WMO (Geneva, May 1979) to give its formal approval to the World Climate Programme (WCP). In effect, Congress made official the final stages of the preparatory work which had begun after Seventh Congress (Geneva, 1975) and occupied a lot of attention on the part of the Executive Council* and of the Secretary-General and his staff. However, with regard to the WCP, the formal business of Eighth Congress, though free from serious conflict, called for careful and detailed consideration because it was essential to decide on policy, organization and administrative matters. Congress appreciated that the programme would not have a short time scale, but would probably still be in existence, subject to appropriate modifications as progress was achieved, at the end of the century and for many years afterwards.

The preceding section with its brief account of most of the topics covered at the World Climate Conference may serve to illustrate that the WCP has a range that is extraordinarily wide and that the research and applications must be tackled in great depth. Eighth Congress therefore took the view that WMO must seek the co-operation of many other UN agencies and other international organizations as indispensable support for the programme. Congress accepted the proposal that the WCP should have four components:

- World Climate Data Programme;
- World Climate Applications Programme;
- World Climate Impact Study Programme;
- World Climate Research Programme.

Congress requested the Executive Council* with the help of the Secretary-General to arrange for the establishment of an intergovernmental and inter-agency panel to provide overall co-ordination of the four programmes and related activities. The

* See note on p. 8.

Executive Council*, holding its thirty-first session immediately after Eighth Congress, agreed that an early decision should be made regarding the responsibilities for the World Climate Programme. Whilst WMO would be the responsible Organization for the programme as a whole, the Executive Council envisaged and subsequently confirmed that responsibilities would be undertaken as follows:

- WMO would be responsible for the Data and Applications Programmes;
- The United Nations Environment Programme (UNEP) would be responsible for the Impact Study Programme;
- WMO, jointly with the International Council for Scientific Unions (ICSU), would be responsible for the Research Programme.

Among the many other international organizations involved are Unesco, FAO, the World Energy Council, the International Atomic Energy Agency and the Consultative Group on International Agricultural Research.

At the highest level, WCP co-ordination is ensured through the meetings of the Executive Heads of Agencies involved. The Executive Heads of WMO, UNEP, ICSU, Unesco, and other interested organizations, or their representatives, meet approximately at annual intervals.

Some information about the four components of the WCP is given below.

World Climate Data Programme

The acquisition and processing of climate data are essential for successful implementation of the other three programmes. The requirement is for the development of vast amounts of meteorological, oceanographic, hydrological and other pertinent geophysical data. The World Weather Watch, WMO's global system for observing and exchanging weather information in real time, is a main contributor of current global meteorological data. Oceanographic data are supplied mainly by the WMO/IOC Integrated Global Ocean Services System (IGOSS) and by other observing systems including drifting buoys and satellites. The programme is also concerned with investigating all possible sources of climatological and related data and with assisting national data centres in exhaustive searches for historic data. Countries are also assisted in improving their systems for climate data management, including building national and regional data banks. The co-ordination of a global network of climatological stations is another objective.

Developments in data acquisition and exchange give much encouragement. Of special importance are satellite data, giving global coverage, and coupled oceanic and atmospheric data sets. Organizations co-operating with WMO in this programme include UNEP, Unesco, and the Committee on Climate Change and Oceans (SCOR/IOC).

World Climate Applications Programme (WCAP)

It was shown above in paragraphs (*f*) to (*k*) that a knowledge of climate and its characteristics is an essential element in national and regional economic and social development. There are many examples of applications of climate information and knowledge in sectors such as those related to land-use planning, food production, energy, water resources, transport and environment as well as the combat or control of the effects of drought and desertification. The planning, design and operation of airports, industries, urban areas and individual buildings, recreational facilities and flood-control schemes are particularly sensitive to climate conditions—as are the many aspects of human health—including the impact of pollution. Close collaboration between climatologists and the various users is called for in defining the requirements for climatological information. Important aspects of the programme are the development and exchange of knowledge, methods and techniques between nations. The exploiting of modern computer and communications technology in handling and disseminating information and data based on conventional observations as well as remote sensing, e.g. from satellites, has opened up promising new opportunities for cost-effective use of climatological applications.

World Climate Impact Study Programme (WCIP)

Only rudimentary techniques for the study of the consequences of increasing the concentration of carbon dioxide and other foreign ingredients in the atmosphere were available at the time (1979) of setting up the World Climate Programme. The purpose of the Impact Study Programme is thus to assemble multi-disciplinary teams to develop methodologies for impact studies. The context is global and the time scale probably very long. Within this framework there is scope for regional studies, e.g. with regard to impacts on agriculture and on water-management facilities. The effects of climate change may be found to depend primarily on the nature of the socio-economic system that has to accommodate to the climate changes. The Impact Study Programme must therefore include studies of the vulnerability of different kinds of society and of land resources and must also aim to provide estimates of their resilience under stress.

The lead role in this programme was accepted by UNEP, with strong support promised not only by WMO but also by other international organizations. ICSU's Scientific Committee on Problems of the Environment (SCOPE) had already embarked on a review of the methods of climatic impact assessment. The reports of this committee will make a valuable contribution to the programme as will any reports from the Climate Research Programme to be carried out by WMO/ICSU.

World Climate Research Programme (WCRP)

The main objectives of the Climate Research Programme are to determine:

- To what extent climate can be predicted;
- The extent of man's influence on climate.

These questions had already received a lot of attention from research workers, notably in the WMO Research and Development Programme, in the WMO/ICSU Global Atmospheric Research Programme (GARP) and in various ICSU activities. It was widely agreed that co-ordination of all research efforts would be desirable and it was readily agreed by WMO and ICSU that there should be one World Climate Research Programme. The programme would be conducted by a Joint Scientific Committee (JSC) which was established for the formulation of a research strategy and for overall co-ordination at the international level. For the inception of the WCRP, a great advantage was the experience already gained by the GARP in research into the general circulation of the atmosphere and in drawing up requirements for global data in order to extend the period for which reliable weather forecasts could be produced. Two major GARP projects were particularly significant, the GARP Atlantic Tropical Experiment (GATE) in 1976 and the first Global

* See note on p. 8.

Weather Experiment, which took place throughout 1979. The latter, in particular, gave rise to an enormous quantity of data covering the whole globe and involving data acquisition by means of satellites, balloons, ships, drifting buoys and aircraft as well as by the established weather stations of the WMO Global Observing System. Although the GARP experiments as well as other associated research studies enabled the WCRP to begin in favourable circumstances, it must not be overlooked that research into global climate and its prediction is one of the most complex tasks ever undertaken by teams of scientists. It would be unwise to expect rapid success. The Joint Scientific Committee for the WCRP has summarized as follows the achievements required to meet its objectives of determining (*a*) whether climate change can be predicted; and (*b*) man's influence upon climate:

- To improve our knowledge of global and regional climates, their temporal variations, and our understanding of the responsible mechanisms;
- To assess the evidence for significant trends in global and regional climates;
- To develop and improve physical-mathematical models capable of simulating and assessing the predictability of the climate system over a range of space and time scales;
- To investigate the sensitivity of climate to possible natural and man-made stimuli and to estimate the changes in climate likely to result from specific disturbing influences.

Statement issued by WMO on atmospheric carbon dioxide

At its thirty-first session (Geneva, May–June 1979) immediately after Eighth Congress, the Executive Council*, as already recounted, dealt with a host of items aimed at the implementation of the World Climate Programme. Within this context it also approved the issue for publication of a statement on atmospheric carbon dioxide.

The climatic events of the 1970s, which caused a number of disasters, and were most injurious to the populations involved, had aroused much public interest, and, to a great extent, anxiety about the possibility of climatic changes that might have undesirable consequences for humanity. This interest on the part of the general public was renewed and enhanced by the World Climate Conference, which was widely reported in the world press and on radio and television. Numerous questions about climate arose persistently but the dominant one was about atmospheric carbon dioxide and other greenhouse gases. It appeared that meteorologists and associated scientists in most if not all countries needed guidance as to the nature of the problem posed by the burning of fossil fuels with regard to the manner in which the problem was viewed by research workers. It was obviously appropriate that any statement on the subject should be issued by WMO with its high standing in the atmospheric sciences and in view of its position as a Specialized Agency of the United Nations.

The WMO statement was headed "Information material for answering queries on atmospheric carbon dioxide and possible climate change due to its continued increase". The statement contained a wealth of facts and speculations which admirably served the purpose of its distribution. It will suffice to select a small number of aspects discussed in the statement.

It was estimated that if the world's consumption of fossil fuels were to continue to rise at its existing rate (in 1975), the atmospheric carbon dioxide concentration would by about the year 2030 reach twice the pre-industrial level (in the early part of the nineteenth century). Carrying the extrapolation still further, a quadrupled concentration was estimated to be reached before the year 2100. Estimates such as these are likely to be reasonably accurate over a decade or two, but there are inevitably some uncertainties attaching to projections for a long period ahead. For example, population increase and hence energy consumption per head may slow down. Furthermore, substitutes for fossil fuels may be increasingly used. On the other hand, continued deforestation would reduce the extraction of carbon dioxide from the atmosphere, an effect that would also occur if for one reason or another the ocean became less efficient as a sink for atmospheric carbon dioxide.

If a still longer time scale is considered, and if all the recoverable fossil fuel, most of which is coal, were burned in the next few centuries, the concentration of carbon dioxide would rise to approximately eight times its pre-industrial level. There would then be negligible fresh input to the atmosphere, but, because of carbon dioxide's slow rate of removal from the atmosphere, it would take about one thousand years for the concentration to fall from eight times to four times its pre-industrial level. So, if the global climate were adversely affected by ever-increasing carbon dioxide in the atmosphere, the resulting climate would be almost permanent, subject to other natural or man-made influences.

With regard to temperature effects, it was estimated in the statement that a doubling of carbon dioxide concentrations could lead to an increase in the global average surface temperature of 1 to 3°C and that this effect might occur before the year 2050. It was also suggested that if a warming attributable to carbon dioxide increase were sustained for a sufficiently long period, of the order of several centuries, the major ice sheets of the polar regions might partially disappear, with important consequences for the sea-level.

The first decade of the World Climate Programme

Since the WCP was officially approved by WMO's Eighth Congress in 1979, it may seem somewhat premature at the time of writing this account (1987), to put "the first decade" of the programme in the heading of this section. Nonetheless, this licence with the time scale can be explained by recalling that from 1975 onwards, in Seventh Congress, in the Executive Council*, and among the constituent bodies of WMO and in other international bodies such as ICSU and Unesco, there was much discussion and no little preliminary planning about the compelling necessity for a World Climate Programme.

The early stages of any scientific and technical programme are inevitably concerned with setting up the organization, defining responsibilities, securing the collaboration of all interested parties and preparing, as a basis for critical examination, outline plans of tasks to be undertaken and of the appropriate structure of committees, expert panels and working groups. Within this context the World Climate Programme may with some justification be regarded as a special case. The programme itself is of almost unparalleled complexity; the five physical components—atmosphere, oceans, ice masses, land surface and biosphere—demonstrate that the programme is multidisciplinary to a considerable extent. The data deficiencies and data requirements are immense; the research problems must be tackled, however unyielding they may appear to be, and the applications and the impact sectors of the programme, which incorporate economic and social factors, present challenging problems of basic and applied research with objectives aimed at improving the welfare of mankind. Such are the scientific and technical problems and the range of difficulties that it seems no exaggeration to suggest that the first decade marked effectively the early stages of a programme that must be expected to

* See note on p. 8.

continue well beyond the lifetime of the participants in the World Climate Conference. It is all the more gratifying, therefore, to report that the World Climate Programme and its four components are being pressed forward with commendable vigour, enthusiasm and initiative under the overall supervision of Congress and its Executive Council.

Under the general arrangements for the programme, a separate department, the World Climate Programme Department, was established in the WMO Secretariat in March 1982. It carries responsibility for the administrative aspects of the implementation of two components of the programme—Data and Applications—and also for general co-ordination of the WCP. The Impacts component is being implemented by UNEP and the Research component is organized by the WMO/ICSU Joint Planning Staff.

In order to review the current status of climatological work in different regions a number of technical conferences have been held. (Among them were three WMO Regional Conferences in Africa, Asia and South America held in 1980–1983 and devoted to regional aspects of the climate-change issue.) Valuable benefits have been derived in providing information about the WCP and in assisting in the development of national and regional programmes.

It has been recognized that many countries need to develop basic capabilities in climate services. These countries require help with regard to the treatment of climate data, including quality control and processing, and also with regard to the capability of applying the data to critical areas such as food production, water and land and the efficient utilization of energy. Advice must also be given where necessary so that national Meteorological Services know what action to take with their governments in the event of climate variations or changes which are likely to affect important economic and social sectors.

In May 1983, the Ninth World Meteorological Congress reviewed the first four years of the implementation of the World Climate Programme and adopted plans for programme development in conjunction with approving the WMO First Long-term Plan for 1984–1993.

In the following two years an important international conference was prepared with the active participation of WMO. The conference was originally called merely "the Conference on CO_2 Assessment" and that is how the thirty-sixth session of the Executive Council (1984) referred to it, requesting that the conference should produce reports on:

(*a*) The carbon cycle;

(*b*) The effect of CO_2 on the climate and the detection of CO_2-induced climate changes; and

(*c*) Climate impacts on terrestrial and marine ecosystems.

The Council also stated that, in addition, the conference should make a statement on the current capability to make predictions of future climate as affected by CO_2 and other factors and propose action to develop further the methodology to determine socio-economic impacts on CO_2-induced climate change.

The conference subsequently received the title "WMO/UNEP/ICSU International Conference on the Assessment of the Role of Carbon Dioxide and of Other Greenhouse Gases in Climate Variations and Associated Impacts", and was held 9–15 October 1985 in Villach, Austria.

The conference, which is often referred to as the Villach 1985 Conference, provided a consensus statement on the probable magnitude of climate warming and its implications. Conclusions in the statement, which is still largely valid and widely used, can be briefly summarized as follows:

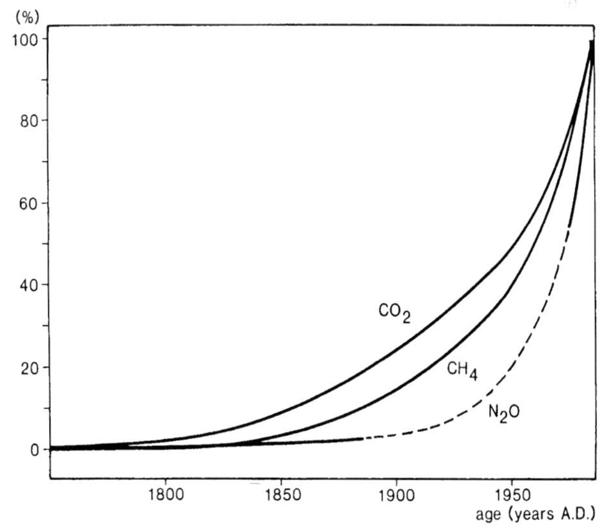

Relative increase of the atmospheric CO_2, CH_4 and N_2O concentrations from 1750 to 1985. (Development of N_2O is not yet known in detail.)

(*a*) If present trends continue, the combined concentrations of CO_2 and other greenhouse gases would be radiatively equivalent to a doubling of CO_2 from pre-industrial levels possibly as early as the 2030s;

(*b*) Increases of the global mean equilibrium temperature for a doubling of the atmospheric CO_2 concentration, or equivalent, may be between 1.5 and 4.5°C;

(*c*) Global warming of 1.5°C to 4.5°C would lead to a sea-level rise of 20–140 centimetres.

The Villach Conference was a milestone event in the development of international activities related to the climate-change issue. Its outcome was noted by many international organizations and national authorities, and in particular by the Tenth World Meteorological Congress, which urged that a concerted effort by WMO, UNEP and ICSU be undertaken to increase understanding and narrow the range of uncertainties in expected climate change. Congress further asked the Executive Council, in co-operation with the governing bodies of UNEP and ICSU, to monitor continuously the implications of increasing concentrations of greenhouse gases and their effect on the global climate and related issues and to arrange appropriate mechanisms to undertake further development of scientific and other aspects of greenhouse gases. At its session following Congress the Executive Council took a crucial step to comply with the Congress directive: it requested that the Secretary-General, in co-ordination with the Executive Director of UNEP, explore and, after appropriate consultation with members of the Executive Council, establish an *ad hoc* intergovernmental mechanism to carry out internationally co-ordinated scientific assessments of the magnitude, timing and potential impact of climate change. This mechanism—the WMO/UNEP Inter-governmental Panel on Climate Change (IPCC)—came into being in November 1988 with an objective of producing by the end of 1990 a comprehensive report that would include three basic parts:

- Assessment of available scientific information;
- Assessment of environmental and socio-economic impacts of climate change; and
- Formulation of response strategies.

Tenth Congress and the subsequent session of the Executive Council approved another important action: the organization of a second World Climate Conference that would review all aspects of the WCP, with emphasis on the economic benefits of climate applications, especially in developing countries, and would also provide a status report on the scientific understanding of climate change. Later, arrangements were made to co-ordinate work of the IPCC and preparations for the Second World Climate Conference (SWCC) (scheduled for November 1990), so that the first scientific presentation of the IPCC report could be made at the SWCC.

In promoting all these activities WMO responded to a demand made by its Members and engaged in even more extensive and intensive activity in one of the most serious global problems which humanity faces at present—and, unfortunately, will probably face in forthcoming decades.

These efforts of WMO and other collaborating international agencies received high recognition by the UN General Assembly which, at its 43rd session, adopted Resolution 43/53, "Protection of global climate for present and future generations of mankind". The resolution endorsed action taken on the establishment of the IPCC and welcomed the convening of a Second World Climate Conference. It was considered "that activities in support of the World Climate Programme... be accorded high priority by the relevant organs and programmes of the United Nations system".

This recognition and an endorsement of the World Climate Programme at the highest United Nations level should not be regarded as an award; rather, it was a challenging demand to increase and consolidate international activities on climate and climate-change issues.

International responsibilities of WMO were identified by the WMO Executive Council at its forty-first session (June, 1989) as follows:

> WMO's responsibility is to provide the authoritative scientific information and advice on the condition and behaviour of the global atmosphere and climate and the factors that affect them.

The Council further stated that:

> To carry out this responsibility, WMO will actively support and encourage the development of environmentally sound policies by the international community to respond to climate change, and encourage its 160 Members in co-operative efforts and appropriate programmes within their own countries.

Programmes concerning greenhouse gases and climate change, the atmospheric and hydrological environments, support to IPCC, preparation of the Second World Climate Conference, an active public information programme, and a programme for information exchange among Members—these are high priority areas where WMO has to increase its efforts based on WMO scientific and technical programmes including the WCP.

As far as the WCP is concerned, the programme has successfully proved its viability and merits. Important achievements can be claimed for all programme components.

The World Climate Data Programme is of extreme importance to the other three components, which are almost wholly dependent upon well-organized, high-quality data resources. The tasks assigned to this programme have, therefore, a degree of urgency and it is to be noted that an outline plan was approved by the Executive Council* at its thirty-second session in June 1980, within a year of the adoption of the WCP by Congress. The plan was put into final form during 1983 after consideration of comments by Member countries of WMO and by the Advisory Committee on the World Climate Applications and Data Programmes which had been set up in 1982 by the Executive Council at its thirty-fourth session. The WCDP implementation plan was updated for inclusion in the WMO Second Long-term Plan adopted by Tenth Congress in 1987 for the period 1988–1997.

It was realized from the outset that the WCDP would have to be concerned with a large variety of climate and climate-related data. It was therefore accepted that a primary aim should be to ensure the timely availability and exchange of reliable data necessary to describe the world climate and the evolutions of climate. The task is complicated since meteorological, oceanographic, hydrological and geophysical data are included and, for each category, several data-acquisition systems have been developed separately. A comprehensive information system is accordingly needed and this is to take the form of a data referral system which will provide references to the sources of data, i.e. to the centres where data sets are stored. As a first priority, planning has been concentrating on data from climatological stations and hydrological stations since these form a major source of data for research and applications. The referral system, which has acquired the acronym INFOCLIMA, has been making good progress with the willing co-operation of Member countries and a series of INFOCLIMA publications has begun.

A project has also been formulated to promote the transfer and exchange of information on the processing and management of climate data, using computers and software packages that can be brought into service by all countries. The acronym for this project is CLICOM (Climate Computing). It is a flexible system and, in addition to its primary purpose, it can cope with data rescue, user services (through periodical data summaries and statistics) and a variety of user applications. As a result of the project about 100 countries are equipped with CLICOM systems. Within a few years, virtually all Members of WMO will have CLICOM systems installed. Data rescue deals with the discovering of neglected data in danger of being irretrievably lost as the papers decay and crumble. It is a worthwhile effort to obtain access to such data wherever possible and to preserve the records on microfilm or other suitable medium. The Data Rescue (DARE) project initiated in Region I (Africa) will be expanded to cover South and Central America and South-East Asia.

Another important project in the data programme is concerned with Climate System Monitoring (CSM), which aims to gather information on the state of the climate system and on the occurrence of anomalies or trends. As the system develops, it should be practicable to advise countries of any climatic trends in their regions and to prepare for world-wide distribution of regular reviews of the global climate with explanatory articles on the interrelationships among the major components of the climate system.

It is evident that the data programme has been making impressive progress. The number of data centres exceeds one hundred and these are located in seventy or more countries. The gathering and organization of data in accordance with INFOCLIMA are gradually becoming more comprehensive. In addition, CLICOM and the CSM projects are proving well worth while and are receiving full support.

As a further development of the WCDP, the Climate Change Detection Project (CCDP) has been initiated in collaboration with other agencies. The main outcome of the project should be:

(*a*) Reliable assessment of climate trends and variability;

(*b*) Data assimilation into homogeneous gridded data sets.

Relevant activities include establishment of global and regional climate baseline data sets by dedicated meteorological

* See note on p. 8.

centres and initiating action to project reference climatological stations.

The World Climate Applications Programme covers a wide range of human activities. The priority areas are food production, water resources, energy and urban and building climatology. An important step, taken in response to a request from the Conference of Ministers of the Economic Commission for Africa (ECA), has been to study the incidence of drought in Africa. Requirements are also expected to increase with regard to human health. The plan of action for the programme was adopted by Ninth Congress (Geneva, 1983); the updated plan was included in the WMO Second Long-term Plan for 1988–1997 adopted by Tenth Congress. Most of WMO's Technical Commissions are involved and other organizations participating in appropriate aspects are FAO, Unesco, WHO, the United Nations Centre for Human Settlements (HABITAT), the Consultative Group for International Agricultural Research (CGIAR) and the World Energy Conference. It has proved convenient to organize the programme into four sub-programmes: WCAP—Food; WCAP—Water; WCAP—Energy; WCAP—Other applications.

The food sub-programme has been engaged in extending and improving long-established activities. The application of climate information to food production has always been prominent in the work of WMO's Commission for Agricultural Meteorology but, as so often happens in applications of scientific and technical knowledge, there are always improvements to be made, especially in applications of new technology, and faulty procedures to be corrected. Additional impetus was given to this work by the World Climate Conference, where the need to take into account the variability of climate and the possibility of climate change was emphasized.

In the food sector also, important agro-climatic studies on a sub-regional scale are under way in collaboration with FAO and Unesco. As with the data programme, a referral system is being developed in order to make more widely available information on current practical techniques. The system, known as the Climate Applications Referral System—Food (CARS—Food) should prove of increasing value to all countries. (The problems of drought and desertification are dealt with more extensively under the Agricultural Meteorology Programme with which all activities within WCAP—Food are closely co-ordinated.)

The aim of the water sub-programme is to help Member countries serve with increasing efficiency the variety of social and economic requirements which depend on the management and control of water resources. Activities are undertaken in conjunction with those of the Hydrology and Water Resources Programme of WMO and in collaboration with Unesco and other international organizations. In this sub-programme there is ample scope to extend knowledge on the relationship between climate and water resources and to investigate the impact of climate variability and climate change upon water resources.

In the energy sub-programme there has been considerable activity, no doubt partly because this area of interest is relatively new compared with activities in food production and water resources. A number of short-term missions have been carried out by experts to assist national Meteorological Services in the application of climatological information in the solution of energy problems. Much attention has also been given to questions of nuclear energy, solar energy and wind energy. The World Energy Conference has been playing a valuable part in promoting expert discussion on the development and peaceful uses of energy resources. The requirements for information and advice in this sub-programme are many and varied and in consequence a section on Energy has been included in the Climate Applications Referral System (CARS).

New interests have also been developing in urban climatology, notably in planning for rapidly growing cities in tropical areas. The Tropical Urban Climate Experiment (TRUCE) represents one of the specific measures taken in this regard. The main long-term objectives of the experiment are to produce a data base on the characteristics of tropical urban climates, to implement co-ordinated research programmes, to develop models capable of stimulating the urban climate system, and to promote the use of climate information in building and urban design.

Like urban and building climatology, human health has shown signs of becoming a growth area in the Applications sub-programme since there would be immense implications for whole populations if any major likelihood of climatic change were to arise. A major climate application relates to the development of objective bioclimatic criteria to be used in planning human settlements.

The World Climate Impact Studies Programme is the responsibility of UNEP, which accepted an invitation from WMO to take on this role in view of the close connection between the programme and the environmental assessments of UNEP. A plan of action for the WCIP was approved by the Governing Council of UNEP in May 1980, less than a year after adoption by the WMO Congress of the World Climate Programme. The plan of action has placed special importance upon the assessment of the political, economic and social consequences of climate changes resulting from industrial processes. Equal stress is laid upon efforts to reduce the vulnerability of food production to climate change and to the development of methods for assessing the impacts of climate variability and climate change.

Implementation of the plan is supervised by a Scientific Advisory Committee created by UNEP and regular meetings are held with representatives of the four components of the WCP. As a follow-up to UNEP's plan of action, the Impact Studies Programme has so far been conducted in three phases.

The first phase concentrated on the development of ways of assessing climate impacts and on measures to render food production less vulnerable to climate variability. Proposals were also drawn up for the study of impacts on other sensitive sectors such as water and energy resources, health and human settlements. The second phase was in part a continuation of the first and additional measures were included on the side of organization, in particular, placing on a secure basis the arrangements for co-ordination with other components of the WCP. A study was also initiated on the use of satellite data to quantify parameters relevant to climate impacts. Various international bodies are involved in a project entitled International Satellite Land-Surface Climatology (ISLSCP). The first two phases covered the period 1981–1985 and a third phase of the WCIP began in 1986. In this phase the activities of earlier phases are being continued and extended and projects are to be undertaken with certain countries to incorporate national climate impact studies into the work of the WCIP.

The first decade of the impact studies programme may be described as largely experimental. There was a lack of suitable methodologies for impact studies, almost a new field of research until the World Climate Conference was held. It was therefore a task of high priority to devise, test and subsequently publish impact methodologies. This work has led to an increased understanding of the response of the atmosphere to changing concentrations of greenhouse and other gases and the stage seems to have arrived for incorporating international and regional policy considerations into climate impact studies.

The World Climate Research Programme was ready for development as soon as the WCP was approved by Congress in 1979. The research programme is in many respects a sequel to the Global Atmospheric Research Programme (GARP) or, perhaps more specifically, corresponds to GARP's second objective. A great deal of experience in global studies was gained in the Atlantic Experiment (GATE) and in the first GARP Global Experiment (FGGE). The overview papers at the World Climate Conference already contained mature and positive ideas as to the contents of a World Climate Research Programme.

Although what needed to be done was well understood, the problems of the global climate are so complex (and, perhaps in some aspects, not solvable in a useful way), that it took some four years of intensive discussion before the scientific plan for the WCRP became available for publication. Involved in the discussions were the WMO/ICSU Joint Scientific Committee and its working groups and also the Joint Planning Staff. Finally the plan was approved by both WMO and ICSU and published by WMO (reference WMO/TD–No. 6) on behalf of the two Organizations.

The plan for the WCRP is a most impressive document which describes the overall strategy and sets out the action to be taken in each of three streams. The first stream is concerned with the relatively short term and is designed to examine the feasibility of predicting large-scale weather systems over periods of one to two months. In effect, the first stream will explore the ultimate limit of predictability associated with specific initial states of the atmospheric circulation. The second stream increases the generality of the approach and is based on models representing the interactions of the global atmosphere, tropical oceans, sea ice and the land surface. By this means, the aim of the second stream is to predict the variations in the heat transport by ocean currents and the resulting variations of atmospheric pressure systems over periods ranging from several months to several years. The third stream extends the time scale still further. Its objective is to study the causes of long-term climate trends so as to assess the potential response of the global climate to natural or human influences over periods of several decades.

The aims in each stream define research programmes which need to be pursued for a very long time if measurable success is to be achieved. The strategy and its accompanying plan seem admirably constructed to enable the scientific community to reach a position where it can confidently advise governments on the likely impact of climate changes upon the economic and social fabric of nations. The research programme is an extremely strong challenge to scientists of many disciplines. The objectives in view represent an equally strong challenge to the United Nations and to governments to ensure that teams of scientists and adequate resources are made available to the programme.

It is of interest to list some of the projects to be carried out under the plan of action. The titles are sufficiently descriptive and in some cases an acronym is in use:

- Global climate analysis and model development;
- Research on climate processes;
- Study of the Tropical Ocean and Global Atmosphere (TOGA);
- World Ocean Circulation Experiment (WOCE);
- Study of Climate Forcings;
- Study of Global Change.

The World Climate Programme—looking to the future

The World Climate Programme is of great importance to all nations. The first decade, which is the subject of this historical account, should be regarded as the foundation period during which the strategies for each component are drawn up and plans of action prepared in all possible detail. It is plain for all to see that strategies and plans have been produced for all four components by expert scientists and their associates with due responsibility, knowledge and wise judgement. It is clear that the first decade of the WCP has been spent constructively and that everything possible has been done in co-ordination and collaboration, in planning, in assembling working groups, subdividing the components to render the tasks somewhat more straightforward and logical, to give promise that, if success is attainable, it will be achieved. It seems reasonable, therefore, to set aside the historical aspect and to take a look towards the future and consider the prospects for progress of each component.

The Data programme has a special responsibility as the other three components of the WCP are essentially dependent on adequate data over the whole range of their activities. It seems likely that the data component will proceed with encouraging success. The rescue of old data and their transfer to permanent form on microfilm or other facility is a well-ordered process in which many years' experience has been gained. Fresh data of high quality become available each day from observing stations throughout the world and from satellites. There can be no doubt that, in accordance with the international arrangements organized by WMO, the data centres will maintain a high level of efficiency.

The Applications and the Impact Studies components can be considered together since they both function through the application of environmental data to practical problems such as food production and the effects upon society of climate change and variability. The plans of each of these two components are imaginative and will be pursued with scientific responsibility. Full advantage will be taken of improved data resources and of any progress made by the research component. The prospects for good progress in the applications and the impact studies components of the WCP seem favourable.

There remains the Research component and here the outlook is extremely difficult to assess. It can be said confidently that highly competent scientists have agreed on a strategy and plan of action that must have the greatest chance of success. But climate is probably unique in its complexity. Some scientists of long experience in climate research are known to consider that it may not be possible to predict climate change or to estimate the extent to which mankind may cause the climate to change. Other scientists, also of long experience, view the problem in two subdivisions. First, there is the problem of identifying different states of the atmosphere if subject only to the numerous interactions within the climate system. Second, there is the problem of how the picture would change as a result of variations in the boundary or external conditions of the climate system. In the case of the first subdivision it is accepted that progress will be extremely difficult. The second subdivision depends on the capability of climate models to represent the long-term equilibrium state of the climate. If the models can meet this requirement, they can be used to trace changes in the system as it responds slowly to changing boundary conditions.

It would not be sensible to speculate further. If the research programme is to make good progress, it will be the result of sustained, dedicated effort and far-ranging co-operation of scientists in many disciplines and of operational services which monitor the global environment.

CHAPTER XIII

EDUCATION AND TRAINING PROGRAMME

Introduction

Education and training are regarded as the most important and far-reaching of the preparatory steps which help to determine the efficiency of a Meteorological Service. In any professional scientific organization it is essential that new entrants should be equipped with the basic educational qualifications appropriate to the grade to which they are appointed. In a meteorological organization the main subjects are mathematics and physics and knowledge of these subjects at the required level equips new staff to undergo training in meteorological duties. These duties consist primarily of the application of scientific knowledge to the great variety of operational and research activities that are to be found in meteorological work. As staff of a Meteorological Service may have a career which extends over thirty years or more it is evident that recruitment and training must be recognized as serving vital, long-term objectives concerned with efficiency and scientific enthusiasm.

When WMO's predecessor, the International Meteorological Organization, was formed in 1873, meteorological activity existed in relatively few countries and the science of the atmosphere was in a condition that might be described as slowly developing. At that time and for the next several decades efforts were focused on observing and measuring the properties of the atmosphere; instruments and techniques were experimental and not suitable for widespread, routine use by people of limited qualifications. At the same time the leaders of IMO were looking forward to the organization in an increasing number of countries of networks of observing stations which, taking advantage of the recently invented telegraph, would permit the rapid exchange of comprehensive weather reports made at prescribed times. With such arrangements it would be practicable for co-operating countries to plot and analyse synoptic charts depicting the movement of pressure systems over a large area of the globe. These preliminary steps were identified as essential for the development of weather forecasting on a sound scientific basis. Disasters involving shipping on the high seas had long emphasized to governments and their peoples the need for meteorological services that would give advance warning of adverse weather along sea routes and in coastal waters. There was also at the time a growing recognition that reliable weather forecasts would serve many other interests besides shipping.

It is now well known that whereas the requirements for gale warnings covering sea areas provided the initial incentive for the establishment of national Meteorological Services, each with networks of observing stations, it was the growth of aviation in the early years of the present century which provided the strongest stimulus for the extensive development of meteorological observing stations and for the investigation and rapid improvement of forecasting techniques. The growth of national Meteorological Services and their integration, through the agency of IMO, into a co-ordinated international organization was a gradual process because there were many technical and practical problems to solve. One of the most important, arising from the steadily increasing number of observing stations, was the need for suitable instruments, for guidance on carrying out observational routines, including visual observations of cloud and visibility, and for general agreement on standards to be maintained in order to safeguard the efficient production of accurate and reliable meteorological data. Provision had to be made for the transition from the days in the 1870s when the few meteorological stations were operated by highly qualified scientists engaged in research to a time when there would be numerous observing stations where staff of less advanced qualifications were employed.

It was quickly appreciated by IMO that among the basic requirements for an efficient meteorological organization, an indispensable one was concerned with the education and training of staff. The introduction of accurate, robust instrumentation and the provision of clear guides on methods of observation in accordance with approved procedures would be ineffective if meteorological staff were not educated and trained to appropriate scientific standards in conformity with their appointed grades and duties.

IMO therefore gave high priority to education and training and when, in 1951, it was replaced by the intergovernmental World Meteorological Organization, the new body had additional reasons for continuing to stress the importance of education and training. In the years up to the late 1930s the expansion of Meteorological Services had been dictated largely by the requirements of aviation. During World War II and in the years immediately following, it was widely recognized that meteorology and climatology could make valuable contributions to many sectors of industry and commerce as well as to aviation. The responsibilities of Meteorological Services were clearly to become of much wider scope, a consideration which further emphasized the importance of education and training. As a result the WMO Convention approved by governments incorporated an article specifying that WMO would aim to encourage research and training in meteorology and to assist in co-ordinating the international aspects of such research and training.

In this field WMO proceeded cautiously at first. The Member countries of the world-wide Organization were primarily responsible for deciding the educational qualifications required by their own staff and for the training courses in meteorology for the different grades. Another factor which gave signs of presenting WMO with numerous problems was that, since World War II, many colonial territories had achieved or were about to achieve independence and their Meteorological Services would need to become self-contained after years as a subsidiary organization of another, more developed country. These new independent countries would become full Members of WMO and might be expected to seek advice and material assistance not only in training but also in many other aspects of an independent Meteorological Service, including its infrastructure.

WMO, itself a new Organization but richly endowed with tradition and experience inherited from IMO, could therefore anticipate wide-ranging problems affecting the scientific standards and efficiency of Meteorological Services in different regions of the world. Of all these problems education and training would be among the most fundamental. WMO, with its status as an official inter-governmental body, was well placed to recognize the requirements and difficulties—

some of long standing, others inevitably associated with newly emergent nations—that confronted the various Meteorological Services in different regions of the world. Soon after its formation WMO had become a Specialized Agency of the United Nations and was able to establish collaborative arrangements with other organizations throughout the UN family, including the UN Technical Assistance Programme. In the implementation of this programme, mostly with countries that had recently gained their independence, WMO was involved at an early stage in providing experts to advise in all aspects of organization and training in meteorology. In 1952 the first WMO mission was sent to Israel to give a course to meteorologists on microclimatology. In 1954 WMO experts gave training courses to meteorological assistants on observing procedures and on the coding and transmission of weather reports. At a higher level, the first WMO fellowship for study at a university was awarded in 1953.

The planning of WMO's role in education and training

At first therefore WMO's activities in the field of training were on a modest scale and for the most part consisted of providing experts to visit certain countries to advise on training generally and to give courses that were urgently required. However, in due course it became clear that WMO would have to expand its work in training considerably and that the most pressing responsibilities were arising as a result of political developments in Africa. The Secretary-General accordingly presented a report to the twelfth session (1960) of the Executive Council*, explaining that the demands upon WMO for support in the various sectors of training were growing and were already exceeding the resources of the WMO Secretariat. He proposed, and the Executive Council agreed, that the Secretariat should draw up a comprehensive long-term plan for training activities, including those under the UN Technical Assistance Programme, to be undertaken by WMO. While work was proceeding, WMO received an urgent request from the newly independent Member countries in Africa for immediate assistance in training for all meteorological grades from observing staff up to professional forecasters. The Executive Council considered the commitment at its thirteenth session (1961) and gave the Secretary-General authority to take action on the following lines:

* To appoint an expert to prepare an overall plan for the training activities that WMO should assume responsibility for on behalf of Member countries;
* To carry out a survey of the meteorological requirements of newly independent States of Africa;
* To consider the needs of the Secretariat for staff and facilities to enable WMO to meet its responsibilities for training.

The Secretary-General invited Professor J. Van Mieghem (Belgium), a distinguished scientist with a high reputation in international meteorology, to prepare a draft training plan and to provide advice on special aspects such as training in Africa and the structure of the Secretariat.

Professor Van Mieghem viewed the whole subject of training in a general perspective and within this scope gave special attention to the problems in the developing countries of Africa and Asia. He produced three reports covering the following aspects:

* The professional training of all grades of meteorological staff in the less-developed countries;
* Plan for meteorological training in Africa;
* Establishment of a Training Section in the WMO Secretariat.

These three reports provided the foundations of the WMO Education and Training Programme for the decade which followed. At the same time, a number of important events were taking place which had implications for training in meteorology and these were taken into account in both planning and projects.

Course at a meteorological training school in Afghanistan

The most important of these events concerned the collaboration among UN Agencies, which were aware of the importance of education and training for implementing development plans all over the world. With other Specialized Agencies WMO joined a sub-committee set up by UN to ensure co-ordination of the various training programmes which to some extent contained overlapping elements. An important task of this sub-committee was to assess manpower requirements in the different countries and to give guidance on education and training. As a member of the sub-committee WMO was able to provide estimates of meteorological requirements for inclusion in overall plans.

* See note on p. 8.

Another event with a bearing on the formulation of training policy was the request made by the Executive Council* to the presidents of the WMO Technical Commissions to put forward a list of topics that they would like to see included in a comprehensive training plan. WMO's responsibility for meteorological training received recognition and at the same time was extended when the International Civil Aviation Organization asked WMO to advise on the meteorological content of courses to be given at its Regional Training Centres. These courses were designed for staff working at civil airports whose duties called for a background knowledge of such aspects of weather as fog, mist and strong gusty winds.

This section has been largely concerned with the years 1950–1961 and it may be said that this period was notable for a more mature understanding of WMO's responsibilities in education and training and for a sudden upsurge of interest in these fields, accompanied by a strongly growing demand for WMO's support from an increasing number of countries.

Professional training in meteorology

Training is a subject in which all professionals are involved, at least initially, as students but in which relatively few participate as instructors or as directors of policy. Nevertheless, almost everyone is prepared to claim expert authority on training and to argue forcefully as to what a syllabus should include and what should be omitted. Not surprisingly, therefore, the Van Mieghem report aroused a great deal of controversy. The report was comprehensive. It proposed four categories of staff from university honours graduates (Class I) down to weather observers (Class IV) and a detailed syllabus was presented for each grade. The ensuing discussion among the Member countries of WMO ranged very widely and was prolonged. Some commentators suggested that each syllabus was unnecessarily advanced and too detailed. Others supported the academic nature of the four courses on the grounds that meteorology was rapidly evolving from a comparatively elementary science using largely empirical techniques to a sophisticated branch of mathematical physics using complex theoretical methods and requiring observational data of the highest quality. It was argued that training programmes should reflect these developments but there were many who considered that somewhat lower standards should be required for the generality of staff in each grade, leaving the more advanced studies to those who were qualified and selected to carry out basic research.

Thus the proposals contained in the training report were controversial and also stimulating. It was particularly gratifying that such a report should arouse widespread interest and elicit comment from the majority of WMO Members. In 1962 the Executive Council* discussed the report and agreed that it contained a great deal of valuable information and advice concerning the technical nature and content of meteorological training courses. However, because of the divergence of opinion among WMO Members, the report was accorded the status of an advisory document and after some revision, based on the comments that had been made, it was circulated as a WMO *Technical Note* (No. 50). As a result the report became widely known and acquired considerable prestige and authority. Its guidance proved of great value in determining the technical organization of Meteorological Training Centres which were established in Nairobi, Lagos, Kinshasa and Cairo with the help of funds provided by the UN Development Programme.

Whilst the descriptions of the categories(Classes I to IV) provided valuable and indeed authoritative advice concerning the field of education, recruitment and training of personnel, they could not be regarded as a static set with a long-term validity. The reason for this was that the functions of national Meteorological and Hydrological Services were continually expanding and becoming more exacting as the users of meteorological information developed their own scientific and technical procedures. It was necessary, therefore, to take into account the increasing role of atmospheric and certain earth sciences in a wide spectrum of activities covering the biosphere, energy and social, economic and industrial problems. Furthermore, each Meteorological or Hydrological Service had a dual role—namely, to carry out national commitments and also to participate in the different international programmes of WMO and of other governmental and non-governmental agencies. Other factors to be allowed for were mainly of an administrative nature, such as the way staff grading systems differ from one country to another and the variety of training programmes in operation with the national Services.

WMO realized, therefore, that a primary objective of co-ordinating and standardizing meteorological, hydrological and related activities would be well served by the provision of guidance on education and training. In addition, WMO accepted that this guidance must be kept under review and brought up to date from time to time. The latest review carried out by the EC Panel of Experts on Education and Training was published, after approval by the Executive Council of WMO, in February 1987. It describes in valuable detail, for each class, the level of education required for entry, the numerous duties to be performed and the training to be provided, including the formal course of instruction and practical work and the subsequent period of on-the-job training.

The panel offered a note of encouragement in pointing out that training and work experience at each level should provide adequate preparation for advancement to the next higher grade. The panel also noted that, especially in reference to Class I and Class II personnel, requirements have broadened to include graduates not only in mathematics and physics but also in other natural sciences, such as chemistry, biology and economic and social sciences.

Guidelines for education and training

One of the most valuable practices of WMO has been the publication and distribution to all Members countries of guidebooks for the various sectors of meteorology and its applications. These *Guides* are among the so-called mandatory publications of WMO and as such have a special status. The objectives are to ensure the world-wide adoption of technical and other procedures and the observance of standards above specified minimum levels in all branches of meteorological work. *Guides* so far issued by WMO deal with subjects such as meteorological instruments and methods of observation, agricultural meteorological practices, hydrological practices, and marine meteorological services. A *Guide* is not accepted for general issue until the whole subject of concern has been discussed widely and in depth and a text is finally approved by all Members in the interests of international co-ordination. A *Guide* carries substantial authority.

The *Technical Note* on training, which was issued in 1963, was widely utilized by developing countries and indeed each syllabus prescribed for the various grades (Classes I, II, III and IV) exercised a profound influence on the training courses of all Members of WMO and in regional training centres. Although the *Technical Note* had been issued with no more than advisory status, it soon became necessary for WMO to consider whether it should be accepted as a *Guide* in the mandatory series of WMO publications. It was realized that such a step would have

* See note on p. 8.

important implications for all Meteorological Services since formal recognition would thereby be given to the classification of meteorological personnel and to the corresponding syllabus for each grade.

This question was considered by the Executive Council* at its eighteenth session, in 1966. There was no question about the value of the *Technical Note* but there was no prospect of unanimous agreement to raise its status from advisory to mandatory. A major barrier was that in applications of meteorology, e.g. to agriculture, and also in hydrology, training courses could vary from one region to another because of differences in climate and it seemed inadvisable therefore to attempt to draw up a universal, mandatory syllabus for any of the four classes of meteorologists. However, the Executive Council decided to seek the views of all WMO Members. The task was entrusted to the EC Panel of Experts on Meteorological Education and Training which was directed to consult the Technical Commissions of WMO as well as the Meteorological Services in the Member countries.

It was anticipated that the panel would produce a comprehensive report covering both basic and specialized fields. Expectations were fully realized and the panel produced detailed programmes for training in what may be described as core subjects such as synoptic meteorology, climatology, aeronautical meteorology and hydrometeorology. It was also foreseen that training courses would be required in the near future in relatively new fields such as atmospheric chemistry, computerized data processing, environmental meteorology and generally in applications to a country's economic and industrial developments.

Evidently the EC panel worked with profound knowledge and mature judgement. The great value of its report was immediately recognized and brought the Executive Council* to the outstanding question—whether to confer the status of *Guide* upon the report on education and training. The problem, in effect, focused on a consideration as to the wisdom of modifying the criteria required to provide a *Guide* with its mandatory nature. There was no question that, taking training programmes as a whole and appreciating their very wide range, they could not be subject to hard-and-fast rules as in the case, for example, of the *Guide to instruments and methods of observation* and the *Guide to climatological practices*. The Executive Council therefore decided at its twenty-first session (1969) that the material produced by the panel should be published with the title *Guidelines for the education and training of meteorological personnel*. In order to avoid misunderstanding the text of the publication was preceded by a note to say that the contents were of an advisory character and did not have the standing of a *Guide*.

The *Guidelines* have proved remarkably successful and have been put into practice to a great extent by a majority of the WMO Members. A revised and enlarged second edition was published in 1977 when the title was extended to include operational hydrology. In 1984 the third edition was issued. These successive editions were required partly in order to take into account the incorporation of hydrology into the formal responsibilities of WMO, as recorded in the Convention, and partly in the light of more advanced technologies and new fields of specialization.

Organization of WMO's training activities

In the early years of WMO's existence as an intergovernmental organization and as a Specialized Agency of UN, training was accorded a high level of importance while assessments were made of the responsibilities and requirements which Members would wish to place upon WMO and also of the advice and practical support which WMO should offer to Members carrying out their own training programmes. Specific questions were dealt with as they arose and relationships were established with other UN Agencies which had interests in training as part of regional and national development projects. Apart from the WMO Secretariat itself, the main components of WMO playing a part in the field of training were the Technical Commissions, which dealt with problems in their own specialized fields, and the Regional Associations.

As the training interests of WMO grew to substantial proportions and the range of activities widened, the Executive Council* in the early 1960s began to examine the need for formal, long-term arrangements for exercising oversight of WMO's training efforts, for advising on individual questions, whether of an advisory nature or of practical consequence, and for assuming responsibility for matters which might affect training policy. Discussions in the Executive Council resulted in 1965 in the formation of an EC Panel of Experts on Meteorological Training. Over the years this body has functioned very successfully in supervising the implementation of policies issued by Congress and by the Executive Council. In its responsibility over the whole WMO field of training, the panel is also in a position to make recommendations to the Executive Council on major plans and projects.

The main tasks of the panel, in conjunction with the Secretary-General, were to advise and assist in the training of staff in national Meteorological Services, especially those in the developing countries, and to conduct studies into various aspects of the organization of training, e.g. facilities, conferences, seminars and training centres.

Education and training are subjects which are apt to give rise to controversial questions. Very soon the panel found itself considering some exceedingly complex problems including the classification of meteorological staff, contents of training courses and suggestions for the issue of WMO certificates to those who successfully attended approved training courses. The panel also studied the possibilities of establishing regional training centres on a long-term basis and the arrangements for their supervision. From the outset the panel showed a determination to recommend speedy, decisive action when urgent matters arose, some of them of a regional rather than a universal nature. The creation of a Chair of Meteorology at the University of Costa Rica was initiated by the panel, with beneficial results in a region where the profession of meteorology was greatly in need of encouragement. Another example of the panel's practical outlook was the formation of a mobile unit which visited Panama and other countries of Central America for the purpose of training Class IV meteorologists, the operators of the networks of observing stations.

A stimulating factor in the work of the panel was that its recommendations to the Executive Council* were a signal for prolonged discussion and, at times, dissension. Even so, the Council attached great value to the activities of the panel and noted favourably that it was interpreting its terms of reference in the widest possible way. The panel was thus authorized to evaluate the standard of training provided in WMO projects and seminars and to monitor the work of the instructors.

The panel has been well served by its chairmen. Professor Van Mieghem, author of reports which contributed effectively to training policy, was an obvious choice for chairman when the panel was set up in 1965. He was succeeded in 1971 by Dr Alf Nyberg (Sweden) who served in that capacity until 1979, when Dr R. L. Kintanar (Philippines) became chairman. The high prestige accorded the panel is demonstrated by the fact that Dr Nyberg and Dr Kintanar each served an eight-year term as President of WMO.

The organization of WMO training activities was completed by the establishment of a training section in the Secretariat. Such a section was, of course, indispensable if the work of the panel

* See note on p. 8.

were to make significant progress beyond the discussion and planning stages. In the early years of WMO education and training responsibilities were borne by the Technical Co-operation Division and by the Technical Division. The former was engaged in planning training projects for approval and financing by the United UNDP. The Technical Division had an advisory role in support of Member countries organizing their own training programmes. Eventually, in 1964, a small unit consisting of a scientific officer and a secretary was set up in the office of the Secretary-General to deal with all training matters under the personal supervision of the Secretary-General.

This unit was the nucleus around which a more substantial effort from the Secretariat became available for all aspects of training. There was indeed a quite rapid expansion of training activities and action was taken to reorganize the Research Division into a Research, Education and Training Division. Finally in 1976 it became necessary to establish a separate Education and Training Department within the WMO Secretariat.

The remarkable growth of the efforts devoted to training in the WMO Secretariat were in large measure a response to the demands raised by Member countries, notably the less-developed countries, for advice and support in all aspects of training. Help was required at all levels from Class IV up to Class I meteorologist, and in the case of some countries arrangements had to be made for staff to go abroad for study at universities or in the training colleges of other Meteorological Services. The successful work in education and training carried out by the Secretariat demonstrated the imagination, drive and enthusiasm of Dr H. Taba (Iran), who was in charge for some 12 years of the unit which evolved into a Department of WMO. One of his successors as head of the Department was Professor G.O.P. Obasi (Nigeria), formerly Professor of Meteorology in the University of Nairobi, who was later appointed Secretary-General of WMO. Some of the major achievements of the Secretariat in the field of education and training are described in the following sections.

Meteorological training in the instruments enclosure of the Civil Aviation School, Casablanca (Morocco)

Training in Africa

The earliest plans for training in Africa were based on a more or less theoretical and subjective assessment since there was little in the way of information of experience to serve as a guide. However, these plans, which were modified in the light of comments from directors of Meteorological Services in Africa, proved successful in practice and provided useful background material when training plans in other areas were in preparation.

The main problem in Africa in the early 1960s was seen to be the professional training of Class I and Class II personnel in all the developing countries. One important factor was the language situation which resulted from the existence over long periods of English and French colonial territories. It was therefore proposed that professional training should be provided by the creation of departments of meteorology in two universities, one in a French-speaking country and one where the language in use was English. It was also envisaged that training centres would be required on a regional rather than a national basis for Class III and Class IV staff and for specialized training in meteorological instruments.

At first progress was very slow because the supplies of suitably qualified recruits to the national Services was hardly more than a trickle. For example, the Department of Meteorology in the University College of Nairobi had only one student for its inaugural course but in a short while it became a flourishing and highly successful centre of studies.

An exercise in pragmatism had to be carried out in order to ensure an adequate flow of trainees to fill the courses that had been arranged. For entry to Class I training it was decided to relax the qualifications for admission and to accept candidates who, though without a degree in mathematics or physics, had acquired reasonably equivalent knowledge in their employment and had demonstrated good potential. With regard to Class II personnel, the remedy appeared to be to extend the courses to a duration of 30 months so that carefully chosen students could receive instruction in the appropriate sciences and then go on to study the applications to the atmosphere. These very practical measures were approved in advance by the directors of the national Meteorological Services and proved to be most successful.

Training in the South American region

Among the areas of the world most urgently in need of training programmes so that national Meteorological Services could be adequately staffed, the South American region was second only to Africa. A fortunate circumstance was that Argentina possessed excellent training facilities for Class I personnel in a university meteorological department. After viewing the problems WMO concluded that meteorological departments were required in two South American universities, one Spanish-speaking and one Portuguese-speaking. It was also proposed that each national Meteorological Service should establish a training centre for Class IV, i.e. weather observing staff. Looking further ahead, the need was recognized for refresher courses and associated seminars when the staffing of Meteorological Services had improved.

Outstanding help in the realization of these plans was given very generously by Argentina. In 1958 it opened its Class I training courses at the University of Buenos Aires free of tuition fees to students from other countries of Latin America. A similar arrangement for the training of Class II personnel was made in 1963. One of the main advantages for Latin American students was that in Buenos Aires they received instruction in their own language.

The next step was to tackle the training problems of the English-speaking countries and territories of the Caribbean region. An Institute of Meteorology was established in Barbados and courses were provided for Class II, III and IV staff in general meteorology and in certain specialized fields.

The efforts of WMO in dealing with the urgent training requirements in Latin America and the Caribbean met with considerable success. Of particular value, and indeed indispensable, was the organization of fellowships which enabled selected personnel to travel abroad in order to attend meteorological courses at universities of other suitable establishments.

WMO fellowship programmes

The United Nations Development Programme was instrumental in providing funds for the training of staff recruited by the Meteorological Services of developing countries. Fellowships were made available in this way but mainly for specialized and postgraduate courses and not for studies leading to a university degree. In the newly independent, developing countries there was a shortage of candidates who would benefit from such fellowships. At first recourse was had to sending WMO experts to various countries to provide tuition on an *ad hoc* rather than an organized basis. This procedure met with success for a time but, as expected, it soon proved to be inadequate since the training requirement had increased and, in addition, the practical problems became intractable. By 1960 there was already an urgent need to train meteorological staff on a much larger scale. In order, therefore, to encourage well-educated young people to become professional meteorologists, WMO introduced a system of long-term fellowships covering attendance at universities, beginning with undergraduate studies in mathematics and physics as a preliminary to a course in meteorology. Fellowships of this type were not eligible for UNDP support and so a fresh source of funds had to be found. A decision was therefore made to set up a New Development Fund (NDF). WMO introduced the scheme in October 1965 and, although the number of fellowships was necessarily limited, some very valuable results and experience were obtained.

In 1968 WMO took a further step by launching the Voluntary Assistance Programme* (VAP), of which the primary objective was to give practical help to national Meteorological Services to enable them to carry out their commitments to the World Weather Watch Programme. The necessity for technical equipment of all kinds was fully recognized but it was also appreciated that adequate numbers of trained staff were essential. Long-term fellowships were therefore included in the VAP scheme and a number of donor countries responded generously, in some cases offering courses of study for up to five years. Even under these conditions some nations were unable to produce students with the necessary grounding in mathematics and physics. Once again the meteorological community showed resourcefulness and offered to provide preparatory courses to bring candidates up to the required level. In some cases there was then an additional problem because some of the students needed to learn the language of the host country providing the courses. Ideally, candidates should receive training in their native language and, since this was not always possible, a number of long-term fellowships were not used. The problem was discussed at Fifth Congress (Geneva, 1967) and, in order to ensure that funds would be available for the full utilization of long-term fellowships, the decision was made to allocate US $500 000 for the purpose from the WMO regular budget. Such a step involved sacrifices or shortfalls in other important activities, but Congress felt that the training of staff for the Meteorological Services of developing countries should be accorded very high priority. The arrangement has been continued by subsequent Congresses so, on aggregate, the Members of WMO have been making a substantial contribution to the system of long-term fellowships.

The United Nations Development Programme has been a generous source of funds for the education and training of meteorologists but, as pressure on UNDP resources increased, an important difficulty emerged. Requests for UNDP assistance have to be made by governments and WMO has always given every help to any government submitting an application on behalf of its Meteorological Service. However, the large increase in development programmes made governments face the reality that UNDP finances, though large, were not unlimited. The question of priorities therefore had to be considered by all developing countries and in a number of cases it was found that other requirements were considered of greater urgency than those of the Meteorological Service. The results were that meteorological training was delayed at a time when it needed to be accelerated and that existing training facilities were not being used to their full capacity. The problem was a frustrating one because unless a Meteorological Service could recruit and train an adequate complement of staff, the country's development programmes in industry, commerce and social affairs would be handicapped because the contributions required from its Meteorological Service would be of limited efficiency and comprehensiveness.

WMO raised this very important problem with UNDP at a meeting in 1969 and it was agreed that if WMO drew up a training project and demonstrated the need for special provision to be made for fellowships, UNDP would try to find a way of meeting the requirement. It appeared UNDP contemplated an exceptional arrangement to provide funds for a scheme consisting wholly of fellowships. WMO therefore carried out a survey among the countries in Latin America and in the Caribbean. As a result WMO submitted proposals to UNDP for a "special fund project" for meteorological training, including 20 fellowships for Class I personnel and 40 for Class II grades. The scheme was classified as a regional project, thus freeing the individual countries from having to meet any of the costs from their own programmes. The proposals were approved by UNDP and, in a most constructive and far-reaching initiative, were extended to cover a wide range of projects. With regard to training and education, the response was excellent. In the first year no fewer than 47 fellowships were awarded for study in the universities of Buenos Aires, Costa Rica and Rio de Janeiro and at the Caribbean Institute of Meteorology in Barbados. Encouraged by this success, UNDP approved a similar project for African students to receive fellowships for attendance at the East African Institute for Meteorological Training and Research.

It was realized that the fellowship scheme involved some degree of uncertainty. The students attended a wide course of study

* Now Voluntary Co-operation Programme

in mathematics and physics as well as in meteorology and they in consequence qualified to seek work over an extensive scientific field. There was therefore the risk that those who successfully completed their studies with the help of fellowships might seek more lucrative employment than the national Meteorological Service could offer. WMO therefore considered it advisable to keep a record of the subsequent history of fellowship students and was greatly encouraged to find that well over 90 per cent returned to and remained with their country's Meteorological Service.

The WMO fellowship scheme has had a relatively large number of sponsors, namely UNDP, the Voluntary Assistance Programme*, funds-in-trust, the regular budget of WMO, and bilateral support. These sponsors have contributed substantially to providing developing countries with university-trained meteorologists as well as staff qualified in such special fields as aeronautical meteorology, agrometeorology, marine meteorology, in applications to hydrology and in meteorological instrumentation. Fellowships have long been recognized as an essential means of assisting countries in establishing Meteorological Services of high efficiency and with staff of good scientific standing.

Expert missions

Many other methods have been used for the training of meteorologists in developing countries. One was the use of experts to organize training courses in the countries themselves in the form of classroom tutorials, on-the-job training or, of special importance, the training of national instructors. By this means WMO provided or organized the training of more than 1 300 personnel in the first decade of the Technical Co-operation Programme. However, from about 1960, the training requirement was so great that additional methods had to be resorted to, resulting in the establishment of regional training centres and university chairs.

Seminars have also proved their worth in contributing to further and refresher training. Their primary purpose is to communicate theoretical understanding and practical techniques covering the most recent advances in meteorological research. The first WMO seminar was held in the Caribbean in 1956 and in the next 20 years some fifty regional seminars took place. According to their nature these seminars were supported or co-sponsored by other international organizations, notably UNDP, Unesco, FAO and ICAO. WMO was quickly on the scene with seminars on the interpretation of satellite data, on the development of World Weather Watch and on the role of meteorology in economic planning on global, regional and national scales. The report of one of these seminars so interested the UN Economic Commission for Africa (ECA) that in 1969 it adopted a resolution calling upon its Members to assist in the implementation of the World Weather Watch.

An important objective of WMO was that all countries should be able to assume responsibility for their own training. It was always assumed that countries would train Class IV staff with limited assistance from WMO. The Executive Council** decided that regional seminars should be arranged for instructors who would then be qualified to give courses to Class II and Class III trainees. The first of these seminars was held in Cairo in October 1966. The scheme has met with considerable success.

Mobile training units

From the creation of WMO in 1951, training activities concentrated largely on courses for personnel in Classes I, II and III, i.e. with staff of professional or near-professional status. It had been taken for granted that each country would train its own Class IV staff where duties included making routine observations of the weather, plotting synoptic charts and preparing climatological records. However, by the time of Fifth Congress in 1967 it was learned that some developing countries were experiencing difficulty in organizing training courses for their Class IV staff. Accordingly, the Secretary-General made arrangements for the establishment, as a temporary measure, of mobile training units which would visit selected countries for periods of about six months. Their task was primarily to give Class IV instruction but the expert in charge of each unit was also prepared to give any desired advice on the organization of a national Meteorological Service.

Regional Meteorological Training Centres

A logical development of WMO's Educational and Training programme was the establishment of Regional Meteorological Training Centres (RMTCs). These were intended to fill important gaps in training facilities. Thus a centre would not be set up in a Region unless a sufficient number of Members declared, through their Regional Association, that deficiencies existed which could not be catered for adequately within the Region. From about 1962 it was realized that a regional organization would provide an efficient means of assisting national Services in training meteorologists to the standard required in each of the four classes or grades. The provision of staff and equipment for a centre was expensive, beyond the means of developing countries, and it was necessary to seek the financial support of UNDP. For this purpose a centre had to be properly organized with terms of reference, criteria for the appointment of instructors and with detailed plans of the courses to be provided. It was also specified that the EC Panel of Experts on Meteorological Training with the assistance of the WMO Secretariat would supervise the operation of each centre that was established. Since 1965 RMTCs have been set up in Kenya (Nairobi), the Caribbean (Barbados), the Philippines (Quezon), Brazil (Belém), Egypt (Cairo), Algeria (Oran), Argentina (Buenos Aires), India (Pune) and Niger (Niamey). By the end of 1987 more than 10 000 students had satisfactorily completed courses of instruction at these centres.

These centres have been doing very valuable work contributing to the efficiency of national and international meteorology. Their existence has been abundantly justified. The centre in Niamey should be selected for special mention since its primary purpose is training in agrometeorology and operational hydrology. It was set up in 1975 as a result of the disastrous drought which affected the Sudano-Sahelian region of Africa. Nearly twenty countries, all liable to experience severe drought, have sent students for training at the centre, many of whom were awarded fellowships. The training has formed part of a comprehensive programme organized by the United Nations with the help of relevant Specialized Agencies to bring relief to the drought-stricken areas and to undertake a variety of measures designed to prevent, or at least to mitigate the potentially drastic consequences of, any further experiences of widespread drought. For the training programme which bears the title AGRHYMET the executing agency is WMO and FAO was appointed associate agency. In recent years the African drought has spread well beyond the Sahelian zone and the surrounding countries have intensified their efforts to have a sufficient number of trained staff available to cope with any emergency that might arise.

Training for research

In 1970 WMO and the International Association of Meteorology and Atmospheric Physics (IAMAP) jointly arranged a

* Now Voluntary Co-operation Programme

** See note on p. 8.

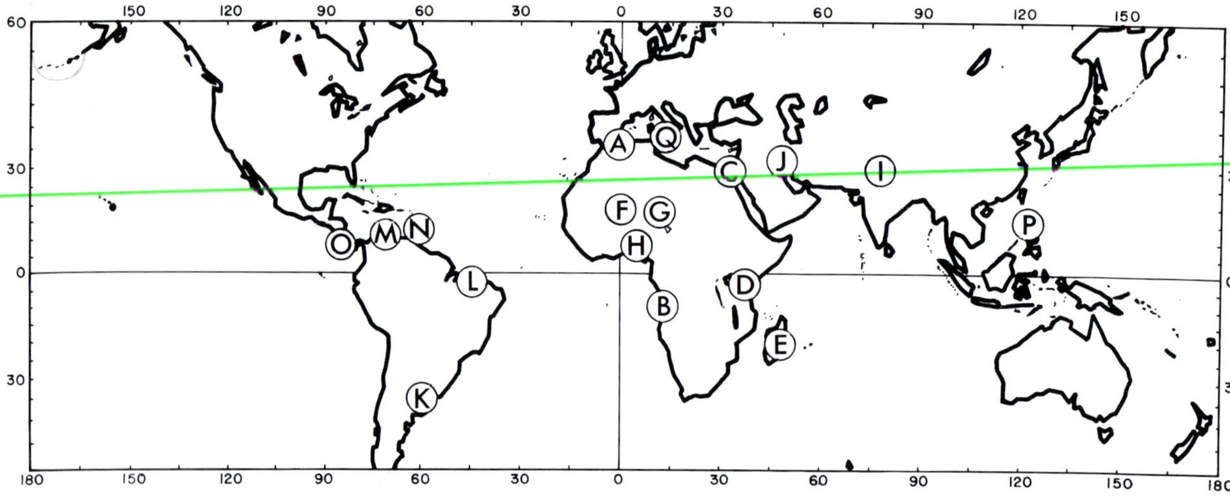

Regional Meteorological Training Centres

	Country	Name of Centre	Region
A	Algeria	Hydrometeorological Institute for Training and Research, Oran	I
B	Angola	Regional Meteorological Training Centre, Mulemba	I
C	Egypt	Regional Training Centre for Instruments Specialists, Cairo	I
D	Kenya	Institute for Meteorological Training and Research, and the Department of Meteorology, University of Nairobi, Nairobi	I
E	Madagascar	Etablissement d'Enseignement Supérieur Polytechnique (EESP), University of Madagascar, Tananarive	I
F	Niger	African School of Meteorology and Civil Aviation (EAMAC), Niamey	I
G	Niger	Regional Training Centre for Agrometeorology and Operational Hydrology and their Applications (AGRHYMET), Niamey	I
H	Nigeria	Nigerian Institute for Meteorological Research and Training, Lagos	I
I	India	Telecommunication and Radiometeorological Training Centres, New Delhi, and Training Directorate, Pune	II
J	Iraq	Regional Meteorological Training Centre, Baghdad	II
K	Argentina	Department of Meteorology, University of Buenos Aires, and the Training Centre, National Meteorological Service, Buenos Aires	III
L	Brazil	Department of Meteorology and Geophysics, University of Pará, Belém	III
M	Venezuela	Department of Meteorology and Hydrology, Central University of Venezuela, Caracas	III
N	Barbados	Caribbean Meteorological Institute, and the University of the West Indies, Barbados	IV
O	Costa Rica	Meteorology Section, School of Physics, University of Costa Rica, San José	IV
P	Philippines	Department of Meteorology and Oceanography, University of the Philippines, and the Philippine Atmospheric, Geophysical and Astronomical Administration (PAGASA), Quezon City	V
Q	Italy	International School of Meteorology of the Mediterranean, Erice (Trapani)	VI

symposium on higher education and training for Class I and Class II staff, the two grades which are principally concerned with meteorology as the science of the atmosphere and with the applications of meteorology to such disciplines as agriculture and hydrology and to operational services in aviation, shipping and so on. The purposes of the symposium were to broaden and strengthen meteorological knowledge, to discuss problems in fundamental and applied research and to illustrate how these problems are studied and how new techniques may result. Meteorologists from 30 countries attended the symposium, which was held in Rome. The Executive Council* considered that the symposium was most successful and worth while and accordingly decided that a similar one should be held every four years as an essential element in the Education and Training Programme.

Another scheme with similar objectives to these symposia consisted of an association of higher training with the World Weather Watch Programme (WWW). The World Meteorological Centres (WMCs) located in Melbourne, Moscow and Washington and also the various Regional/Specialized Meteorological Centres (RSMCs)** in the WWW system are forecasting centres of the highest scientific levels and are fully equipped with computers and associated facilities to participate in basic and applied research. It was a logical step therefore to consider whether and to what extent advantage could be taken of the resources of the centres to provide further training for forecasters in particular. In formulating the scheme it was necessary to take into account that world and regional centres operate to strict time schedules which must have overriding priority. However, it was noted that most of these centres are able to devote a proportion of their time to research into numerical weather prediction and into the further development of the Global Data-processing System. It appeared therefore that the type of training which could be supplied whilst causing minimum disruption would be familiarization and on-the-job practical training.

The scheme was implemented to a limited extent in the late 1970s and early 1980s and there was excellent co-operation at the centres involved. The subject was reviewed by the Executive Council at its thirtieth session (Geneva, 1984) and it was confirmed that the WMCs and RMCs had made invaluable contributions to the WMO training programme. The Executive Council invited centres taking part in the scheme to concentrate on training for research, notably in applications to operational meteorological responsibilities.

* See note on p. 8.
** See note on p. 21.

Granting of certificates by WMO

When training courses were first arranged by WMO there was no intention to award certificates to trainees who had successfully completed a course. It seemed to be sufficient that a student had attended a course and returned to his national Service; and that WMO would send to the director of the Service a report setting out the syllabus covered and an assessment of the student's performance. This procedure was accepted for a few years and then pressure began to be exerted for each trainee to be presented with a certificate after completing a course and reaching the prescribed standard. The question gave rise to considerable debate and in the end the Executive Council* decided at its nineteenth session (Geneva, 1967) that graduates of the Regional Meteorological Training Centres should be given certificates on the completion of their courses. It was also decided that the certificates should contain a summary of the syllabus and the extent, or level, to which each subject was studied.

Library and publications for training

One of the important problems facing WMO in regard to training was the lack of textbooks and other material in developing countries. The Secretary-General, with the approval of Fifth Congress (Geneva, 1967), therefore decided that a training library should be set up in the Secretariat. The objective was to install in the library not only textbooks but also audio and visual aids, interesting series of weather maps and meteorological cross-sections. Within a few years the library was well stocked with a variety of training material including some 65 films which were available for issue on loan to national Services. The library has provided a valuable service for instructors and fellowship students on their visits to the WMO Headquarters in Geneva. The Executive Council*, as a further step, authorized the Secretariat to purchase additional textbooks for issue on long-term loan to Meteorological Services which could not obtain such books by other means. In more recent years from 1980 onwards the library has acquired computer equipment specifically designed for use in training programmes. By this means valuable extensions have been made to the support which the Secretariat provides for regional and national training institutions.

While the library was expanding and adding to the range of its available services, the Education and Training Department in the Secretariat carried out a critical examination of the publications that had been assembled and considered whether they satisfied fully the requirements of instructors and trainees, especially in developing countries. It was known that at some long-established training colleges instructors had prepared lecture notes for distribution to trainees attending the various courses ranging from Class I to Class IV. Moreover, at some of the colleges booklets had been compiled containing meteorological problems together with guidance on step-by-step approaches to practical solutions. There was no question that material of this kind, in typescript notebook form, was a great aid to students in private study to consolidate the knowledge imparted by lectures and tutorials. The Secretariat therefore gathered as many examples as possible of notes and problem papers and, by careful selection, set about the production of WMO training manuals. The first one published, in 1966, was for Class III personnel and similar manuals for other classes followed at commendable speed.

The initiative regarding manuals of lecture notes and problems was quickly followed by another. Consultations with instructors and students had indicated that existing textbooks, though very helpful, did not exactly fulfil all requirements. This deficiency was, of course, mainly a question of orientation. A textbook normally aims to serve the maximum number of teachers and pupils but in doing so, usually very successfully, it may fail to give adequate treatment to the special interests of a group of students and their instructor. After considerable discussion it was decided that there was a requirement for a publication or series of publications somewhat more refined and more elaborate than anything attempted earlier. The objective was described as a compendium of lecture notes and the first of its kind, provided for Class IV personnel, was published in 1970 in two volumes. Volume I, entitled *Earth science*, was descriptive of the human environment; Volume II had the informative title *Meteorology, meteorological instruments and methods of observation*. The two volumes were published in English, French and Spanish. Later in 1970 a compendium of lecture notes for Class III meteorologists was published; it consisted of 450 pages. Further volumes covering marine meteorology and agricultural meteorology were published in 1976 and 1982 respectively, again for Class III and Class IV, and an up-to-date compendium on meteorological instruments was issued in 1986.

The appearance of these publications, which catered for staff in Class III and Class IV grades, soon gave rise to demands for similar volumes for use by meteorologists in Classes I and II. Plans were accordingly made and the services of outstanding scientists were enlisted to write the required books. The first volume, published in 1973, dealt with dynamic meteorology and a succession of publications followed. It was a considerable achievement on the part of the Education and Training Department to organize the preparation and publication of a series of volumes tailored closely to the needs of staff in the various national Meteorological Services. After the first one on dynamic meteorology, the following volumes had been issued up to the end of 1986:

- Physical meteorology (1973);
- Synoptic meteorology (1973);
- General hydrology (1977);
- Aeronautical meteorology (1978);
- Marine meteorology (1979);
- Tropical meteorology (1979);
- Hydrometeorology (1984);
- Air chemistry and air pollution (1985);
- Workbook on numerical weather prediction for the tropics (1986).

International Education Year

The General Assembly of the United Nations designated the year 1970 as International Education Year and invited Member States and Specialized Agencies to make the year one of increased effort in education and training. WMO responded to a pleasing extent with action which may be summarized as follows:

- "Meteorological education and training" was adopted as the theme for World Meteorological Day in 1970;
- Seven world or regional conferences and symposia were held during the year on topics concerned with meteorological education and training;
- Intensive efforts were made by means of booklets, pamphlets, articles in the press and in broadcasts to stress the importance of meteorology in the economic and social development of all countries.

---

* See note on p. 8.

Outlook for the long term

In essence, education and training are fairly routine operations illuminated from time to time by flashes of new knowledge. Sound knowledge has a permanent status and is constantly in use. For example, Boyle's Law, discovered some three centuries ago, will continue to figure prominently in the thermodynamics and dynamics of the atmosphere. The Training and Education Department of WMO and the training institutions of national Meteorological and Hydrological Services have acquired vast experience in all aspects of training, have been ready to innovate and have learned the lessons of faulty judgement. New challenges have been found, as when remote sensing by satellite made its appearance and had to be incorporated into the syllabus of almost all training courses. There would, however, be little or no surprise if the contents of a course showed hardly any variation from one year to another.

Nevertheless, much work of major importance remains to be done in the field of education and training, especially as there is a continuing expansion of the applications and techniques of meteorology and operational hydrology to economic developments in all countries. The first priority must be to assist developing countries to acquire adequate numbers of trained staff able to carry out all the varied tasks in meteorology and hydrology. An important aim should be to improve constantly the content and presentation of each course so that the students attending it would receive a somewhat better training than those on the preceding course. There is no end to this important work since each year new staff have to be recruited to offset retirements and other departures. Training should also be regarded as a recurring process with advanced courses and refresher courses forming an integral part of the careers of all members of staff. The organization of education and training, including fellowships, has long been dependent upon generous financial provision from UNDP, the Voluntary Co-operation* Programme, funds-in-trust schemes and bilateral arrangements. It is very necessary that these sources of finance should continue while any requirement remains. Sponsorship is the more readily provided if there is clear evidence of efficiency and it therefore seems probable that highly specialized courses, involving a small number of trainees, would be subject to regional rather than national arrangements. For this and other reasons it would be advisable to maintain the existing Regional Meteorological Training Centres and perhaps add one or two more.

It is evident that the workload on the Education and Training Department is likely to increase substantially—with, as a result, a strengthening of its various facilities, including the training library and up-to-date equipment as well as books. Training activities should be extended, possibly in a liaison capacity at first, towards users of meteorological and hydrological information, forecasts and other advice. There is much progress to be made in improving the services provided to customers and in developing new services as the requirements and problems of users are better understood. There would appear to be a twofold responsibility: to increase the internal efficiency of the training effort and to assist recipients of meteorological and hydrological information to increase their own efficiency as improvements are made in the services provided.

Demonstration of training aids and equipment

Since it was created in 1951, WMO has carried out an enormous amount of work, much of it freshly conceived, in the field of education and training. It has received the warm appreciation of all countries that have been helped to develop efficient Meteorological and Hydrological Services of high scientific standing. The importance of education and training is widely recognized and the responsibility and commitment of the Organization in this respect are unlikely to diminish.

---

* See note on p. 116.

CHAPTER XIV

TECHNICAL CO-OPERATION PROGRAMME

General review of the sources of assistance

It will be clear from the preceding chapters that full and friendly international co-operation is essential in meteorology—essential both to ensure overall progress in the scientific understanding of the atmospheric processes, and also to ensure that the practical benefits derived from such progress become available to all countries of the world. Difficulties experienced in any country in fulfilling its allotted role in WMO programmes are thus of international, as well as national, concern. This was in fact recognized as far back as the time of the Vienna Congress in 1873, when Buys Ballot presented a proposal for "the formation of an International Fund for the establishment of meteorological observatories on islands and distant parts of the Earth's surface". The time was not however ripe for the approval of such a far-sighted proposal. Indeed, it was not until the creation of WMO as a Specialized Agency of the United Nations, some three-quarters of a century later, that such arrangements for mutual assistance and co-operation between the countries of the world became possible on any significant scale. This was because the Organization, as a Specialized Agency, could participate in the assistance programmes of the United Nations; and also because, as a governmental body, WMO itself could now initiate appropriate action, if so decided by the Member countries. In fact, it seized both opportunities—and to good effect.

Taking first the UN sources of such assistance, it is important to note that one of the basic aims of the United Nations—and one which is indeed incorporated in its Charter—is to assist and encourage economic and social development in all countries. By the time WMO had become a Specialized Agency, the UN Expanded Programme of Technical Assistance (EPTA) was already in full operation, being funded on a voluntary basis by its Member countries.

The First Congress of WMO (1951), in a resolution outlining the general policy of the Organization, made clear its wish to participate in EPTA and, in another resolution, specifically authorized the WMO Executive Council* to take such steps as it considered appropriate for the participation of WMO in that programme. The Executive Council* took prompt action on this matter—action which was endorsed four years later by Second Congress (1955). WMO's participation in this programme continued to develop and did so with highly beneficial results. Thus a resolution of Third Congress (1959) noted the "valuable technical assistance already rendered by WMO through its participation in UNEPTA" and decided "that every endeavour should be made to increase WMO participation". It is interesting to note that by this time, the scale of assistance was such as to have made it difficult for WMO to meet the administrative and operational services for the projects for which it was now responsible. Third Congress drew attention to this difficulty. The UN Technical Assistance Board, for its part, readily recognized the problem and appropriate financial steps were soon taken to assist WMO in providing these services.

The overall success of EPTA, and the need to finance projects on a larger scale than had been previously foreseen, led to the establishment in 1958 of a new UN programme, the Special Fund, for helping the developing countries. WMO was not slow to see the significance of this decision and in the following year (1959), Third Congress adopted a special resolution on this subject. This resolution expressed the hope that WMO could participate in the activities made possible by the Special Fund and requested the Executive Council* and the Secretary-General "to co-operate fully with the Managing Director of the Special Fund". The outcome was a further increase in the scale of assistance rendered by WMO.

Thus in 1963, Fourth Congress "noted with satisfaction the steady increase in scope and magnitude of the Organization's participation in the United Nations technical co-operation programmes and expressed the hope that this favourable development would continue".

In 1962, yet another UN programme was established—this time for giving assistance in public administration by providing experts for operational and executive functions (OPEX). WMO was able to participate in this programme also. A year later, Fourth Congress approved such participation and urged increased WMO participation in it.

Thereafter the three UN programmes (EPTA, Special Fund and OPEX) continued to develop and in due course UN technical co-operation activities became a major factor in the overall role of UN and the Specialized Agencies in world affairs. In 1972, these programmes were combined into a single comprehensive programme—the United Nations Development Programme (UNDP). Since its inception, the UNDP has functioned in a highly effective manner, as has also WMO's participation in it. The subsequent four-yearly Congresses of WMO have duly recorded the constantly increasing participation of WMO in the UNDP.

The nature and scale of WMO's participation in the UNDP and in the UN programmes which preceded it are dealt with later in this chapter.

Turning now to WMO's own programmes in this field, it is significant to note that even at First Congress, the need for some means of assisting Member countries in their meteorological activities was recognized. Thus, despite the need for strict financial stringency in those early days, a token sum of one thousand US dollars was included in the regular budget for technical assistance to developing countries. Four years later (1955), WMO created the Operational and Technical Development Fund and a somewhat greater, but still relatively modest, sum of US $9 600 was included in the regular budget for this purpose. This was, however, increased by Third Congress (1959) to US $60 000 and this figure was maintained by Fourth Congress (1963). At this stage, important developments were taking place in the overall role of WMO—developments which had direct repercussions on the scale and scope of the technical co-operation activities of the Organization. These led to a decision of major importance being taken by Fifth Congress (1967), namely to create the WMO Voluntary Assistance Programme. To understand the full significance of this decision, some further explanation is necessary.

* See note on p. 8.

Expansion of Meteorological and Hydrological Services in Ecuador.
Left: *Installation of instruments;* right: *setting up a raingauge at an agrometeorological observatory*

As is described in Chapter III, following the advent of outer space technology and greatly improved computer techniques, WMO introduced in 1964 a new and highly sophisticated operational programme: the World Weather Watch. It soon became clear that it would not be appropriate to rely solely on the UNDP for providing the assistance which this new programme would require. It was equally recognized that the needs in this case went far beyond the relatively modest programmes which WMO had itself operated with funds from the regular budget of the Organization. Thus, in 1968, WMO established its own Voluntary Assistance Programme, which was later renamed the Voluntary Co-operation Programme (VCP).

The VCP was initially approved to facilitate the full implementation of the World Weather Watch Programme and the granting of long-term fellowships. Within the WWW Programme, each Member country is required to perform certain functions on a routine operational basis, the total of such functions comprising the programme as a whole. With the introduction of the VCP, it became possible for a Member to receive assistance, when necessary, in its efforts to fulfil its responsibilities. Such a Member is required to prepare a project which, if implemented, would enable it to perform its allotted task. The project is generally drawn up with the help of the Secretariat and is then submitted to the Executive Council for endorsement. It is then circulated to all Members with a view to finding one (or more) able and willing to assist in its implementation. The assistance may be either financial or in the form of direct practical aid, such as the supply of equipment and the arrangements for its installation in the country concerned. The training of local personnel in the use of the equipment is also frequently necessary. When a mutually acceptable project has been developed, the details are drawn up as two separate agreements: the first between WMO and the donor country; the second between WMO and the recipient Member. Several donor countries may combine to support a project and a single project may be designed with a view to benefiting several recipient countries. The implementation of a programme of this kind evidently requires a high degree of co-operation between all countries involved, with WMO serving as the focal point.

Although, as already indicated, the VCP was originally designed to facilitate the implementation of the WWW Programme, specific reference was made in the programme to the award of long-term fellowships—an acknowledgement that the WWW (or indeed any other meteorological programme) cannot be fully successful without adequately trained staff in national Meteorological Services. As will be seen later, the types of project which could be implemented under the VCP were later considerably extended to cover many other programmes and activities of WMO.

The introduction of the VCP system of technical co-operation brought with it many advantages. In the first place, the donor country is able to make a contribution to a specific project which it feels is important, and to do so either by a specific contribution to the fund VCP(F) or by direct practical assistance in the form of equipment and services, such contributions constituting the VCP(ES). An important feature of the VCP is that all the arrangements are made through the good offices of WMO, so ensuring that the project, when completed, will enable the receiving country to be one step nearer to meeting its responsibilities under the approved WMO global plan. Yet another advantage is that the role of the WMO Secretariat is essentially that of giving technical advice without direct financial involvement. It is moreover a system from which all Members benefit by virtue of the dependence of all on the global effort.

Mention must also be made of another form of assistance in the technical co-operation programmes—namely, funds-in-trust

(FIT) projects. Under this form of assistance, WMO may carry out a project which is financed by a national government—either the government of the country in which the projects will be carried out, or the government of another country which wishes to support the project. Apart from the material assistance provided to satisfy urgent requirements, the FIT system has also been of great value in financing expert missions and fellowships.

Fields and types of assistance

The fields in which assistance has been given in the various programmes mentioned are of course prescribed in the rules applying to each. As far as the UN programmes are concerned, all such fields are directly related to economic development and may be appropriately presented as follows, this being the form given in the rules governing the UNDP itself:

(*a*) Physical resources (agriculture, forests, fisheries, water, minerals);

(*b*) Industry;

(*c*) Infrastructure (general and regional development planning, public administration and technical services, public utilities, housing, building and physical planning); and

(*d*) Human resources (population, education and training, manpower policy, health, social welfare).

In all these fields meteorological factors are involved and, as already explained, WMO has played its part in providing the necessary assistance.

As regards WMO's own technical assistance programmes, the main aim has of course been to facilitate implementation of the various WMO programmes. Assistance was therefore initially given in many fields, albeit on a relatively modest scale. The importance of the WWW Programme, however, led to the creation of the VCP, which was very successful and which soon developed into WMO's major programme in this field. Its success led later to its extension to many other WMO programmes.

In both the UN and WMO programmes, the types of assistance given may be classified under three broad headings:

- Expert missions;
- Fellowships;
- Supply of equipment and related services.

In some cases, the individual projects involve just one of these types of assistance. In other cases, however, two or all three types of assistance may be financed from a different source (i.e. WMO or UNDP). In all such cases, full co-ordination of the two forms of assistance was ensured by the Secretariat from the outset. Indeed, arrangements were made for the UNDP and the VCP to participate jointly in certain projects. A special advantage of this particular form of co-operation was that some non-technical, but very relevant, problems relating to the use of the currencies of the developing countries concerned could in many cases be satisfactorily resolved.

Benefits derived by Members

It is of course not possible to describe in full detail the benefits which Members have derived from the technical co-operation activities described above. There is, however, no doubt that these activities have constituted an essential element in the success story of the World Meteorological Organization as recorded in the other chapters of this book. The technical co-operation activities grew steadily throughout the period under review. After a very modest start, expenditure for the last four-year period for which figures are available was in excess of US $86 million. A clearer picture of the scale and importance of these activities will, it is felt, be derived from the following descriptions of some selected projects which comprise both national and inter-country activities.

Selected projects

The summary descriptions which follow include five inter-country technical co-operation projects, which have been undertaken either wholly or partly within four of the six WMO geographic Regions, and ten country projects undertaken in five of the six Regions. A description of an important funds-in-trust project is also included.

INTER-COUNTRY PROJECTS

Region I: Hydrometeorological Survey of the Catchments of Lakes Victoria, Kyoga and Albert (1967-1972)

The unprecedented rise in the level of Lake Victoria in the period 1961–1964 stimulated an interest in the subject of lake-level fluctuations and the factors that cause these fluctuations. At the same time, the lake system, including Lakes Kyoga and Albert (later Mobutu Sese Seko), offers a natural storage of considerable developmental significance, and problems associated with storage and regulation in these lakes have always been of considerable interest to all the countries in the Nile Basin.

The availability and need for data for design of such control measures in the Upper Nile in Uganda, Kenya and Tanzania and the potential for agricultural development of these water resources were evaluated by a joint WMO/FAO mission in 1963.

A water-level recorder in use (Hydrometeorological Survey of the Catchments of Lakes Victoria, Kyoga and Mobuto Sese Seko)

The mission recommended that a hydrometeorological survey should be undertaken for the collection and analysis of hydrometeorological data of the lake catchments, in order to study the water balance of the Upper Nile. The data collected and the study were expected to assist the countries in the planning of water conservation and development, and to provide the groundwork for intergovernmental co-operation in the storage, regulation and use of the Nile.

Five of the nine riparian countries (Egypt, Kenya, Sudan, Tanzania and Uganda) requested the assistance of UNDP in this study, and the project became operational in May 1967, with headquarters at Entebbe, Uganda. In the later stages of the project, in 1972, the area of study was extended, by the participation of Rwanda and Burundi, to virtually the whole of the 395 000 km^2 catchment of the Upper Nile to Nimule, on the Uganda-Sudan border.

To ensure the all-important task of co-ordinating the participation of the governments, a Technical Committee was established consisting of representatives of the five (later seven) participating countries, with the Director-General of the East African Meteorological Department as a co-opted member. A project co-manager was appointed by the Technical Committee, to work directly alongside the WMO project manager.

The project plan of operation included an extensive programme of installation of new data-collection stations and for upgrading of many existing stations. A total of 25 new first-order meteorological stations, 200 ordinary raingauges, 60 hydrological stations, and ten lake-level recording stations were installed, as well as seven index catchments, to give adequate representative cover to the project area. In particular, several islands in Lake Victoria were instrumented for the first time; an automatic weather station was installed on the central island. Aerial surveys were undertaken of flat lake-shore areas liable to significant changes with variation in lake-water levels, and a hydrographic survey of Lake Kyoga was carried out for definition of a reliable surface-water elevation-area-capacity relationship.

Data for the project area were collected in a computerized data bank, and analytical procedures derived for the various parameters involved in the water balance of Lakes Victoria, Kyoga and Albert. Preliminary estimates of water balance of the three lakes were worked out, laying a foundation for further refined studies; a mathematical model of the lake catchments was prepared. Specific hydrological reports were also prepared, for the Kafu basin in Uganda, the Niondo Basin in Kenya, and the Kagera Basin in Tanzania, Rwanda and Burundi, as an aid to formulation of water-development proposals in the catchment areas. The Kagera report formed a basis for the UNDTCD Kagera River Basin Development project, implementation of which commenced in 1971.

The successful conclusion of this large and complex project marked an important milestone in promoting and stimulating the efforts of the riparian countries of the Nile Basin to co-operate in planning for exploitation of the water resources of the Nile.

Region I: Programme for Strengthening of National Meteorological and Hydrological Services in the Sahel, and Establishment of a Regional Centre for Training and Applications in Agrometeorology and Operational Hydrology (AGRHYMET Programme) (1975-1991)

The devastating drought which afflicted the Sahelian countries of West Africa between 1968 and 1972 served as a sharp reminder of the extent to which production in agriculture and livestock rearing, and human activities in general, depend on the hazards of the weather and climate in the Sahel, a zone situated on the boundary of the monsoon rains of West Africa and therefore an area of highly variable seasonal and annual rainfall. The drought, more severe than any recorded in the zone during the previous fifty years, required the mobilization of aid on an unprecedented scale, not only to bring immediate relief, but also to create conditions that would alleviate the consequences of continued or later drought.

The affected countries themselves took the lead in this respect, through the formation of the Inter-State Committee for Drought Control in the Sahel (CILSS), the membership of which comprised Chad, Niger, Upper Volta (now Burkina Faso), Mali, Senegal, Gambia, Mauritania, and, later, Cape Verde and Guinea Bissau. The WMO Sahel Programme (AGRHYMET) originated in an early CILSS resolution requesting that national Meteorological and Hydrological Services be strengthened, and that a Regional Centre be created for training and applications in agrometeorology and operational hydrology. A joint WMO/FAO mission in mid-1974 broadly outlined the needs of the national Services and the centre to meet the CILSS objectives, and based on the mission recommendations the AGRHYMET Programme, comprising a regional project for creation of the Regional Centre at Niamey, Niger, and seven national projects, was approved by UNDP. A national project commenced in Cape Verde in 1978, and in Guinea Bissau in 1986.

With WMO designated as executing agency and FAO as associated agency, the programme had an initial Phase I term 1975–1980, which was extended later to include 1981. Subsequent extensions were made for Phase II (1982–1986) and Phase III (1987–1991). In addition to continued financial support by UNDP, and early support of the UN Sahelian Office (UNSO), the programme has been assisted both financially and in kind by a number of donor countries which, for purposes of input co-ordination and programme review, formed a Co-ordinating and Advisory Committee (CAC). Technical direction of programme activities is provided by a Technical Committee comprising representatives of the national Meteorological and Hydrological and Agricultural Services. Cross-representation is provided in each of the committees, which meet at least annually, and more frequently when necessary. The regulations of these committees and the responsibilities of all agencies, organizations and countries involved are laid down in a CILSS/WMO Agreement.

The first phase of the programme saw the development of necessary infrastructures, at national and regional levels, in the countries through strengthening of Meteorological and Hydrological Services in terms of training of personnel at all levels, the extension and upgrading of observational and data transmission networks, and improvements in data-processing capacity. Activities at the regional level included creation of the Regional Centre at Niamey, with early implementation of training courses at technical (Class III) and engineer (Class II) level. With completion of the physical construction of the Centre, in 1978, installation commenced of the data-processing and telecommunication facilities needed for the collection and interpretation of data observed throughout the Region.

The second phase of the programme, 1982–1986, saw the development of operational activities for the production and distribution, both regionally and nationally, of agrometeorological and hydrological information and forecasts to agricultural planning and extension services, and to farmers and herders; these are designed to help promote an increase in food production and provide warning of potential food shortage. Ten-day agrometeorological bulletins, and monthly hydrological bulletins, were issued by the Regional Centre. Because of delays in postal distribution of the bulletins throughout the Region, a ten-day telex summary was also dispatched. A daily bulletin was issued locally, in Niger, as a pilot exercise for development of national bulletins in each country.

During the second half of the Phase II period, therefore, considerable effort was made to develop the issue of national agrometeorological bulletins. The basic organizational structure for this is the national Multi-disciplinary Working Group, including representatives of the Meteorological, Agricultural, Hydrological and other Services, which is responsible for crop monitoring and for data receipt from the national networks, their analysis, and the issue of bulletins and alerts.

The objectives of the third phase of the programme have been to bring about, so far as possible, a uniform national capability in this regard, and also to consolidate the progress made in the operational sense by increased inputs of agrometeorological data and information, and also improvements in information-communication procedures, in order to ensure the receipt by farmers and herders, as well as by government Services and planning and development agencies, of information that will have a beneficial impact on food production in the Sahel.

Region II: Technical Support to the Regional Typhoon Programme (1974-1981)

The project "Technical Support to the Regional Typhoon Programme" arose out of a call by the UN General Assembly in 1972 for increased co-operation between WMO, UNDP, the Office of the Disaster Relief Co-ordinator, and the future environmental programmes of the United Nations. The project commenced in January 1974 with an initial duration of three years, which was later extended to eight years. Its specific objectives were improvement of meteorological, hydrological and telecommunications networks, the improvement of typhoon forecasting techniques and of arrangements for the timely distribution of typhoon and flood warnings, and the establishment of natural disaster prevention plans, as well as the provision of necessary staff training and the promotion of research activities.

Under the meteorological component of the project a substantial improvement was made in observation and telecommunication facilities. A total of fifteen ground weather-radar stations were installed throughout the Region and arrangements made for proper calibration and maintenance, and also for exchange of radar information. Satellite cloud photography provided on APT receivers, modified in 1978 through launching of the TIROS-N satellites to provide vertical profiles of temperature and humidity as well as surface temperature, was supplemented in the same year by launching of the Japanese Geostationary Meteorological Satellite (GMS), which covered the whole of the typhoon area and for which receiving stations were established in five countries. Telecommunications networks were reinforced by provision of ten additional regional links, and by improvement of national data collection in seven countries of the Region.

Flood-forecasting systems were implemented, under the hydrology component of the project, for the first time in some countries, and extended in other countries, and the component was enlarged to include flood-loss prevention and management evaluations, in seven pilot areas throughout the regions vulnerable to heavy flood damage, and also flood risk evaluation. The project led also to a much greater awareness of the importance of disaster prevention and preparedness, ranging from pre-disaster planning and emergency disaster prevention measures to relief operations and rehabilitation measures.

The success of the project as an exercise in regional collaboration was illustrated by the very extensive bilateral support which it generated within the Region, supplementary to the UNDP financial allocation, and by the basis which it provided for continued regional and international co-operation in support of the Typhoon Committee.

It can be noted that the same UN General Assembly resolution of 1972 resulted in the establishment of the WMO Tropical Cyclone Programme (TCP), the ultimate objective of which was to establish national and internationally co-ordinated systems to ensure reduction in loss of life and damage by tropical cyclones in all storm-prone areas of the world. A description of this programme will be found in Chapter V.

Region II: Regional Development and Applications of Components of the Hydrological Operational Multi-purpose Sub-programme of WMO (HOMS) (1981-1987)

A description of the HOMS programme is given in Chapter IX of the present publication. One important objective of the programme is the adaptation and transfer of appropriate technology required by national Hydrological Services for assessment and management of water resources. The project described had the overall objective of improving the quantity and quality of hydrological data by providing a framework for the transfer to the Asia and South-West Pacific Regions, and among the participating Services within the Region, of hydrological technologies developed elsewhere.

The desirability of making available applied water resource technologies to and within the Region was apparent at the time of conception of the project, when thirteen countries of the Region were listed as having specific existing or planned hydrological and water resources development undertakings.

In its implementation, the project was given a flexible and adaptable structure based upon self-contained but closely co-ordinated sub-projects, each dealing with a particular subject, and each one led by one national agency and carried out in co-operation with other national agencies concerned with the subject, with support provided through a regional co-ordinating centre and WMO. A total of 14 sub-projects were identified for implementation, seven of which were concerned with various aspects of streamflow or flood forecasting, two each with real-time data acquisition, project planning, design and management, and estuarine hydrology, and one with computerized processing of hydrological data. Consultant advice was provided to collaborate with the lead agency in technical adaptation and testing of the technology, and in training of staff. A total of 79 consultancies were provided, the great majority under bilateral co-operation, WMO's Voluntary Co-operation Programme (VCP), or under TCDC arrangements, for which the project provided a very fruitful field of application.

Training was a substantial component of most sub-projects, and was presented through individual fellowships or through on-the-job training workshops organized at a regional centre or by a sub-project lead agency. A total of 29 workshops were organized in the Region during the course of the project, 12 in Bangkok and the remainder at thirteen other centres in the Region, and fellowships were awarded to six countries for participation in a workshop on storage and management of hydrological data, held in Brussels. In summary, nearly 600 hydrologists from more than 20 countries participated in the training programme.

The project was co-ordinated at the global and inter-regional level by WMO, which also organized the planning and monitoring of all project activities. It interacted with other regional WMO projects, notably the projects "Programme Support to the Typhoon Committee", and "Programme Support to the Panel on Tropical Cyclones", and also with national projects in the field of hydrology in Bangladesh, Burma (now Myanmar), India, Nepal and Pakistan.

Region III: Meteorological Training in Latin America (1971-1979)

The countries in Latin America and the Caribbean had gradually developed their Meteorological Services, and by 1969 almost every country had a Service dealing with meteorological activities at the national level. There was, however, considerable variation in their state of development, and in their ability to carry out the responsibilities assigned to them. Increasing demands were made on the Services to provide meteorological information, advice and warnings needed for the national economic development programmes, particularly in the fields of agriculture, water resources, transportation, communication and industry as well as in connection with environmental preservation problems associated with development. Also, the WMO World Weather Watch Programme called for improvement in the observational networks and telecommunications facilities of the national Meteorological Services. A widespread shortage of trained personnel of the junior, middle and professional levels was strongly felt in Latin America as being a handicap to the ability of the Services to meet the above-mentioned demands.

The situation was that the large majority of the countries in Latin America were obliged to send their staff abroad for training at the middle and professional levels. Most of them required financial assistance in the form of fellowships and the governments had the possibility of requesting this type of assistance from UNDP, bilateral sources, or the WMO programmes. The UNDP fellowships which could be obtained from the respective country programmes were, however, specifically destined to provide specialized training to personnel already employed in the Services and who needed to improve and increase their knowledge and skills. The UNDP fellowships were not therefore to be used for a full training of technicians (two years) or for full university courses (four–five years).

The provision which had been approved by the fifth World Meteorological Congress (1967) from the regular budget of the Organization, US $125 000 per year, was enough to satisfy only about six to ten requests per year from all developing WMO Member countries in the world, and in view of the need to ensure geographical distribution, only two fellowships per year, at the most, could be expected for Latin American candidates. Bilateral possibilities were limited and their utilization was complicated still further by the requirement for proficiency in, or to spend a considerable part of the fellowship period studying, a foreign language.

This whole question was raised by WMO at the UNDP Programme Working Group in March 1969. The UNDP Administrator, considering "the difficulty in providing in the various country programmes sufficient fellowships for sending participants to WMO-sponsored training courses", agreed that in a new and unique approach, a fairly large-scale project should be designed consisting entirely of fellowships.

In order to provide the justification required by UNDP before approving the project, WMO conducted a survey among all the countries in Latin America and the Caribbean. Completed in July 1969, this inquiry showed that the Universities of Buenos Aires, Rio de Janeiro and Costa Rica could accept 20, 10 and 15 foreign students per year respectively, for Class I training. The University of Buenos Aires could additionally accept 40 Class II foreign students per year and the Caribbean Meteorological Institute in Barbados could take five foreign students every other year, for Class II training. A Class III training programme then being designed under another project for the six countries in the Central American Isthmus could take 15 students per year.

A further aspect of the survey carried out by WMO indicated that in all Latin American countries and the Caribbean, there was a requirement to train 500 Class I and 500 Class II staff during the following four years. It also indicated that, according to the capacity to absorb new staff, a minimum of 25 Class I and 25 Class II students could and should commence their training abroad in each of the following four years. On the basis of these results and justification, UNDP agreed to finance a regional project entitled "Meteorological Training in Latin America".

The first fellowships were awarded in March 1971 and the last student completed training in March 1979. Under the project, a total of 67 awards were made, 43 of which were for Class I academic studies and 24 for Class II technicians courses. For various reasons 14 of these students could not complete their studies; the majority were Class I students and in general their problems could be attributed to the substantial difference existing in educational systems and academic levels in the various countries in Latin America.

The success of this exercise in regional and inter-regional co-operation was reinforced by the fact that, according to the yearly evaluation surveys carried out by WMO, almost all former fellows are still working in the Meteorological, Hydrological or related Services in their country or origin.

COUNTRY PROJECTS

Region I: MALAWI—Advancement of Hydrological Services in Malawi (1978-1983)

A WMO/UNDP project "Water Resources Assessment of the Lake Malawi Catchment" was undertaken during the period 1972–1981. The present project commenced in January 1978. The earlier hydrological activities had demonstrated the need for additional experienced professional staff to perform duties of an operational nature, for training of Malawian professional and technical staff, and also for review of the status and operation of the network of hydrological stations and for assistance in the processing and publication of hydrological data. Meeting these requirements was the immediate objective of the project. The assistance provided under the project was concurrent with the provision until 1980, by the UK Technical Aid Programme, of the services of a principal hydrologist and a hydrologist/data processor, and also the services of a hydrologist from the earlier WMO project to September 1981, as well as limited-period assistance of a data processor and a hydrologist/water resources expert assigned to a UNDTCD project "National Water Resources Master Plan", which commenced in 1981.

Emphasis was placed upon training of Malawian personnel at both professional and technical levels, there being at this stage no professionally trained Malawian hydrologist and an insufficient number of both senior and junior technicians. The project contributed, together with other bilateral aid sources, to a very intensive programme of overseas fellowship training which provided overall:

- Two fellowships for graduate civil engineering studies;
- One two-year M.Sc. course fellowship in Water Resources Engineering;
- Four fellowships for hydrometric supervisors to attend the WMO Regional Training Course held at Nairobi, May–June 1982;
- A 2½-year water resources technology training programme for one hydrometric supervisor; and
- Three- to six-month courses in water resources technology for two hydrometric assistants.

The project also provided local training, of both on-the-job and classroom types. In particular it ensured the completion of a 2½-year hydrometric supervisors' in-service training course started by the previous project in September 1978.

In regard to the hydrological network, a particular effort was made toward standardization of equipment, and for the establishment of an instrument maintenance/repair workshop in the charge of a trained instrument technician. The network itself, comprising 160 river-discharge, 32 water-level, 62 pan-evaporation, and 35 special-purpose rainfall stations, was considered in general to be adequate in density to the needs for assessment of the water resource.

The data-processing capability of the Hydrology Service received particular attention. A computerized data-processing system was introduced, in collaboration with the UNDTCD project, which became operational in April 1982, and the first steps were taken for initiation of publication of data. Also in the context of data processing, and following a recommendation of the previous WMO project, the study "Assessment of the Water Resources of the Lake Malawi Catchment" was updated, using recently acquired data as well as revised historic data.

Project staff also assisted in the drafting of Water Pollution Control Regulations and in the implementation of a pollution-monitoring programme in and around the cities of Lilongwe and Blantyre.

Region I: BOTSWANA—Strengthening of the Meteorological Services (1984–1991)

The Republic of Botswana covers part of the interior tableland of southern Africa known as the Kalahari Desert. It has an arid to semi-arid climate, and about three-quarters of the country is grazing land. Most of the population is engaged in subsistence farming and livestock breeding. In addition to the general semi-aridity of the country, the main constraint to agriculture is the highly irregular distribution of rainfall in time and space, even in the summer rain season. Annual and monthly rainfall totals are extremely variable from year to year, and heavy storms covering very limited areas are separated by long periods of dry weather.

UNDP approval was given for a project to strengthen the Meteorological Service, particularly through the creation of an Agrometeorology Section. This section would have the responsibility to provide user-adapted climatological and quasi-real-time information to government services and other users, in the form of regular bulletins, crop water-balance and crop production assessments, and information on the actual or potential incidence of pests and disease, as well as rainfall statistics and frequency analyses of important parameters in agrometeorological applications.

Further immediate objectives of the project were to create a Data Processing Section within the Climatological Division, to assist the Meteorology Service in the training of aeronautical forecasters and Class IV personnel, to establish a meteorological instruments workshop, and to expand the meteorological station network to all areas of the country.

The project also undertook a number of studies for the development of operational agrometeorological applications in Botswana. These included an agrometeorological analysis of rainfall data in the region of an Integrated Farming Pilot Project, presenting a survey of the water balance of three major soil types in the region, and testing methods for sorghum and maize production analysis. Another important study established a relation between the FAO water satisfaction index and the related yields of sorghum, for the various provinces of the country.

It can be noted that drought conditions have existed in the country during six successive years, and that Botswana is a participating country in the regional WMO project "Assistance to Drought-stricken Eastern and Southern African Countries in the Fields of Agrometeorology and Hydrology". The objectives of this project are to assist in the timely monitoring of drought in terms of its intensity, geographical extent, duration and impact in Eastern and Southern Africa, and in devising measures to alleviate the effects of such droughts.

Region I: MALI—Agrometeorological Pilot Project (1982–1991)

The Mali Pilot Project is being undertaken in the context of the inter-country AGRHYMET Programme, a description of which is included earlier, as an inter-country project. One objective of this programme, which is directed towards increasing food production in the Sahel region of West Africa, is the development and adaptation of agrometeorological applications in the particular meteorological and climatological conditions of the Sahel.

To this end, a proposal developed from the national AGRHYMET project in Mali for an agrometeorological pilot project, to be undertaken by the Agrometeorology Division of the Mali Meteorological Department, with the objective of identifying in practical terms appropriate agrometeorological methodologies, information and advice which the programme must provide, to the rural community throughout the Sahel. The pilot project commenced early in 1982 with financial support from the Government of Switzerland, in the context of its donor support to the AGRHYMET Programme.

The pilot zone is an area of some 900 km^2 centred on Bankoumana, on the left bank of the River Niger, about 20 km south-west of Bamako. It is within an existing integrated rural development project "Operation Haute Vallée", and includes an irrigated area of some 500 km^2. Crops representative of those cultivated throughout the Sahelian region (such as millet, maize, sorghum and groundnuts) are grown in this zone, as well as certain other crops adapted to irrigation, such as rice and market garden produce.

Project activities are undertaken in collaboration with the Crop Technology Unit of the Sotuba Research Section, and with the International Crop Research Institute for the Semi-arid Tropics (ICRISAT). The project collaborates with the CILSS/FAO Integrated Plant Protection project in entomological and phytosanitary aspects.

Within a pilot area of 100 km^2 four villages were chosen, and in each village four farmers were selected, each cultivating a one-hectare plot divided equally between experimental and control areas. Traditional farming methods are practised in the control areas, and in the experimental areas the farming practice takes account of advice provided by a multi-disciplinary team comprising specialists in the fields of agronomy, meteorology, agrometeorology, crop protection, and information communication. Maize and sorghum are grown, and fertilizer is applied equally in control and experimental areas. An agrometeorological station is located within the area of the four villages and raingauges are installed near each plot, to take account of spatial variation in rainfall.

A comparison of crop yields from the experimental and control areas of each of the 16 plots, averaged over the three-year period 1983–1985, showed a consistently higher yield from the experimental areas, averaging 17% for sorghum and 26% for maize. A comparison of yields from the experimental areas with the average yields for the villages showed an average increase on the experimental areas of 46% for sorghum and again 26% for maize. What has been found particularly encouraging is that, profiting from the increased yields, the farmers now appreciate and put into practice the advice determined by the multi-disciplinary group and issued through the Agrometeorology Division and the agricultural Extension Service. At the same time the project has learned to respect and take into account certain intuitive factors in the traditional farming practices of the area concerned. The project duration was extended in 1987 for a further five years.

Region II: CHINA—Improvement of the Real-time Telemetering and Flood Forecasting System for the San-Men Gorge and the Huayankou Reach of the Yellow River (1981–1985)

The Yellow River, the second largest in China, has long been known as the "harmful" river, because of its propensity for flooding. This results from dyke breaches in the lower reaches, the average occurrence of which was twice in three years. The flood of 1933 affected a total of 3.6 million people, and caused 10 000 deaths. The flood-prone area covers some 250 000 km^2, in which 100 million people live, and is one of the most important industrial and agricultural areas of the country.

The flood plain in the lower reaches is fed by runoff from three areas, including the reach between Sanmexia and Huayankou (known as the San-hua region), which has a catchment area of 41 673 km^2 and with which the project was particularly concerned. This catchment is estimated to have contributed 70% to the maximum peak flow of 22 300 m^3 s^{-1} observed at Huayankou in July 1958, where a possible maximum flow of 55 000 m^3 s^{-1} has been estimated.

The project prepared a plan and design for a real-time telemetering and flood forecasting system for the San-hua reach. A pilot data telemetering system was set up on the Yihe River, based on a network of four hydrometric stations already installed by the Government, and this network was increased to 14 telemetering stations. The relay station network was reinforced, and a flood forecasting centre established at Yhengzhou equipped with a Prime computer and two microcomputers, and with modern rainfall-runoff and flow regulation models. Also, training was given on the installation, maintenance and repair of telemetry and communication equipment and instruments, and on computer operation and hydrological modelling.

It was recognized that further sub-basins in the San-hua reach would have to be similarly equipped for the flood-forecasting system to extend to the whole reach, and that the project activities represent only one step toward an automatic water routeing system for water management projects on the lower reaches of the Yellow River, which will optimize water conservancy and control to ensure maximum possible safety from floods.

Region II: YEMEN ARAB REPUBLIC—Establishment of Meteorological Services (1982–1986)

The Yemen Meteorological Department was established in 1969 under the Civil Aviation and Meteorology Authority (CAMA), as part of the Ministry of Communications. Following accession of the country to Membership of WMO in 1971, a number of phased projects were undertaken with the general aim of establishment of national Meteorological Services. Under the first of these projects, which had a duration of less than two years, synoptic stations were established at the three main airports, namely Sana'a, Taiz and Hodeidah, a training course was initiated for meteorological observers, and a modest fellowship programme was implemented.

A Phase II project followed in November 1974, with an initial term of four years. The main objectives were to render the national Meteorological Service operational in the provision of regular weather information and data to user-interests, to extend the networks of meteorological stations, and to continue and extend the training programme for national personnel. The project assisted also in formalizing the status of the General Department of Meteorology (GDM).

At the time of commencement of this Phase II project, the Government of Saudi Arabia offered extensive assistance to the Yemen Meteorology Service in the provision of upper-air instruments, weather radars, meteorological satellite monitoring systems, and telecommunications equipment, as well as in the construction of buildings and shelters for the housing of equipment and personnel and in the training of personnel. This assistance was defined in collaboration with the WMO project, to avoid duplication, and was implemented during the four-year period 1981–1984. A trust-fund maintenance agreement was signed in May 1980, according to which the Government of Saudi Arabia would be responsible for operating the equipment, and WMO was responsible for equipment maintenance and the selection of engineering and technical staff for this purpose.

In evaluation of the WMO Phase II project in January 1981, it was concluded that in spite of the training provided under the project, and under the regular WMO budget and its Voluntary Co-operation Programme, the number of Yemeni professional personnel was still limited. A Phase III project was therefore approved, for implementation during the period 1982–1986, with emphasis on this training component but less on equipment and other components and on equipment maintenance, this being covered by the Saudi FIT Agreement. The new project provided also for strengthening of the technical capability of the Climatology/Agrometeorology Division in data processing, and in the analysis and publication of data. Improvements were also proposed for the synoptic network and to the organizational structure of the GDM in general.

Training of staff was undertaken locally whenever possible, and a total of 117 persons received classroom training through refresher and familiarization courses or on-the-job training. Overseas courses were nevertheless necessary for professional and specialized training, and the project co-ordinated the processing of a total of 49 fellowships, including 30 under the UNDP/WMO project or other WMO programmes and 17 under bilateral aid from Saudi Arabia. The fellowship programme included M.Sc. and B.Sc. studies in meteorology, B.Sc. studies in electronics and electrical engineering, and Class II forecasting and technician training in various technical fields.

In addition to the extension and upgrading of observational networks, and to the improved infrastructure and capability of the Forecast Centre, to both of which the Saudi assistance contributed materially, progress was made in collaboration with user agencies in defining the optimum network of 15 agrometeorological stations, as well as in studies of important agrometeorological applications.

Better equipped in material and in trained personnel, and with an improved organizational structure, the GDM was considered in 1986 to be functioning effectively and to be providing relevant services to user interests. A further four-year WMO/UNDP project, which commenced in 1987, has the particular objective of developing these advisory services for weather-sensitive economic activities. Mention must be made of the very effective assistance provided to the GDM, through the WMO/UNDP project, of up to ten UN Volunteers, seven of whom were meteorological forecasters, and the remainder equipment maintenance technicians.

Region II: BURMA—Expansion of Meteorological and Hydrological Services (1964–1967)

The monsoon climate of Myanmar (formerly Burma) and the striking variation in the physiographical features of the country make for significant variations in annual rainfall distribution. Although the country is blessed with abundant water, the resource is poorly distributed in time and space. The heavy rains during the south-west monsoon and the torrential downpours associated with sudden storms lead to sustained flooding in the wetter areas and to flash floods in the drier parts, or where steep mountain torrents overflow, causing serious erosion of valuable agricultural land and transport of sediment, which in turn results in silting up of reservoirs and irrigation canals and reduction in the efficiency of navigation channels. During the dry season, on the other hand, scarcity of water becomes a problem over much of the country. The dry zone, a region of great potential fertility and economic importance, regularly suffers acute water shortage, in some years even during the wet season.

Until 1964 hydrometeorological observations were carried out on only a limited scale. Out of a total of 270 rainfall stations, only 54 were directly controlled by the Meteorological Department, and network densities varied markedly from one part of the country to another. These networks were wholly inadequate to meet the data requirements for the planned expansion in economic development, particularly in the fields of agriculture, further hydropower generation, improvements in inland water transportation, and for flood and erosion control. A request was therefore submitted to the UN Special Fund for assistance to the Meteorological Department in the extension and improvement of the hydrological and hydrometeorological station networks and in data processing and analysis, particularly for the formulation of flood forecasts and alerts.

The project commenced early in 1964, with an initial duration of three years. During its course a very substantial increase was made in the density of observational networks, by the establishment of 18 river discharge and 28 water-level stations, together with 17 hydrometeorological and nine meteorological stations, and also 170 rainfall stations, of which 42 were equipped with rainfall recorders. For the efficient operation and maintenance of the reinforced networks, and for carrying out the new responsibilities in the field of hydrology, two regional offices of the Meteorological Department were created, one in Mandalay, in Upper Burma, and the other in Rangoon. Hydrological forecasting centres were set up in these regional centres, which started issuing flood forecasts in 1966 and which became engaged also in special assignments for various government departments and agencies concerned in the extent or the effects of floods. A hydrological laboratory was also established, for sediment analysis and runoff studies, and a fully equipped workshop set up for repair and maintenance of hydrological and meteorological instruments.

Training formed an integral part of project activities. Five courses were held in hydrology for Burmese personnel who would have responsibility for operation of the new stations, or for the processing of data from them, and a series of manuals and handbooks prepared in support of the courses. At the same time a fellowship programme was implemented for specialized overseas training of five local scientists and technicians. The very appreciable increase in activities generated by the project resulted in an augmentation of staff in the Meteorological Department, from 61 in 1964 to 226 in 1966.

Region III: CHILE—Hydrometric and Hydrometeorological Stations (1960–1965)

In order to provide the necessary technical data for feasibility studies of planned major development projects in agriculture and power generation, it was recognized that an extension of the networks of hydrological and hydrometeorological stations would be required with, at the same time, provision of assistance for the corresponding strengthening of the Meteorological and Hydrological Services in Chile. This assistance was provided by an UN Special Fund project for which WMO was designated executing agency.

The design of the extended network of hydrological stations took account of the wide range of climates through the country, ranging from the northern desert to the southern region, which receives rainfalls of 2 000 mm to 8 000 mm. The topography of the latter region offered fine opportunities for hydropower generation, the feasibility of which was further

Stream-gauging in Chile

increased by the narrowness of the country, which facilitates economic power distribution.

To achieve the objective of providing a systematic and comprehensive assessment of country-wide water resources, it was planned that the existing total of 255 hydrological stations, largely located in discrete networks installed for specific irrigation or power projects, should be increased to 390 stations. The project plan of operations also provided for systematic extension of the hydrometeorological station network from 317 to a total of 455 representative stations, comprising two principal and 40 ordinary climatological stations, 370 rainfall/snowfall stations, and 43 special purpose stations.

During the five-year project, in addition to the very extensive field activities for network extension, training was undertaken of the necessary station observers and arrangements were made for the inclusion of meteorological studies in science degree courses at universities in Chile. Instructional material was prepared for both technical school-leavers and civil engineering graduates who were appointed to respective grades in the Hydrology Service, and fellowships were awarded to six senior personnel for specialized studies at overseas institutions. These training activities, together with the provision of water current-meter calibration facilities and sediment and meteorological laboratories as well as the establishment of a data-processing unit, represented a very considerable degree of strengthening of the Chilean Meteorological and Hydrological Services.

Region III: BRAZIL—Hydrology and Climatology of the Brazilian Amazon River Basin (1977–1983)

The poor communications in many areas of Brazil, together with the difficulties arising from the harsh climatic conditions, had hampered the establishment and operation of reasonably dense networks of meteorological and hydrological stations. In line with the Government's objective of reducing regional disparities, the Regional Development Plan for Amazonia (1972–1974) established well-defined priorities for the Amazon basin, including a natural resources survey of minerals, forests, fish, water (hydrology and climatology) and research to develop these resources.

Following a request of the Government of Brazil for UNDP assistance the UNDP Office for Projects Execution (OPE) was designated executing agency, with WMO as associated agency, for a project with the long-term objective of providing, through the acquisition of hydrological and meteorological data, a basis for the detailed planning of all forms of water-resource development.

During the lifetime of the project, which was eventually extended to six years and two months from the original term of three years and four months, a total of 116 new hydrometric stations were installed, together with 197 rainfall stations, the density of the hydrometric network increasing from one station per 27 000 km^2 to one station per 17 000 km^2. The project also assisted in the entry of all collected hydrometeorological data into two existing computerized data banks. During the course of the project, construction of the Tucurui 8 GW hydroelectric power plant required accurate flow forecasts, and so numerical models designed to use available real-time hydrological and climatological information were developed for this purpose. In view of the size of the basin (about 700 000 km^2) and the difficulty of access to it, the decision was made to resort to remote-sensing systems using the GOES meteorological satellite, and the necessary ground receiving station and data-collecting platforms were purchased with project funds. A network of ten telemetering stations was installed and advice was also provided on the application of hydrological models to flow forecasting. This was the first operational application of the GOES system to hydrological forecasting in Latin America.

The project provided an outline of a climatic zonification for the basin, together with a definition of agroclimatic zones for specific crops selected because of their potential economic importance for the future development of the region. A climatological atlas was also prepared for the basin, with data from the period 1960 to 1980. In addition to a number of detailed recommendations relating to maintenance and extension of activities undertaken and systems developed by the project, it was recommended that a regional project should be implemented among all the Amazonian countries, to help transfer the technology and methodological advances achieved through the project, as well as to co-ordinate future action in the fields of hydrology and climatology.

Region V: INDONESIA—Meteorological Applications to Agriculture (1979–1983)

Agriculture is the predominant economic activity in Indonesia. Variations in soil types, topography, and climate lead to great diversity in crops, farming techniques, and agricultural organization. Of particular importance is the rice industry, and during the 1950s and 1960s the Government imported varying amounts of rice each year to meet food requirements or to maintain a ceiling on domestic prices.

The development objective of the project was to achieve more efficient protection and increased production of food crops, through the use of meteorological/climatological data, and to this end a number of specific activities were undertaken, in collaboration with the Meteorological and Geophysical Agency, the Department of Communications, and the Department of Agriculture:

- Magnetic tape recordings were made of rainfall data for 2 945 stations, for a minimum period of ten years, and 1 117 for stations for a minimum of 30 years;
- Development was undertaken of an adequate statistical base of agricultural data, for agrometeorological modelling;
- Computer programs were written to determine probabilities of occurrence of dry and wet days and of rainfall amounts expected at selected probabilities;
- Proposals were made on ways to improve the efficiency of irrigation methods in Indonesia and for the promotion of farm storage in rain-fed areas;
- Analyses were carried out of hydrometeorological data for the development and management of water resources for agricultural purposes;

Recommendations made as a result of these activities were taken into account in the formulation of a follow-up project, the immediate objectives of which were: to demonstrate the technical and economic feasibility of microclimate modification techniques for improving food production in areas of Indonesia prone to food shortages; to develop an operational agrometeorological advisory service; and to provide postgraduate training in agrometeorology. This project commenced in 1983, and was effectively completed in 1987, except for completion of certain fellowship training courses.

PROJECT GUATEMALA 83/009: Establishment of a Hydrometeorological Data Bank in Guatemala

It was felt that the computerized data systems should be improved, especially since the national Meteorological Service formed part of a larger organization known as INSIVUMEH (National Institute for Seismology, Vulcanology, Meteorology and Hydrology). The Government therefore requested UNDP assistance in the implementation of a project aimed specifically at the establishment of a meteorological, hydrological and agrometeorological data-processing centre at the INSIVUMEH in order to be able to provide meteorological information for application to the Government's various programmes.

The second objective of the project was the training of personnel in different areas of activity of the INSIVUMEH, specifically in data processing. The project started in January 1984 and ended on 31 December 1986. The main result of the project was the installation of a modern data-processing centre. Computer programs were also set up and enabled the improvement of the quality control of data through the application of appropriate procedures. Another of the important achievements of the project was the development of technical co-operation among developing countries which enabled the exchange of technicians from Guatemala with others from Costa Rica and Panama. Personnel were trained on the job with the assistance of the WMO experts and consultants.

The general assessment was that all the project objectives were met satisfactorily. Although not all the computer programs have been installed very good progress has been achieved using a modular system which will enable the easy introduction and expansion of new facilities and handle data-processing requirements in the fields of seismology, underground water and the administrative sector.

It is important to note that as a result of this project and the recommendations made, UNDP has approved a new project to consolidate achievements to date. One of the major elements of this second phase of the project is the application of meteorological and climatological data to agriculture.

FUNDS-IN-TRUST PROJECT

Region I: Improvement of Meteorological Services in SADCC Countries (1985)

Sharing the same global geographical location, and also to a large extent the same ecological conditions and degree of economic development, the nine southern African countries of Angola, Botswana, Lesotho, Malawi, Mozambique, Swaziland, Tanzania, Zambia, and Zimbabwe are grouped in an association named the Southern African Development Co-ordination Conference (SADCC), the purpose of which is to promote co-ordinated regional development in the fields of transport and communications, energy, food, agriculture and natural resources, industry and trade, mining, and tourism.

The existence of such an association of countries facilitates technical assistance project implementation on a sub-regional basis, and in particular the implementation of funds-in-trust projects, providing an effective and convenient basis for input and management of project resource requirements from the point of view of both the donor country and the executing agency.

The widespread drought in both southern and eastern Africa during the 1980s highlighted a need for an infrastructure and methodology to ensure the timely monitoring of drought in terms of its intensity, geographical extent, duration, and impact on food production and water resource management. Measures conceived to meet these objectives were:

- The development of national data bases to study systematically the climate and the hydrological regime and its relation to food production, and for water resource assessment and development planning;
- The development and exchange of techniques and methodologies to be utilized on an operational basis by the national Meteorological and Hydrological Services for input into early warning and management systems;
- The dissemination of agrometeorological and hydrological information in a suitable format to be used by national authorities and farmers;
- The development of national and regional infrastructures for the training of personnel; and
- The development of data networks, data-exchange procedures and archiving.

Fundamental to implementation of these measures in the nine SADCC countries was a full and detailed knowledge of the capacity of the individual national Meteorological Services to undertake and support these measures, and a study was therefore proposed of the present status of the Services, and in particular of requirements for improved collection, transmission and analysis of meteorological data. The study was implemented during the period June to September 1985 by a joint Finnish/WMO mission, on the basis of funding deposited under a funds-in-trust arrangement with the Organization by the Finnish International Development Agency (FINNIDA).

It was the overall conclusion of the mission that there existed a need and great potential for the development of meteorology in the SADCC region; the application of meteorological products could also be intensified for the further development of agriculture and water management. A main task of the mission was therefore to prepare an implementation plan for improving Meteorological Services in the region.

This plan, comprising one regional and nine national components, formed the basis for the FINNIDA/WMO Funds-in-Trust Meteorology Project, undertaken in collaboration with the Southern African Transport and Communications Commission (SATCC) during the period 1987 to 1989 with a budget of US $4 million. The objective of this project is to enable the Meteorological Services of the SADCC countries to participate actively in the forthcoming regional UNDP/WMO project "Drought Monitoring in Eastern and Southern Africa", by providing necessary data to the Harare Drought Monitoring Centre and by utilizing products from the Centre for increase in food production and avoidance of weather-related agricultural losses.

A second phase of the FINNIDA/SATCC/WMO project is to be implemented during the period 1989–1992. It will provide for continuation of Phase I activities in the participating countries, particularly in the areas of telecommunication of observed data and of computerized data processing, and also for improved capability for transfer and receipt of data through the Regional Telecommunication Hub, Lusaka, and through the National Meteorological Telecommunication Centres.

The project activities described have led to the adoption of meteorology by SADCC as a new area of co-operation in southern Africa, and to regular, institutionalized meetings of the directors of Meteorological Services of the SADCC countries.

Concluding remarks

The inter-country projects described above serve to illustrate the scope and the need for both inter-country and also regional co-operation for the observation, analysis and forecast of meteorological and hydrological conditions, and for the solution of problems which arise in both respects. It has been a primary objective in the design of WMO technical assistance projects to ensure this co-operation wherever possible, at inter-country and regional levels, and occasionally at the inter-regional level.

In the practical sense, also, inter-country projects offer an efficient means of providing development assistance in the case where a common problem exists and no viable solution is possible, for either physical or economic reasons, on an individual country basis. A case in point is the provision of training opportunities for national personnel at Regional Training Centres, established under WMO technical co-operation projects.

It is true also, however, that by reason of the varying state of evolution of the Meteorological and Hydrological Services in individual countries, and consequently in their need for technical assistance, and also because of particular physical problems which may exist within national boundaries, the majority of WMO technical co-operation projects have been undertaken on a country basis. Responding to the priority needs of developing countries, these projects have frequently had the basic objective of strengthening the national Meteorological and Hydrological Services, both in trained personnel and in infrastructural provisions, particularly in the extension and upgrading of observational networks, and the provision of facilities for processing of the data so essential to the development of the countries concerned.

CHAPTER XV

REGIONAL PROGRAMME

Introductory remarks

The Regional Programme is largely composed of the activities of the Regional Associations whose tasks include the promotion of the execution of the resolutions of Congress and the Executive Council in their respective Regions. The Regional Associations study, from the regional point of view, the scientific and technical programmes of the Organization; co-ordinate, as necessary, the relevant implementation activities undertaken by the Members concerned; and recommend to Congress and to the Executive Council measures to assist Members, particularly in developing countries, in the implementation of the programmes.

The Regional Programme is closely associated with, and derives considerable benefit from, certain activities of regional importance within other WMO programmes.

The Regional Associations have all been active since their establishment by First Congress (1951). The progressive development of WMO's programmes has naturally led to a corresponding growth in the regional activities, which were therefore given a separate identity by Seventh Congress (1975) under the title "Regional Programme". The meteorological activities in the Antarctic, co-ordinated by a working group of the Executive Council (see page 155) because of the fact that the southern limit of the southernmost WMO Regions is latitude 60°S, are also included in the Regional Programme.

The work of the Regional Associations

The six Regional Associations and the Regions covered by them are mentioned on page 154, while the geographical limits of each Region are shown in the map on page 137. The functions of these regional bodies are specified in Article 18 of the Convention, reproduced as Appendix A. As stated on page 152, the presidents of Regional Associations are members of the Executive Council and present the views of their respective Associations to the Council as well as to Congress.

The Regional Associations hold their sessions generally once every four years, questions related to all the major programmes of WMO being on their agenda for study and co-ordination from a regional point of view. The Regional Associations are also required to take action on specific questions referred to them by Congress and the Executive Council.

The noteworthy aspects of the work accomplished by the Regional Associations are described hereafter.

WORLD WEATHER WATCH (WWW)

The resolutions of Congress adopting the World Weather Watch Plan for the four-year period following each Congress session have contained a request to Regional Associations and to the Commission for Basic Systems (*a*) to recommend, *inter alia*, detailed programmes and procedures, as necessary, for the implementation of the Plan and (*b*) to keep the Plan under review in the light of Members' changing requirements and developments in science and technology. The Regional Associations have accordingly played an important role in the implementation plans of WWW in their respective Regions, thereby contributing to the overall implementation of the Plan at the global level.

Global Observing System (GOS)

The establishment of regional basic synoptic networks of surface and upper-air observing stations with their observing programmes had been a traditional responsibility of the Regional Associations even before the inception of the WWW Programme. These networks have an important place in the WWW Plan, constituting the main element of the surface-based sub-system of the GOS. They have been drawn up by the respective Regional Associations and reviewed at each of their sessions in the light of requirements.

Global Data-processing System (GDPS)

The Regional Associations have kept under review the adequacy of the output products of the Regional Meteorological Centres (RMCs) as well as of the data needed by these Centres and have proposed necessary steps to remedy deficiencies. Mention may be made of the inclusion in the WWW Plan of an additional RMC in Region II in 1981 as recommended by Regional Association II. The adoption of regional coding procedures for reporting data required by Members within the Regions has been a regular task of the Associations.

The WWW Plan adopted by Tenth Congress (1987) provides for the strengthening of the GDPS at the regional level through Regional/Specialized Meteorological Centres (RSMCs) with geographical and/or activity specialization. Three centres with activity specialization in tropical cyclone analysis, forecasting and related advisory services were designated by the Executive Council in 1988. As requested by the Council, the Regional Associations concerned will have the task of keeping under review the services provided by these centres in the light of requirements. Plans for the establishment of an RSMC in South-east Asia (at the initiative of ASEAN) for weather prediction services and provision of special products, and of another RSMC in the South Pacific for the provision of specialized products for tropical cyclone warnings, are now being studied in collaboration with the Regional Associations concerned (see pages 136 and 167).

Global Telecommunication System (GTS)

An impressive achievement has been the planning of the regional meteorological telecommunication networks, keeping their performance under constant review, and their improvement through upgrading of circuits, inclusion of additional circuits in the plans and the introduction of new techniques. The credit for this activity goes to the Regional Working Groups on Telecommunications for their intensive and sustained efforts over many years.

Operational WWW Systems Evaluation (OWSE)

A new form of activity included in the WWW Plan and Implementation Programme adopted by Tenth Congress (1987) is the Operational WWW Systems Evaluation (OWSE), the ultimate aim of which is to improve the implementation of WWW. Following the first OWSE, in the North Atlantic, Region I (Africa) has been chosen for the second OWSE, which will concentrate on the evaluation of the potential of the data-collection and retransmission missions of METEOSAT with a view to improving national and regional telecommunications in Africa. The Regional Association for Africa has been taking, through its Working Group on WWW Systems Planning, Co-ordination and Implementation a leading role in planning and carrying out this complex exercise, which is expected to have a duration of about two years, commencing in 1989.

Role of Regional Associations in intercomparisons of meteorological instruments

The Regional Associations have been given a role in the procedures developed by CIMO for intercomparisons of certain meteorological instruments to ensure compatibility and uniformity of data. All Regional Associations have designated their Regional Standard Barometers and laid down instructions concerning the comparison of barometers at synoptic stations with the national standard barometers and of national standard barometers with a regional standard barometer. These instructions are included in the *Manual on the Global Observing System* (Regional Aspects). Regional Associations have also designated Regional Radiation Centres for comparison of radiation instruments within the Regions with standard instruments maintained at these centres, as well as comparison of Regional Standards with International Standard Instruments at the World Radiation Centre. Regional Radiation Centres of all WMO Regions have, in fact, participated in one or more of the International Pyrheliometer Comparisons, of which six have taken place so far at the World Radiation Centre, Davos (Switzerland). This activity has been making a valuable contribution to the availability of reliable radiation data which are needed for the assessment of solar energy potential as well as for scientific investigations, including some within the WCRP, for the study of the role of solar radiation in atmospheric processes.

Tropical Cyclone Programme

The regional activities under the Tropical Cyclone Programme are mainly concerned with the development of co-ordinated operational systems among groups of Members in distinct geographical areas subjects to tropical cyclones, with a view to improving their forecasting and warning services through concerted efforts. These activities are carried out through five regional tropical cyclone bodies, namely:

- ESCAP/WMO Typhoon Committee;
- WMO/ESCAP Panel on Tropical Cyclones for the Bay of Bengal and the Arabian Sea;
- RA I Tropical Cyclone Committee for the South-west Indian Ocean;
- RA IV Hurricane Committee;
- RA V Tropical Cyclone Committee for the South Pacific.

Operational plans have been drawn up by the regional bodies which review them periodically, in order to ensure the most effective co-operation between the relevant countries in strengthening the cyclone warning services in the area concerned. Regional Associations I (Africa), IV (North and

Above: *The President of WMO, Dr R. L. Kintanar, receiving the Sasakawa/UNDRO Award from Ambassador Yoshio Hatano of Japan on behalf of the ESCAP/WMO Typhoon Committee, July 1988*

Left: *Standard solar radiation instrument for Regional Association III, at the Buenos Aires Regional Radiation Centre*

Central America) and V (South-West Pacific), which have established regional tropical cyclone bodies as indicated above, play major roles in the relevant activities. Regional Associations II (Asia) and V (South-West Pacific) are closely concerned with the work of the two joint WMO/ESCAP bodies mentioned above.

The regional activities under the Tropical Cyclone Programme will be enhanced by the Regional Specialized Meteorological Centres already mentioned in the section dealing with the Global Data-processing System.

World Climate Programme

REGIONAL CLIMATIC ATLASES

The preparation of *Regional Climatic Atlases*, following a decision of Second Congress (1955), represents an early activity in the field of climatology within Regional Associations through their working groups or rapporteurs. *Climatic Atlases* for Europe, South America, North and Central America and Asia, consisting of maps of first priority as recommended by CCl, were published in 1970, 1975, 1979 and 1981 respectively, with support from Unesco and, in the case of the *Atlas* for Asia, also from UNEP. The first three *Atlases* mentioned above were published jointly by WMO, Unesco and Cartographia (Budapest), and the fourth jointly by WMO, Unesco and the USSR State Committee for Hydrometeorology and Control of the Natural Environment. The valuable collaboration of the Meteorological Services of the Hungarian People's Republic and of Argentina, the Voeikov Main Geophysical Observatory, Leningrad, and of Brock University, Canada, which undertook the final preparation of the maps contained in the *Atlases* of their respective Regions, should be gratefully acknowledged.

Additional sets of maps for the *Atlases* for Europe, South America and North and Central America have since become available and are awaiting arrangements for their publication.

The preparation of a *Climatic Atlas for Africa*, a revised and improved version of the *Atlas* published in 1961 as a joint project of two technical bodies in Africa outside WMO, but with the collaboration and assistance of the Regional Association for Africa, has been in progress within the Association.

It may also be mentioned that a project for a *Climatic Atlas and Compendium of Climatic Statistics*, now being implemented by ASEAN, is of special interest to the Regional Association for the South-West Pacific.

CLIMAT AND CLIMAT TEMP REPORTING STATIONS

The monthly CLIMAT and CLIMAT TEMP data from selected meteorological stations throughout the world form part of the global climatological data needed for both applications and research. The stations in each Region are designated by the Regional Association concerned, which keeps the network under review in the light of requirements.

Education and training

Presidents of Regional Associations have been requested by Congress to ensure that problems related to education and training are given adequate attention in the Regions. In order that Members may fulfil their responsibilities in this field, at both the national and international level, their requirements have been considered by the Associations concerned, which have made appropriate recommendations to the Executive Council. The Regional Meteorological Training Centres which have been set up over the years and now occupy a prominent place in the Organization's education and training activities (see page 117), have been designated by the Executive Council to meet the requirements of the Regions in accordance with the recommendations of the Regional Associations.

The recommendations of the Regional Associations are also taken into account by the Executive Council in approving regional technical conferences and training seminars (see the following section).

Regional projects

The Regional Programme is closely linked to, and is supported by, the activities of the Organization aimed at assisting Members, particularly in developing countries, in carrying out their tasks at the national and regional levels. Activities in this category include (*a*) regional projects for the transfer of knowledge and proven methodologies through training seminars, workshops and technical conferences and (*b*) regional surveys and co-ordination of Members' implementation plans on a regional or sub-regional scale for particular aspects of a WMO programme. Many such activities have been undertaken on the recommendation of Regional Associations, which have a general responsibility in this regard, as indicated in the opening paragraph of this chapter. The associations have also been requested by Congress to recommend the necessary measures to support Members in their implementation and operation of the WWW Plan. Projects as indicated above were initially carried out under technical co-operation activities financed by UNDP and its predecessors. Provisions in the regular budget of the Organization as an additional source of financing were introduced by Seventh Congress (1975).

The Regional Programme has benefited greatly from the assistance provided to Members under the Training Programme (Chapter XIII) and the Technical Co-operation Programme (Chapter XIV). Examples of regional projects aimed at strengthening national Meteorological and Hydrological Services or in support of WMO's main scientific and technical programmes are given below, indicating the wide range of subjects covered by these projects.

REGIONAL TECHNICAL CONFERENCES, TRAINING SEMINARS, WORKSHOPS, SURVEYS

Development of national Meteorological and Hydrological Services

- Technical Conferences on (*a*) Management of Meteorological and Hydrological Services; (*b*) Role of Meteorological Services in Economic Development; and (*c*) Role of Hydrology and Hydrometeorology in Economic Development

WORLD WEATHER WATCH

Global Observing System (GOS)

- Seminars/workshops (*a*) for network inspectors; (*b*) for meteorological instrument technicians; (*c*) on the management, processing, interpretation and methods of utilization of meteorological satellite data; and (*d*) on the use of meteorological radar

Global Data-processing System (GDPS)

- Seminars on archiving, storage, quality control and retrieval of data
- Implementation co-ordination meetings

Global Telecommunication System (GTS)

- Survey missions on the status of meteorological telecommunications and to develop recommendations for their improvement
- Seminars on management, operation and maintenance of meteorological telecommunications and for training of meteorological telecommunications personnel
- Implementation co-ordination meetings

TROPICAL CYCLONE PROGRAMME

- Seminars on tropical cyclones, their forecasting techniques and warning systems

RESEARCH AND DEVELOPMENT

- Seminars/Workshops on (*a*) Tropical meteorology; (*b*) numerical weather prediction and its different aspects such as NWP in the tropics and routine use of NWP products; (*c*) modern weather forecasting; (*d*) synoptic analysis and forecasting in the tropics; (*e*) local weather system prediction; and (*f*) background air pollution monitoring

APPLICATIONS OF METEOROLOGY

Agricultural Meteorology

- Seminars/workshops on (*a*) agrometeorology and its different aspects, such as agrometeorological observations and their utilization, plant protection, tropical agrometeorology, agrometeorology of semi-arid zones and humid tropics; (*b*) drought and desertification problems; and (*c*) meteorology and the desert locust

Aeronautical Meteorology

- Seminars on synoptic analysis and forecasting for meteorological services to aviation

MARINE METEOROLOGY AND ASSOCIATED OCEANOGRAPHIC ACTIVITIES

- Seminars on Marine Meteorological Services including services to coastal activities

HYDROLOGY AND WATER RESOURCES

- Seminars/workshops on (*a*) field methods and equipment; (*b*) hydrometeorological instruments, methods of observation and establishment of hydrometeorological networks; (*c*) water resource assessment; (*d*) Methods of hydrological forecasting; (*e*) flood forecasting; and (*f*) methods of forecasting heavy precipitation and associated floods

Working groups and rapporteurs

The implementation of WMO programmes at the regional level calls for many detailed and sometimes continuous studies of their regional aspects. Working groups and rapporteurs have played an important role in this task as will be seen from the following examples.

The WWW Plan has well-defined regional requirements, yet Regional Associations have found it necessary to establish

Regional Training Seminar for national instructors from Regional Associations II and V (Jakarta, Indonesia, 18-28 September 1989)

working groups to study the means of meeting these requirements. Perhaps the best example is provided by the Regional Working Groups on Telecommunications, which, over the years, have developed the regional meteorological telecommunication networks, kept them under constant review to ensure their efficient operation and improved them progressively through the introduction of new techniques in a cost-effective manner.

Working groups and rapporteurs have been engaged in developing regional activities within the other major WMO programmes, depending on problems and priorities in the individual Regions. The Regional Associations have been requested by Congress to give particular attention to regional aspects of the World Climate Programme. The development of WCDP and WCAP activities, such as climate data management, preparation of climate data sets, climate applications in fields such as energy (methods of presenting data for use by energy technologists) and urban and building climatology, has received special attention. Within the Applications of Meteorology Programme regional working groups have carried out studies to provide practical assistance to Members in the establishment or expansion of their agrometeorological services, and in the use of agrometeorological information for crop production and protection in general, and in particular for a wide variety of crops which are important for the economy of individual Regions. In the field of hydrology and water resources, working groups have studied regional problems relating to the adequacy of networks of hydrological observing stations, transmission and collection of data, establishment of data banks and hydrological forecasting.

Co-operation with regional organizations

Co-operation with the regional bodies of the UN and with other governmental and non-governmental regional organizations is of particular importance to the Regional Programme in view of the wide applications of meteorology and hydrology and the role of these bodies in fostering regional co-operation for economic and social progress. The description of WMO's collaboration with the five regional Economic Commissions of UN—namely, ECA, ECE, ECLAC, ESCAP and ESCWA—given in Chapter XVI includes examples of co-operative projects in the different WMO Regions. Mention should also be made of the growing co-operation with the Association of the South-East Asian Nations (ASEAN) in their projects for the establishment of a Regional Specialized Meteorological Centre for the provision of weather forecasts and for the preparation of a Climatic Atlas and Compendium of Climatic Statistics (see pages 133 and 135).

Regional Offices

Three Regional Offices within the WMO Secretariat provide support to the Regional Programme. They are: (*a*) the Regional Office for Africa, located at Bujumbura (Burundi); (*b*) the Regional Office for the Americas, located at Asunción (Paraguay); and (*c*) the Regional Office for Asia and the South-West Pacific, located at the WMO Secretariat in Geneva. Each of these Offices is headed by a Regional Director who is assisted by one or two professional Staff.

The Regional Offices for Africa and the Americas have evolved from the posts of the Regional Representatives for Africa and for Latin America, which were included in the WMO Secretariat in Geneva in 1964 and 1968 respectively. The former post was approved by Fourth Congress (1963) in response to a recommendation of the third session (1962) of the Regional Association for Africa to deal with the special problems calling for advice and assistance to Members in the Region. A similar post for Latin America was subsequently approved by Fifth Congress (1967). The term "Regional Offices" for Africa and South America was also adopted by Fifth Congress, while Seventh Congress (1975) changed the title "Regional Representative" to "Regional Director".

Seventh Congress (1975) considered the question of transfer of the Regional Office for Latin America from Geneva to a location within the Region as requested by the Regional Association for South America at its fifth (1970) and sixth (1974) sessions, but postponed a decision on the matter on account of budgetary limitations. Congress, however, decided that the question of transfer of the Regional Offices for both Africa and Latin America should continue to be studied. The financial difficulties in transferring the latter office were later resolved as a result of the kind offer made by the Government of Paraguay to host the office in Asunción. At its thirtieth session (1978) the Executive Council* approved the transfer of the Office to Asunción, where it has since been located.

As regards the Member countries to be served by the Regional Office in Paraguay, Seventh Congress (1975) agreed with a request of Members from the Caribbean area that it should also assist the other developing countries in Region IV (North and Central America). Subsequently, Eighth Congress (1979) decided that the Office should serve all the countries in Regions III and IV. Ninth Congress (1983) changed the name of the Office to "Regional Office for the Americas" to reflect its area of responsibility.

Eighth Congress (1979) endorsed the recommendation of the Regional Association for Africa at its seventh session (1978) urging the transfer of the corresponding Regional Office from Geneva to a suitable location within the Region. Steps were taken to this end and the Organization accepted the kind offer of the Government of Burundi to host the Office at Bujumbura, where the Office was transferred in 1981.

The Regional Office for Asia and the South-West Pacific was set up in 1979, initially under the title "Regional Office for Asia" upon a recommendation of the Regional Association for Asia at its sixth session (1975) and in compliance with the directive of Seventh Congress (1975). (This Office could not be established earlier on account of budgetary difficulties.) Subsequently, Eighth Congress (1979) decided that this Regional Office should serve not only Region II, but also Region V. Ninth Congress (1983) later changed the title of this Office to "Regional Office for Asia and the South-West Pacific".

The responsibilities of the Regional Offices have been laid down in detail by Congress, which has also reviewed them regularly. They include provision of advice and assistance to Members of the Regions concerned in consultation with the presidents of Regional Associations with a view to achieving maximum compliance with the relevant decisions of the

* See note on p. 8.

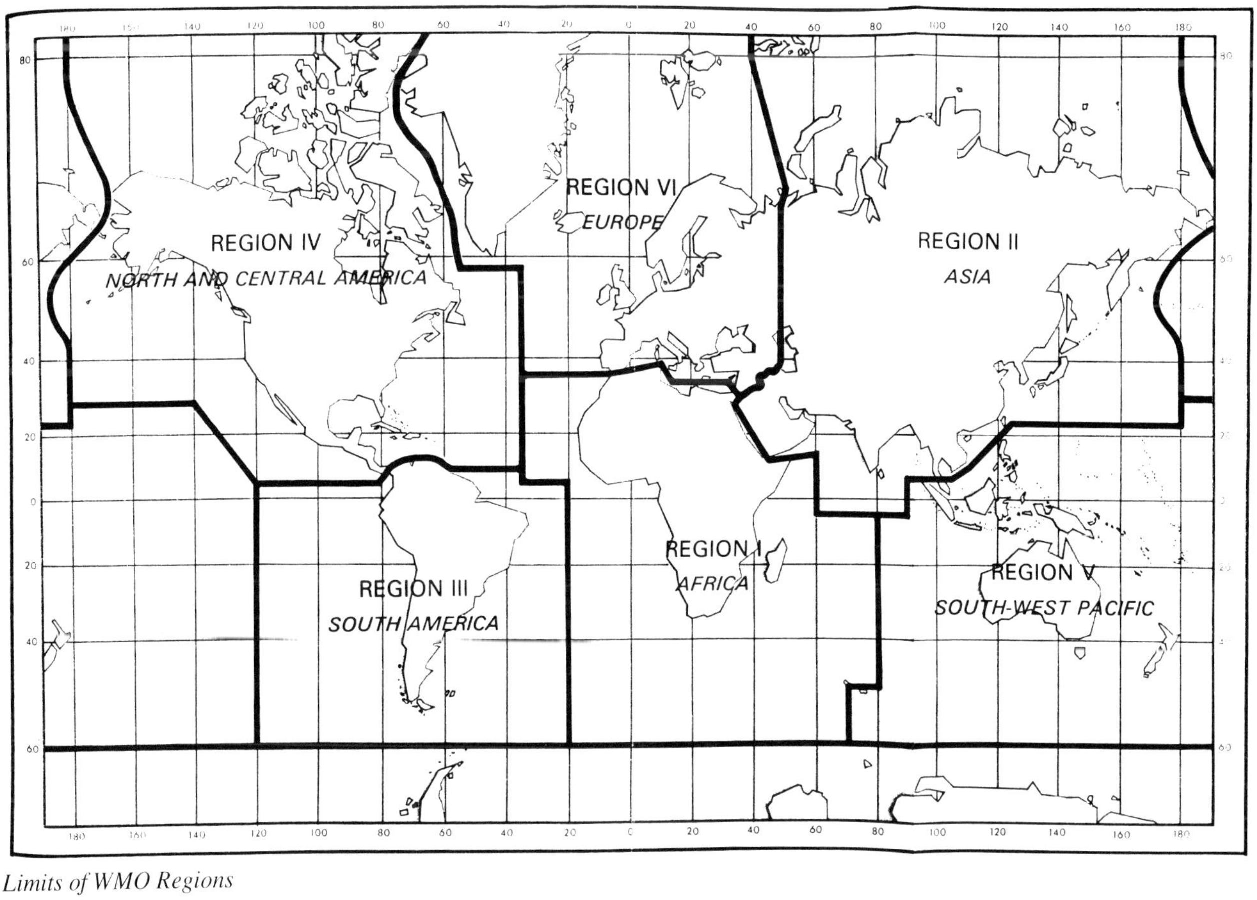

Limits of WMO Regions

Organization, appropriate assistance to the presidents of Regional Associations and liaison with the regional bodies, both within and outside the United Nations system. It may be added that the work of the Regional Offices has been noted by each Congress, which has expressed its satisfaction with the results of their activities.

Antarctic meteorology

The meteorological observing programme in the Antarctic, initiated during the International Geophysical Year (1957/58), has since then continued with additional stations, the present network comprising about 35 stations. The co-ordination of meteorological activities in the Antarctic was undertaken initially by the Executive Council*, which established a working group for this task in 1964; this arrangement has continued since then. As the boundaries of WMO Regions do not extend beyond latitude 60°S, the working group assumes a co-ordinating role which would normally belong to a Regional Association.

The EC Working Group on Antarctic Meteorology is composed of members nominated by countries which are Parties to the Antarctic Treaty and of experts nominated by other countries having active meteorological programmes in the Antarctic. The working group has kept under review the activities in the Antarctic relating to WMO programmes, particularly the WWW, and made recommendations to the Executive Council. As a result, the WMO *Manuals on the GOS, GDPS* and *GTS* include in respect of the Antarctic, as for the six WMO Regions, appropriate material concerning the basic synoptic network, the network of CLIMAT and CLIMAT TEMP reporting stations, arrangements for the collection and transmission of meteorological data, coding procedures, and requirements for processed information to be exchanged within the Antarctic and with centres outside the Antarctic.

* See note on p. 8.

Concluding remarks

The Regional Programme aims at the implementation of the scientific and technical programmes of WMO in the Regions through appropriate action at the national and regional levels. It includes activities to ensure that the Members concerned receive the necessary assistance and support to enable them to take part in these programmes and that their implementation is carried out in close co-ordination between Members. Co-operation with the Regional Economic Commissions of UN and other intergovernmental regional bodies is also an important feature of the Regional Programme.

The wide scope of the Regional Programme and the activities at the national and regional levels undertaken through this programme have contributed to a greater awareness among the countries of the contribution of meteorology and operational hydrology to human activities, the development of the national Meteorological and Hydrological Services and a strengthening of regional co-operation in the implementation of WMO programmes.

CHAPTER XVI

COLLABORATION WITH THE UNITED NATIONS SYSTEM

It has been explained that the principal reason which led to the creation of the World Meteorological Organization to take the place of the International Meteorological Organization was the desire to have an organization of greater strength, an intergovernmental body, to enhance further the application of meteorology to various human activities, and to work in close collaboration with the United Nations and other appropriate organizations associated with it. It is therefore natural that one of the major policy decisions adopted by the First Congress of WMO (Paris, 19 March–28 April 1951) was that the closest cooperation should be established between WMO on the one hand and UN and its other Specialized Agencies on the other, in all fields of common interest, a policy which has been reaffirmed over the years by subsequent sessions of Congress.

AGREEMENTS AND WORKING ARRANGEMENTS WITH ORGANIZATIONS IN THE UNITED NATIONS SYSTEM

Agreement with the United Nations

The Charter of the United Nations lays down that "the various Specialized Agencies shall be brought into relationship with the United Nations", while the WMO Convention states, reciprocally, that "the Organization shall be in relationship to the United Nations".

Among the main organizational tasks undertaken at the First WMO Congress was the approval, pursuant to the above-mentioned provisions, of the draft of an Agreement between UN and WMO, governing their relationship. This Agreement came into force on 20 December 1951, the date on which it was approved by the UN General Assembly during its sixth regular session, thereby conferring on WMO the status of a Specialized Agency responsible for all questions which fall to it under its Convention. The Agreement provides, among other things, for reciprocal representation at certain meetings, exchange of information and documents, submission by WMO of an annual report on its activities, consideration by WMO of recommendations addressed to it by UN including compliance with any request from UN for the furnishing of special reports or studies, and for WMO's participation in the co-ordination of the activities of Specialized Agencies and those of the United Nations. ECOSOC is specifically mentioned in this connection as the body responsible for such co-ordination. It is also the body which considers the annual reports of the Specialized Agencies (and therefore WMO's annual report). Further information about the co-operation of WMO with the United Nations and its various organs in accordance with the foregoing principles, as well as the results of great practical value obtained from such cooperation, is given later in the present chapter.

Working arrangements with Specialized Agencies

The First WMO Congress authorized the Executive Council* to establish such working arrangements with other international organizations as would be desirable in the best interests of WMO. Formal Working Arrangements with the following Specialized Agencies, setting out the principles upon which mutual co-operation should be based, came into force in the period 1952 to 1960, upon approval by the Executive Council and the respective governing bodies of these organizations:

- Food and Agriculture Organization of the United Nations (FAO);
- United Nations Educational, Scientific and Cultural Organization (Unesco);
- World Health Organization (WHO);
- International Civil Aviation Organization (ICAO);
- International Telecommunication Union (ITU);
- International Maritime Organization (IMO).*

By a similar procedure, working arrangements with the International Fund for Agricultural Development (IFAD), which came into being as a Specialized Agency in 1977, were established in 1981.

Agreement with the International Atomic Energy Agency (IAEA)**

Following the establishment in 1957 of the International Atomic Energy Agency (IAEA), a formal Agreement between WMO and this organization came into force in 1959, upon its approval by two-thirds of WMO Member States by correspondence in that year and by the General Conference of IAEA in 1958. The highly beneficial results of WMO's collaboration with IAEA are described later.

WMO'S COLLABORATION WITH ORGANIZATIONS IN THE UNITED NATIONS SYSTEM

General remarks

Collaboration with organizations in the United Nations system is not only an obligation on the part of WMO in view of its status as a Specialized Agency and of its formal agreements and working arrangements with them, but it also provides an effective means of the application of meteorology and hydrology, through concerted efforts. Such collaboration may be in the form of co-operative projects in which WMO is a joint partner, or through provision by WMO of its specialized knowledge and experience. Similarly, assistance and advice are provided to WMO by other organizations.

The major aspects of WMO's relations with organizations in the UN system and a brief account of the activities in which collaboration between WMO and other organizations has

*See note on p. 8.

*In 1982, the Inter-governmental Maritime Consultative Organization (IMCO) was renamed as the International Maritime Organization (IMO).

**IAEA is an organization within the United Nations system, but not a Specialized Agency.

played a role are described below, taking first the United Nations and its bodies and organs, and thereafter each of the other organizations in turn. Some of the activities resulting from collaboration with other organizations are described in more detail in previous chapters dealing with the relevant scientific and technical activities.

It may be appropriate at this stage to say a few words on co-ordination of efforts which becomes important when several organizations are involved in a particular activity. Arrangements for necessary co-ordination in such cases exist under the auspices of the United Nations, and WMO takes part in these arrangements.

UNITED NATIONS GENERAL ASSEMBLY

Peaceful uses of outer space

The evolution of WMO's activities relating to the meteorological aspects of the utilization of outer space has been described in detail in Chapter III. Suffice it to say here that the Organization was quick to realize that the launching of the first artificial Earth satellite, SPUTNIK-1, from a base in the Soviet Union on 4 October 1957, a remarkable event, held a great promise for meteorologists. Some seven months later, the Executive Council* at its tenth session (Geneva, 29 April–17 May 1958) and thereafter Third Congress (Geneva, 1–28 April 1959) took important decisions** with a view to keeping under review the uses of outer space for meteorological purposes as well as keeping the Members informed of developments in this field. A particularly important decision was that WMO should collaborate with the United Nations and other appropriate organizations in carrying out relevant studies. About a year later, the first TIROS experimental weather satellite was successfully launched in the USA on 1 April 1960, transmitting pictures of great interest to meteorologists.

Important developments in the field of outer space were also taking place within the United Nations. In December 1958 the General Assembly, faced at its thirteenth session for the first time with the question of outer space, set up a Committee on the Peaceful Uses of Outer Space. The task of the committee was, *inter alia*, to review international co-operation in this field and to study the means for giving effect to relevant programmes under UN auspices. WMO extended its collaboration to this committee in view of the benefits of the use of outer space for meteorology and of the WMO decisions mentioned above.

Against the background described above, on 20 December 1961 the sixteenth session of the General Assembly unanimously adopted Resolution 1721(XVI) entitled "International co-operation in the peaceful uses of outer space". The resolution had a profound impact on WMO's work as it marked the beginning of a process which led to the establishment of the World Weather Watch (WWW), now widely known as the basic programme of the Organization upon which, to a great extent, many of its other programmes depend. There were four sections in the resolution which dealt with aspects of exploration and the use of outer space for peaceful purposes. One section was exclusively devoted to atmospheric sciences, weather forecasting and related questions. Extracts from this section are reproduced below.

The General Assembly

1. Recommends to all Member States and to the World Meteorological Organization and other appropriate Specialized Agencies the early and comprehensive study, in the light of developments in outer space, of measures:

   (*a*) To advance the state of atmospheric science and technology so as to provide greater knowledge of basic physical forces affecting climate and the possibility of large-scale weather modification;

   (*b*) To develop existing weather forecasting capabilities and to help Member States make effective use of such capabilities through regional meteorological centres;

2. *Requests* the World Meteorological Organization, consulting as appropriate with the United Nations Educational, Scientific and Cultural Organization and other specialized agencies and governmental and non-governmental organizations, such as the International Council of Scientific Unions, to submit a report to its Member Governments and to the Economic and Social Council at its thirty-fourth session regarding appropriate organizational and financial arrangements to achieve those ends, with a view to their further consideration by the General Assembly at its seventeenth session.

WMO was thus given prime responsibility by the United Nations General Assembly for undertaking the necessary studies on a subject which held great promise, and for making proposals for action at the international level. The steps taken by WMO to prepare the report called for in the resolution and the intense activity which took place thereafter within the Organization for several years, culminating in the establishment of the World Weather Watch, are described in detail in Chapter III.

As requested, WMO submitted its report to ECOSOC and, by Resolution 913 (XXXIV), adopted unanimously, ECOSOC expressed "its appreciation to WMO for its forward-looking approach to the advancement of atmospheric sciences" and transmitted the WMO report to the General Assembly at its seventeenth session. The action taken by the General Assembly was embodied in Resolution 1802 (XVII), adopted unanimously on 14 December 1962. The text of this resolution is reproduced below.

The General Assembly,

1. *Notes with appreciation* the prompt initial response of the World Meteorological Organization to the request of the General Assembly, as embodied in Resolution 1721 (XVI), that it report on a programme to advance atmospheric science research and to develop improved weather forecasting capabilities in the light of developments in outer space;

2. *Calls upon* Member States to strengthen weather forecasting services and to encourage their scientific communities to co-operate in the expansion of atmospheric science research;

3. *Recommends* that the World Meteorological Organization, in consultation with other United Nations agencies and governmental and non-governmental organizations, should develop in greater detail its plan for an expanded programme to strengthen meteorological services and research, placing particular emphasis on the use of meteorological satellites and on the expansion of training and educational opportunities in these fields;

4. *Invites* the International Council of Scientific Unions through its member unions and national academies to develop an expanded programme of atmospheric science research which will complement the programmes fostered by the World Meteorological Organization;

5. *Invites* United Nations agencies concerned with the granting of technical and financial assistance, in consultation with the World Meteorological Organization, to give sympathetic consideration to requests from Member States for technical and financial assistance to supplement their own resources for these activities, including the improvement of meteorological networks;

6. *Requests* the World Meteorological Organization, following its Congress in April 1963, to report to the Committee on the Peaceful Uses of Outer Space, and to the Economic and Social Council at its thirty-sixth session, on steps taken relating to these activities.

---

*See note on p. 8.

**Resolution 14 (EC-X)—Participation of the WMO in the investigation of the atmosphere by artificial satellites; and Resolution 28 (Cg–III)—Meteorological data from artificial satellites

Within about four months of this further encouraging sign from the General Assembly, the Fourth WMO Congress (Geneva, 1–27 April 1963) approved the concept of the World Weather Watch and accordingly took a number of steps for the preparation of a detailed plan.* In compliance with General Assembly Resolution 1802 (XVII), WMO then submitted a second report with the decisions of its Fourth Congress to ECOSOC at its thirty-sixth session (1963). ECOSOC expressed "its appreciation to the WMO for its prompt and thorough planning to make maximum use of the opportunities offered by the availability of data from meteorological satellites".

Intensive and concerted activity which then took place within WMO led to the preparation by its Secretariat of the WWW Plan which was adopted by the Fifth WMO Congress (Geneva, 3–28 April 1967).** In September 1967, the UN Committee on the Peaceful Uses of Outer Space expressed great appreciation of the WWW Plan, which it considered to be an excellent demonstration of the practical benefits which could be derived from the peaceful uses of outer space.

Some twenty years have elapsed since the adoption of the first WWW Plan and a vastly improved version of the original Plan is in operation as a result of the progressive incorporation of the striking scientific and technological developments which have taken place during this period. The new generation of professional meteorologists all over the world, making use of the WWW Plan and its products on a round-the-clock basis, may not be aware of the important role played by the United Nations General Assembly in the establishment of the WWW.

Peaceful uses of atomic energy

WMO's active interest in the meteorological aspects of the peaceful uses of atomic energy began in 1956 with the establishment by the Executive Council*** of a panel of experts to study this subject so as to enable the Organization to provide necessary advice to the United Nations and to other organizations as well as to its Members.†(Experts were subsequently designated by UNSCEAR and IAEA on this panel in 1962.)

The panel was largely responsible for carrying out WMO's early and extremely important work in this field, as described in the following account. The first United Nations International Conference on the Peaceful Uses of Atomic Energy, in which WMO participated, had then already taken place in Geneva in 1955, while steps for the establishment of the International Atomic Energy Agency were well advanced, the draft statute of the Agency having been circulated to governments by UN in 1955. The Administrative Committee on Co-ordination had established a Sub-committee on Atomic Energy and the Secretary-General of WMO had participated in its first session in December 1955. The General Assembly had also established, in December 1955, the United Nations Scientific Committee on the Effects of Atomic Radiation (UNSCEAR) with a mandate which included the study, in collaboration with the Specialized Agencies, as appropriate, of the world-wide contamination of the environment by radioactive debris. As the mechanisms controlling the movement of radioactivity in the atmosphere involve many meteorological problems, the need for close collaboration between UNSCEAR and WMO was self-evident. UNSCEAR was very active right from the beginning and WMO collaborated closely with this body, notably through participation at its sessions. Particular mention may be made of the seventh session (January 1960), which included a three-day Symposium on the Meteorological Aspects of World-wide Fallout, organized by WMO, which later published (1961) the full texts of the symposium papers as *Technical Note* No. 43 under the title "Meteorological factors influencing the transport and removal of radioactive debris". A similar symposium was also arranged during the thirteenth session (1964) of UNSCEAR and the papers presented on that occasion were likewise published (1965) as WMO *Technical Note* No. 68 under the title "Meteorological aspects of atmospheric radioactivity".

The close co-operation between UNSCEAR and WMO led to the adoption on 27 October 1961 by the UN General Assembly at its sixteenth session of a resolution* in which it "invited WMO, in consultation with IAEA and UNSCEAR, to examine urgently the feasibility of extending the present meteorological reporting system to include measurements of atmospheric radioactivity and to ensure day-to-day exchange of this information by means of meteorological telecommunication systems". The plan prepared by WMO in response to this request was considered by the General Assembly at its seventeenth session which, in a resolution** adopted on 20 November 1962, "commended WMO for its prompt and effective response and authorized its implementation after further consultations with UNSCEAR". The plan was thereafter completed on the basis of the latter's views and approved by the Fourth WMO Congress (April 1963).*** For technical reasons, however, it was necessary to arrange with the WMO Members making the required observations for the exchange of data by air mail upon request from other Members desiring them, but this would not detract from the objectives of the plan—namely, improving the global picture of atmospheric radioactivity and assisting UNSCEAR in its work. At its eighteenth session (1963) the General Assembly noted the plan as approved by WMO Congress and invited WMO to proceed with its implementation.† Subsequently, by Resolution 2258 (XXII), adopted on 25 October 1967, the General Assembly commended WMO for "its work in carrying forward the scheme for monitoring and reporting levels of atmospheric radioactivity".

Tropical cyclones

Following the world-wide concern at the series of tropical cyclone disasters which struck some countries in Asia in 1970, causing exceptionally heavy loss of life and extensive material damage, the General Assembly at its twenty-fifth session adopted on 16 December in the same year a resolution†† which included the following operative paragraphs:

> *Recommends* to the World Meteorological Organization that it take, if necessary, further appropriate action for mobilizing capable scientists, technologists and other pertinent sources from any or all nations towards obtaining basic meteorological data and discovering ways and means to mitigate the harmful effects of these storms and remove or minimize their destructive potential;
>
> *Calls upon* Member States to exert efforts within their means to implement fully the World Weather Watch Plan of the World Meteorological Organization.

*Resolution 21 (Cg–IV)—Analysis of the world weather system

**Resolution 16 (Cg-V)—World Weather Watch

***See note on p. 8.

†Resolution 22 (EC–VIII)—Participation of the World Meteorological Organization in international developments on the peaceful uses of atomic energy.

*UN Resolution 1629 (XVI)—Report of the United Nations Scientific Committee on the Effects of Atomic Radiation

**UN Resolution 1764 (XVII)—Report of the United Nations Scientific Committee on the Effects of Atomic Radiation

***Resolution 25 (Cg-IV)—Measurement of atmospheric radioactivity

†UN Resolution 1896 (XVIII)—Report of the United Nations Scientific Committee on the Effects of Atomic Radiation

††UN Resolution 2733 (XXV)—International co-operation in the peaceful uses of outer space

The responsibilities of WMO, and of its predecessor IMO, in regard to tropical cyclones have for decades been well known to maritime countries and to mariners alike. The organization of a system whereby basic data are collected from sea areas to facilitate the detection of cyclones and to trace their subsequent movement, as well as arrangements for the operation of a world-wide warning service, has always been an important activity of both WMO and IMO. Also, in 1968, WMO had sponsored jointly with ESCAP (then known as ECAFE) an Inter-governmental Typhoon Committee with a view to promoting and co-ordinating efforts to minimize typhoon damage in the region.*

At its thirty-second session (1977) the General Assembly adopted another resolution** on this subject containing the following text which concerned WMO directly:

> *Welcomes* the report submitted by the World Meteorological Organization on its tropical cyclone project and the World Weather Watch, in response to General Assembly Resolution 31/8, and notes in particular that the satellite has revolutionized the initial detection of tropical cyclones, that the availability of five meteorological geostationary satellites by 1978 would mean that all tropical areas of the world would be under constant surveillance and that the success of the project depends upon continued and increased committal of essential resources to this programme, and calls upon the World Meteorological Organization to intensify its efforts in this field and to report thereon in accordance with the relevant resolutions of the Assembly.

The scope of the Tropical Cyclone Project has widened considerably over the years. In 1979 Eighth Congress raised its status and it became the Tropical Cyclone Programme.

Marine sciences

At its twenty-first session (1966) the General Assembly adopted Resolution 2172 (XXI) entitled "Resources of the sea", calling upon the Secretary-General of UN to make effective arrangements, in co-operation with other organizations, for an expanded programme of activities in marine science and technology. WMO, which was one of the three Specialized Agencies specifically requested to co-operate in this task, participated actively with the UN Secretariat in the work pursuant to this resolution. Several other resolutions dealing with different aspects of this problem were adopted subsequently, both by the General Assembly and ECOSOC. These resolutions have had a major influence on WMO's activities in marine sciences and oceanography. The Executive Council*** established a Panel on Meteorological Aspects of Ocean Affairs to co-ordinate all WMO activities in this field. A noteworthy role was played by the ACC in securing co-ordinated action by the various organizations on the General Assembly resolutions involving complex issues. Later UNEP also joined forces in some of the tasks and has since been providing considerable support.

Exchange of seismic data in connection with disarmament and related questions

The Group of Scientific Experts to Consider International Co-operative Measures to Detect and Identify Seismic Events, set up by the UN Committee on Disarmament, which reports to the General Assembly, was engaged in a study of the scientific and technical aspects of a proposal for the global exchange of seismological data to assist in the monitoring of nuclear tests. The group sought the collaboration of WMO in a study of the possibility of the regular use of the WWW Global Telecommunication System for the international exchange of the seismic data. On the basis of the examination of the proposal from a technical point of view by the appropriate WMO body (CBS), the WMO Executive Council at its thirty-fifth session (1983) approved the inclusion of seismic bulletins in the global exchange programme on the GTS.

ECONOMIC AND SOCIAL COUNCIL (ECOSOC)

The role of ECOSOC vis-à-vis the Specialized Agencies

The Charter of the United Nations entrusts ECOSOC with the task of co-ordinating "the activities of the specialized agencies through consultation with and recommendations to such agencies and through recommendations to the General Assembly and to the Members of the United Nations". The Charter also states that "ECOSOC may take appropriate steps to obtain regular reports from the specialized agencies" as well as "reports on the steps taken to give effect to its own recommendations and to recommendations on matters falling within its competence made by the General Assembly".

The account which follows shows that ECOSOC has on different occasions considered several matters concerning WMO's activities and also taken action with a view to securing co-ordination of the activities of organizations, thereby providing great encouragement to WMO. An important role in the task of co-ordination is also played by the Administrative Committee on Co-ordination and, as will be seen below, this is another high-level body whose deliberations have also been of great assistance to WMO.

Administrative Committee on Co-ordination

The Administrative Committee on Co-ordination (ACC) was established in 1947 by ECOSOC, to which it reports, in order to avoid overlapping of activities of the intergovernmental organizations related to the UN and to promote the co-ordination of their efforts. The Committee is composed of the Secretary-General of UN, who acts as chairman, and the Executive Heads of the Specialized Agencies and of IAEA. There are others, such as the heads of the various UN organs and programmes, who customarily attend the sessions of ACC and participate in its work.

In 1970, ACC appointed a Functional Group on Human Environment with Dr D.A. Davies, then Secretary-General of WMO, as Convener. Although not exclusively concerned with the preparations for the UN Conference on the Human Environment (Stockholm, 1972), the group gave close attention to the drafting of an important document for this conference on behalf of the UN system.

ACC has established sub-committees to facilitate consultation and co-ordination at the working level. Those which are particularly useful to WMO's activities and in which it takes an active part are the Sub-Committees on Water Resources Development, Marine Science and its Applications (previously called Oceanography), Education and Training, and Science and Technology.

*The General Assembly Resolution 2733 (XXV) was, in fact, in response to a resolution adopted by the third session of the Typhoon Committee (November, 1970), appealing to the United Nations for help in reducing the harmful effects of typhoons and other tropical cyclones. The preamble to the General Assembly resolution noted the role of this committee and the discussions at this session.

**RES/32/196A—Report of the Committee on the Peaceful Uses of Outer Space

***See note on p. 8.

Annual Report of WMO

The UN–WMO Agreement states that WMO "shall submit to the United Nations an annual report on its activities". In accordance with the UN Charter, the report is presented to ECOSOC, which has generally commended WMO for its work and on occasion adopted resolutions to this effect.

In 1972, ECOSOC carried out its annual examination of the reports of the Specialized Agencies in accordance with a new procedure adopted in the previous year, providing for an in-depth review each year of reports of organizations selected for this purpose. WMO was one of the two organizations whose reports were chosen for detailed consideration in 1972. Following the debate in which the Secretary-General of WMO took part, the results of the examination by ECOSOC of the WMO report were embodied in a resolution* adopted unanimously on 28 July 1972. The text of this resolution is reproduced below.

> *The Economic and Social Council,*
>
> Considering that many aspects of WMO programmes have been undertaken in response to requests addressed to it by the United Nations, and in particular, in response to General Assembly resolutions on international co-operation in the peaceful uses of outer space, namely Resolution 1721 (XVI) of 20 December 1961, which led to the World Weather Watch, Resolution 1802 (XVII) of 14 December 1962, which led to the Global Atmospheric Research Programme, and Resolution 2733 (XXV) of 16 December 1970, which led to the Tropical Cyclone Project,
>
> 1. Draws attention of Governments of Member States to the need for recognizing the vital character of their atmospheric and hydrological resources in maintaining progress in economic and social development, including the protection of life and property against weather-related disasters;
> 2. Urges Governments of Member States to support the balanced development of their national services in these fields, thereby ensuring that these services can make their fullest contribution to national economic and social development and, at the same time, implement the various programmes of the World Meteorological Organization, thus contributing to the well-being of mankind.

World-wide network of meteorological stations

A number of questions bearing on the application of science and technology for peaceful ends were discussed in 1961 in ECOSOC at its thirty-second session. On the basis of a report entitled "Main trends of inquiry in the natural sciences, the dissemination of scientific knowledge and its application for peaceful purposes" prepared under the auspices of Unesco, two resolutions were unanimously adopted by ECOSOC on 1 August 1961. It is significant that one of these referred exclusively to WMO's activities.** The text of this resolution is reproduced below.

> *The Economic and Social Council,*
>
> *Considering* the importance of the application of meteorology to food production, safety of life at sea, air transport, assessment and development of water resources and other human activities,
>
> *Recognizing* the fundamental role of the permanent world-wide network of meteorological observing stations in providing the basic information for the full application of meteorology to the above-mentioned activities and also for improving our basic knowledge of the global processes of the atmosphere,
>
> *Noting* that there are serious gaps in the existing networks of permanent meteorological stations particularly in tropical regions and in the southern hemisphere,
>
> 1. *Calls* attention to the urgent need to eliminate these gaps;
> 2. *Invites* governments to take steps individually or collectively to establish meteorological observing stations where serious gaps in the world network exist;
> 3. Supports the efforts of the World Meteorological Organization in establishing a plan for the world network of meteorological stations and in assisting governments in implementing this plan.

* ECOSOC Resolution 1728 (L III), Part C

**ECOSOC Resolution 829B (XXXII)

Water resources

The importance of the development of water resources, both in the context of a growing world shortage of water and for the economic progress of countries, began to be realized within the United Nations early in the 1950s. In 1952 it was decided by ECOSOC that UN should promote and co-ordinate international activities concerning water resource development and keep under review the work of the international organizations in this field. Subsequently, in 1954 ECOSOC recommended*, *inter alia*, that particular attention should be given by governments and the UN organizations to the assembly of hydrological data since there was an urgent need for full knowledge of all water resources. While no single organization was concerned entirely with water problems, a great need was felt for one of the organizations to assume the responsibility for promoting the collection of hydrological data and for standardizing procedures to this end. WMO was considered as the most suitable agency for this purpose and it accepted the responsibility to undertake the work dealing with those aspects which were common to both meteorology and hydrology.

In the light of the continuing action by ECOSOC, the role which WMO was to play in this field was gradually emerging. With a view to overcoming deficiencies in hydrological data and in accordance with a decision of the ECOSOC in 1956, a questionnaire drawn up jointly by the UN and WMO Secretariats on existing Hydrological Services, plans for their extension and conditions for the execution of these plans, was sent by UN to its Members and to those of the Specialized Agencies in 1957. The replies received were analysed by the WMO Secretariat and incorporated in a report which was submitted by the Secretary-General of UN to ECOSOC at its twenty-fifth session (April–May 1958). The report also examined current problems with regard to hydrological activities (on the basis of some other studies undertaken by UN) and outlined some steps which might be taken to meet them. One of the recommendations in the report was that the responsibilities of WMO should be expanded to include hydrology. ECOSOC thereupon adopted Resolution 675(XXV) which stated as follows:

> *Commends* the Secretary-General and the World Meteorological Organization for the report entitled "A preliminary inquiry on existing hydrologic services";
>
> *Notes* the recommendations with respect to the functions of the World Meteorological Organization in the field of hydrology;
>
> *Invites* the World Meteorological Organization to consider the report and to take appropriate action thereon bearing in mind the necessity of avoiding duplication in the work of the United Nations and the specialized agencies.

WMO's policy and programme were considered by the Organization's Third Congress (1–28 April 1959). The main decision of Congress was the establishment of a new Technical Commission for Hydrological Meteorology, which would increase significantly WMO's work in the field of water resources without, however, meeting the above recommendations of the UN completely.

ECOSOC's reaction was, nonetheless, favourable and embodied in Resolution 743A (XXVIII), adopted unanimously at its twenty-eighth session on 31 July 1959, the text which is reproduced in part overleaf:

*ECOSOC Resolution 533 (XVIII)

The Economic and Social Council

Notes with interest Resolution 19 (Cg–III) adopted by the WMO and its decision to establish a new Technical Commission on Hydrological Meteorology;

Commends the WMO for the establishment of the Commission;

Expresses the hope that the Governments of the Members of the United Nations will co-operate fully with the work of the WMO and its new Technical Commission on Hydrological Meteorology.

The subsequent developments within WMO over the years, leading to the establishment of the highly successful Hydrology and Water Resources Programme as one of its major programmes—a programme which has, moreover, given WMO a central place among the organizations of the UN system in the field of water resources—have been described in Chapter IX.

WMO has taken an active part in the Inter-Agency Meetings on Water Resources Development, started in 1954 and continued since then on an annual basis, acting as a sub-committee of the ACC. It has also co-operated closely with the Water Resources Development Centre in the UN Secretariat established by a decision of ECOSOC in 1958. WMO has also collaborated with the Regional Economic Commissions, notably through the joint sponsorship of important projects. It has executed a large number of major projects for the assessment and utilization of water resources, undertaken through UNDP and its predecessor institutions of UN.

Seismic sea-waves

Expressing concern at the loss of life and material damage caused by earthquakes and seismic sea-waves, ECOSOC requested the Secretary-General of UN to seek the co-operation of WMO and Unesco in undertaking a study of ways of reducing such loss and damage to a minimum, including the development and co-ordination of seismological observations and research, and of systems of warnings of seismic sea-waves.*

The question of the organization of an international warning service concerning the movement of tidal waves caused by different meteorological and geophysical phenomena, including underwater Earth tremors (tsunami), had been under study within the WMO Executive Council** since 1957. Following the request from ECOSOC, a Panel of Experts set up by the Executive Council prepared a plan for an international warning service for the Pacific Ocean area for tidal waves caused by earthquakes (tsunami), making use of the meteorological telecommunication network.

World Climate Programme

The developments leading to the adoption of the World Climate Programme by the Eighth WMO Congress (1979) have been described in Chapter XII. In brief, it was the growing world-wide interest in understanding climatic change and variability, as well as the possible impact of human activities on climate, which led to this new and comprehensive WMO programme. The planning of this programme, of a highly inter-disciplinary nature, extending over a few years, required close co-operation with many interested international organizations both within and outside the United Nations system.

On the basis of progress reports submitted by WMO in 1979 and 1980, ECOSOC affirmed its keen interest in, and its continued support for, this WMO programme. For its part, the ACC too showed a great interest and in 1979 stated the need for all organizations concerned to co-operate fully with WMO in the implementation of the WCP.

* ECOSOC Resolution 767 (XXX), adopted on 8 July 1960

** See note on p. 8.

UNITED NATIONS CONFERENCES

A large number of conferences on topics of particular importance have been held from time to time under the auspices of the United Nations with the object of taking concerted action to deal with the problems involved and to consider the role of UN and its associated organizations in this task. WMO has participated in many of these conferences, taking an active role in some of them. These conferences are mentioned below, indicating their importance to WMO and the role it has played.

United Nations Conference on the Human Environment (Stockholm, 1972)

It was natural for WMO, which has a special interest in environmental problems, to have participated actively in the preparation for this conference and in the conference itself. A large number of basic papers and shorter contributions were prepared by WMO as material for preparations for the conference. The principal document submitted by WMO to the conference was entitled "Meteorology as related to the human environment" and contained contributions by eminent meteorologists. WMO was also responsible for the preparation of a consolidated report on the activities and functions of the UN system in relation to the human environment, in co-operation with the organizations and bodies concerned. The ACC Functional Group on the Human Environment, under the chairmanship of Dr D. A. Davies, then Secretary-General of WMO, was closely involved in this work, and the document was approved by the ACC.

A very large number of recommendations emanating from the conference concerned WMO, which took follow-up action, in many cases with the support of UNEP.

United Nations Water Conference (Mar del Plata, Argentina, 1977)

This conference was of direct interest to WMO, which took an active part both in the conference and in its preparation. WMO submitted several documents, some of which were in the form of publications, and was also closely involved in the regional preparatory meetings, convened by the UN Regional Economic Commissions.

Several recommendations of the conference concerned WMO's activities in hydrology and water resource development. The conference also made specific reference to the Operational Hydrology Programme (OHP) of WMO and the role it could play in the implementation of the action plan adopted by the conference. This plan has since formed the basis of much of WMO's own activities in hydrology and of those carried out in co-operation with Unesco and other organizations.

United Nations Conference on Human Settlements (HABITAT) (Vancouver, Canada, 1976)

WMO contributed a background paper entitled "Weather, climate and settlements", one of 11 papers presented by UN and its Specialized Agencies. Two other publications were also prepared, with support from UNEP:

- "Application of building climatology to the problems of housing and building for human settlements" (*Technical Note* No. 150);
- "Weather, climate and human settlements" (*Special Environmental Report* No. 7).

United Nations Conference on Desertification (Nairobi, 1977)
In accordance with General Assembly Resolution 3337 (XXIX), calling for specific studies in preparation for the conference, WMO submitted a background document "Climate and desertification" prepared by an eminent climatologist and also collaborated with FAO and Unesco in the preparation of a world map of desertification.

The Plan of Action to Combat Desertification, as adopted by the conference, included specific references to WMO programmes, while there were recommendations involving fields of direct interest to WMO and of its responsibilities. WMO has been collaborating with FAO and UNEP in the follow-up action on certain items.

United Nations Conference on New and Renewable Sources of Energy (Nairobi, 1981)

WMO's contribution to this conference was in the form of the following two *Technical Notes* which were received with great interest and appreciation:

- "Meteorological aspects of the utilization of solar radiation as an energy source", with an annex comprising 13 world maps of relative global radiation (*Technical Note* No. 172);
- "Meteorological aspects of the utilization of wind as an energy source", with a world chart of estimated wind-energy potential (*Technical Note* No. 175).

The conference adopted a Plan of Action aimed at increased use of new and renewable sources of energy which was taken into account by the WMO Executive Council* in 1982 when it updated the 1976 WMO Plan of Action in the field of Energy Problems.

Other conferences

WMO has also taken part in other United Nations conferences, such as those on the Peaceful Uses of Atomic Energy, Exploration and Peaceful Uses of Outer Space, Science and Technology for Development and Technical Co-operation among Developing Countries.

UNITED NATIONS ENVIRONMENT PROGRAMME (UNEP)

WMO has very close ties with the United Nations Environment Programme (UNEP), which was created by the UN General Assembly in 1972 as the organ to implement the decisions of the UN Conference of the Human Environment, held in Stockholm in the same year and in which WMO took an active part. In an address to the WMO Executive Council* in 1976, Dr M. K. Tolba, Executive Director of UNEP, stated that WMO was the only United Nations Agency whose total mandate could be considered as falling within the environmental field, and UNEP and WMO had a mutual, intimate and continuing interest in each other's activities, which was illustrated by the close and excellent co-operation between them right from the inception of UNEP.

The fields of activity involving mutual collaboration between the two organizations include background air-pollution monitoring, long-range transport of pollutants, water quality monitoring, marine pollution, climatic changes, ozone layer, atmospheric carbon dioxide, desertification, natural disasters (tropical cyclones) and weather modification. UNEP has been providing substantial support, including financial support, to WMO's activities in most of these fields. The total funds committed by UNEP in this respect up to the year 1986 were about US $6.5 million.

With substantial assistance from UNEP, the observational capability of a large number of ozone stations has been upgraded, while new stations have been established in data-sparse areas. With such expansion in the network and study of the resulting data, WMO has issued official statements entitled "Modification of the ozone layer due to human activities and some possible geophysical effects" in November 1975, October 1978 and November 1981.

Following the Vienna Conference (1985) for the Protection of the Ozone Layer, a Conference of Plenipotentiaries (Montreal, 1987) adopted a Protocol setting limits to the production and consumption of substances which deplete the ozone layer. The Protocol entered into force on 1 January 1989 upon its ratification by the necessary number of signatories as laid down in the Protocol, which also provides for a secretariat located at UNEP to undertake certain tasks for its implementation.

As regards atmospheric carbon dioxide, WMO initiated in 1977, in collaboration with UNEP, a Research and Monitoring Project on Atmospheric Carbon Dioxide to study the trend in global concentrations of this gas and their possible impact on climate. The project has since been incorporated into the World Climate Programme.

Other WMO activities which have benefited from UNEP support include the Global Weather Experiment, through the provision of wind observing equipment to ships; the Tropical Cyclone Programme, through (*a*) the provision of tide gauges for the improvement of storm-surge predictions in the Bay of Bengal and (*b*) joint projects on early warning systems in countries in the Bay of Bengal and Arabian Sea areas and in Central America.

REGIONAL ECONOMIC COMMISSIONS

General remarks

WMO has maintained close co-operation with the Regional Economic Commissions established by ECOSOC with the aim of assisting in the economic and social development in their respective regions. The main fields of co-operation cover development of water resources, economic benefits of meteorological and hydrological services, environmental pollution and mitigation of damage caused by tropical cyclones.

Economic Commission for Africa (ECA)

Water resources

A joint WMO/ECA project for the study of major deficiencies in hydrological data in Africa, initiated in 1963, was completed in 1966. The two organizations undertook a joint project for the planning and development of hydrometeorological networks and related services in Africa, WMO providing a hydrometeorologist during 19711975 and two hydrologists for a follow-up project during 1980-1983, all the experts being based at the ECA Secretariat. About 30 countries benefited from the project.

WMO co-sponsored a regional meeting (Addis Ababa, 1986) on socio-economic and policy aspects of water resources management in Africa, organized by Unesco and ECA to highlight Africa's special water problems and to encourage water resource development in the light of the Action Plan of the UN Water Conference (Mar del Plata, Argentina, 1977) in which WMO took an active part, as was described earlier.

* See note on p. 8.

Training seminars and technical conferences

Several regional training seminars and technical conferences have been organized jointly by WMO and ECA, namely on Hydrometeorological Instruments and Methods of Observation and the Establishment of Hydrometeorological Networks in Africa (Addis Ababa, 1967), the Role of Meteorological Services in Economic Development in Africa (Ibadan, Nigeria, 1968) and Hydrology and Hydrometeorology in the Economic Development of Africa (Addis Ababa, 1971).

Resolution adopted by ECA on the WMO World Weather Watch Programme

ECA itself, at its ninth session (1969), adopted a resolution* emphasizing the beneficial impact of WWW on economic development in the region and inviting its Member States to give full support to the implementation of the WWW.

African Centre for Meteorological Applications for Development (ACMAD)

Following the droughts which have affected several countries in Africa, the Secretary-General of WMO took the initiative of proposing to the Conference of Ministers of ECA countries (1984) the establishment of a centre to monitor drought situations and to issue timely warnings and advice to African Governments. The proposal was readily accepted by ECA, which adopted Resolution 528 (XIX)—Meteorological services to combat drought in Africa. The establishment of the centre, known as the "African Centre for Meteorological Applications for Development (ACMAD)", is now at an advanced stage. The centre will occupy an important place within the Applications Component of the World Climate Programme.

Economic Commission for Europe (ECE)

Water problems

WMO participates in the annual meetings of the ECE Committee on Water Problems in Europe, which co-ordinates technical conferences and drafts principles on such subjects as trans-boundary waters, groundwater management and water-use statistics. Where appropriate, WMO contributes its expertise, and makes use of the resulting documentation in its own work. It also takes part in the annual inter-Secretariat meetings held under the auspices of ECE to co-ordinate water-related activities in Europe and at which representatives of interested organizations both within and outside the UN system are present.

Long-range transport of air pollutants in Europe

The problem posed by the transport of substantial quantities of man-made chemicals over large distances and across national boundaries gave rise within ECE to the Co-operative European Monitoring and Evaluation Programme (EMEP) of the Long-Range Transmission of Air Pollutants in Europe, a joint activity of ECE, UNEP and WMO. This co-operative programme, which arose from a decision adopted by the Conference on Security and Co-operation in Europe (Helsinki, 1975), was started in 1978 and WMO is responsible for the meteorological aspects of the programme. Two Meteorological Synthesizing Centres, one located in Moscow and the other in Oslo, undertake the calculation of trans-boundary fluxes for the evaluation of the long-range transport of pollutants and for testing the applicability and performance of mathematical transmission models. Appropriate stations in the WMO Background Air Pollution Monitoring Network (BAPMoN) also participate in the work.

The great practical importance of this co-operative programme needs no emphasis in the context of the wide concern over acid rain. The programme has been instrumental in bringing about a protocol, within the Convention on Long-range Trans-boundary Air Pollution, providing for a reduction by at least 30 per cent of emissions, or trans-boundary fluxes, of sulphur. The Executive Body for the Convention reviews the work carried out under the programme and makes recommendations thereon.

Economic value of national Meteorological Services

In 1967 ECE provided the services of a senior economist to WMO for a study of the assessment of the economic value of a national Meteorological Service, which was undertaken in connection with the planning of the World Weather Watch.

Economic and Social Commission for Asia and the Pacific (ESCAP)*

Mitigation of damage caused by tropical cyclones

A major field of co-operation between WMO and ESCAP relates to the joint WMO/ESCAP* Inter-governmental Typhoon Committee and the WMO/ESCAP Panel on Tropical Cyclones.

The WMO/ESCAP Panel on Tropical Cyclones, established in 1972, owes its origin to Resolution 2733 (XXV) adopted by the UN General Assembly in 1970.

Water resources

WMO's collaboration with ECAFE in the field of water resources began in 1954 with a joint project for the study of the major deficiencies in hydrological data in the ECAFE region. The report on the study, prepared jointly by the WMO and ECAFE Secretariats and submitted to the third ECAFE Regional Technical Conference on Water Resources Development (Manila, 1957), highlighted the inadequacy of the existing networks of hydrological stations in the region and called attention to certain basic considerations to be borne in mind in planning an extension to the networks.

A series of inter-regional seminars was organized jointly by WMO and ECAFE. These were on Hydrological Networks and Methods (Bangkok, 1959), Field Methods and Equipment Used in Hydrology and Hydrometeorology (Bangkok, 1961), Methods of Hydrological Forecasting for the Utilization of Water Resources (Bangkok, 1964) and Assessment of the Magnitude and Frequency of Flood Flows (Bangkok, 1966).

During 1957–1960, WMO participated, together with several other organizations, in the major ECAFE project in the field of flood control and water resources, for the development of the Lower Mekong River Basin. Collaboration was also extended to ECAFE in the preparation of the "Glossary of hydrologic terms used in Asia and the Far East", published in 1957.

Role of meteorology in economic development

A Conference on the Role of Meteorological Services in Economic Development in Asia and the South-West Pacific was organized by WMO in co-operation with ECAFE (Bangkok, 1973).

Economic Commission for Latin America and the Caribbean (ECLAC)**

Water resources

For over ten years, commencing in 1957, WMO participated in the work of the Water Resources Survey Group of ECLA** by providing an expert in hydrometeorology. The work of this group

* ECA Resolution 196 (IX)—World Weather Watch and economic development in Africa

*The Economic Commission for Asia and the Far East (ECAFE) was in 1974 renamed the Economic and Social Commission for Asia and the Pacific (ESCAP).

**The Economic Commission for Latin America (ECLA) was in 1983 renamed the Economic Commission for Latin America and the Caribbean (ECLAC) led to large Special Fund projects in several countries, with WMO as the executing agency, for the improvement of meteorological and hydrological networks.

WMO has been collaborating with the ECLAC Committee on Water set up in 1979 for promoting the implementation of the Mar del Plata Action Plan (see page 144) in Latin America.

Role of meteorology in economic development

WMO organized, in co-operation with ECLA, a Technical Conference on the Role of Meteorology in Economic Development in Latin America (Santiago, Chile, 1970).

Economic Commission for Western Asia (ECWA)*

WMO presented a paper on "Agroclimatic aspects of agricultural resources conservation" to the ECWA Expert Meeting on the Management, Conservation and Development of Agricultural Resources in the ECWA Region (Damascus, 1981).

INTERNATIONAL ATOMIC ENERGY AGENCY (IAEA)

Measurement of atmospheric radioactivity

WMO collaborated with IAEA in establishing a scheme for measuring atmospheric radioactivity and disseminating the data, in response to a request made by the UN General Assembly in 1961.

World-wide distribution of hydrogen and oxygen isotopes in precipitation

WMO extended its collaboration to IAEA in a project, started in 1961, for a world-wide survey of the concentration of hydrogen and oxygen isotopes in precipitation from a global network of stations, under IAEA's responsibility for the analysis of the samples. The data, published regularly by IAEA, have been widely used, both in research and application, by hydrologists, hydrometeorologists and others all over the world. WMO's collaboration with IAEA in the use of isotope techniques in hydrology should also be mentioned.

Siting and operation of nuclear power plants

As early as 1960 WMO published a *Technical Note* on the meteorological aspects of the safety and location of reactor plants, as part of the activities of the WMO EC Panel of Experts on the Peaceful Uses of Atomic Energy. With the installation over the years of more and more nuclear power plants, WMO and IAEA have collaborated in preparing guidance material on relevant problems associated with their siting and operation. In order to provide meteorological and hydrological advice on this subject, WMO prepared, with the co-operation of IAEA, a *Technical Note* in two volumes entitled "Meteorological and hydrological aspects of siting and operation of nuclear power plants" (*Technical Note* No. 170). Part I is devoted to meteorological aspects, while Part II is devoted to hydrological aspects.

IAEA, on its side, has produced safety codes and guides on various aspects of nuclear power plant siting, design and operation. WMO has provided its expertise to IAEA in the preparation of this material.

IAEA Convention on Early Notification of a Nuclear Accident

As early as 1960–61 WMO had published *Technical Notes* dealing with (*a*) the transport of radioactive effluent from reactors both in routine operation and as a result of instantaneous discharge occurring in an accident (*Technical Note* No. 33) and (*b*) the meteorological factors influencing the transport and removal of radioactive debris (*Technical Note* No. 43).

Following the accident at the nuclear power station at Chernobyl (Ukrainian SSR) in April 1986, WMO was among the international organizations invited by IAEA to participate in activities leading to international agreements that might help to mitigate the consequences of an accidental release of radioactive material across national borders. Two Conventions adopted by a special session of the IAEA General Conference (September, 1986) came into force in October 1986; of these, the Convention on Early Notification of a Nuclear Accident is particularly relevant to the responsibilities of WMO. It stipulates that States which are, or may be, physically affected must be notified of any accident and relevant information made available promptly. Meteorological and hydrological data and forecasts needed to predict trans-boundary flows of radioactive material are part of the information to be provided to contact points designated by IAEA. For WMO, the application of this provision involves both scientific and operational aspects. Consultations and joint meetings between WMO and IAEA as well as planning for the establishment of the necessary procedures within WMO commenced promptly, particularly with regard to the details of the information to be provided at different stages to meet the relevant provisions of the Convention and the requirements of data for atmospheric modelling aspects in order to arrive at the best possible predictions. Close and active collaboration with IAEA as well as with other organizations, especially UNEP and WHO, will need to be maintained in this activity.

FOOD AND AGRICULTURE ORGANIZATION OF THE UNITED NATIONS (FAO)

Resolutions of the FAO Conference

The importance of meteorology in the development of agriculture and the consequent need for co-ordination between agricultural and meteorological services at the national level on the one hand, and for close co-operation between FAO and WMO at the international level on the other, have been reflected in two resolutions adopted by the FAO Conference, the policy-making body of the Organization.

As early as in 1954, the seventh session of the FAO Conference adopted the following resolution:

> **Resolution 36—Co-operation between agricultural and meteorological services**
>
> *The Conference,*
>
> *Noting* the work undertaken by WMO in the application of meteorology to agriculture and food production throughout the world,
>
> *Recommends* that Member Governments should establish co-ordination between agricultural and meteorological services at the national level to assure the practical implementation of their programmes of work.

Again, a resolution was adopted by the FAO Conference at its nineteenth session (1977) calling in strong terms for the closest collaboration between WMO and FAO.

Inter-Agency activity in agroclimatology

WMO, FAO and Unesco have together carried out, since 1960, a series of agroclimatological surveys in different regions of the world, with the aim of assessing their agricultural potential and developing methodologies for the application of agroclimatological data for increasing agricultural production. Of the five

* Since 1985 Economic and Social Council for Western Asia (ESCWA)

such surveys carried out so far, three were in accordance with the plans and recommendations of the WMO/FAO/Unesco Inter-Agency Group on Agricultural Biometeorology, established in 1968 at the initiative of WMO.

In order to take follow-up action on the first four surveys, technical conferences with the participation of agronomists and meteorologists from the countries concerned, as well as international experts, were held respectively in Beirut, (1964), Dakar (1971), Nairobi (1973) and Bogotá (1978). Corresponding action in respect of the survey in the South-East Asia Region is being planned.

Locust control

WMO's collaboration with FAO in anti-locust operations goes back to 1960 when it became closely associated with the UN Special Fund Desert Locust Project with FAO as the executing agency and involving some 20 countries. The project had important meteorological aspects, such as climatological studies, organization of meteorological observations and training and preparation of synoptic maps.

The menace of locusts has remained a matter for much concern and at the request of FAO, WMO arranged for the supply of necessary meteorological information for locust control operations in the Red Sea and the Horn of Africa areas in 1978.

FAO Global Information and Early Warning System on Food and Agriculture

At the request of FAO, in 1978 WMO arranged for the provision, by the Meteorological and Hydrological Services concerned, of the necessary meteorological and climatological information as input to the FAO Early Warning System.

UNITED NATIONS EDUCATIONAL, SCIENTIFIC AND CULTURAL ORGANIZATION (Unesco)

WMO's activities are of a scientific and technical nature, while Unesco's programmes embrace the fields of science and education. Naturally, close and effective co-operation has developed between the two organizations, as programmes and activities of common interest have been undertaken by them over the years. While WMO's collaboration with Unesco in hydrology and in the work of the Intergovernmental Oceanographic Commission (IOC) set up by Unesco is well known, examples of co-operation in other areas can also be given. The major aspects of co-operation between WMO and Unesco are described below.

Filmstrips and slides of cloud pictures

A highly successful and particularly useful joint WMO/Unesco project was the production, in 1959, of filmstrips and slides showing cloud pictures for educational purposes. The material comprised (*a*) reproductions, in the form of high-quality colour filmstrips and colour slides, of a set of 76 photographs selected from some 225 plates contained in WMO's *International Cloud Atlas* showing various types of clouds; and (*b*) an explanatory booklet with texts to serve as commentary notes for the lecturer demonstrating the cloud types.

International Indian Ocean Expedition (IIOE)

An International Indian Ocean Expedition (IIOE), sponsored by Unesco and supported by WMO, was carried out during the five-year period 1961-1965 with the participation of some 40 ships from 15 countries for an intensive multi-disciplinary investigation of the Indian Ocean, which included a comprehensive meteorological programme. An International Meteorological Centre, established in Bombay as part of a UN Special Fund Project with the WMO as the executing agency, was in operation during 1962–1965 to provide special meteorological services to the expedition ships, analyse the meteorological data, undertake research on meteorological problems and to provide training facilities for students from other countries. A symposium on the meteorological results of the expedition was held in Bombay in 1965 under the joint sponsorship of Unesco and WMO.

Hydrology

WMO played a very active role in the International Hydrological Decade (IHD), a programme carried out on a world-wide scale during 1965–1974 under the aegis of Unesco. WMO's contribution dealt with scientific and practical aspects of network planning, establishment of stations, provision of guidance material on instrumentation and observational methods, standardization of records, exchange and publication of data, methods of analysis, standardization of terminology, hydrological forecasting and training of personnel.

Oceanography and marine sciences

Intergovernmental Oceanographic Commission (IOC)

WMO's collaboration with Unesco in Oceanography and Marine Sciences takes place through the Intergovernmental Oceanographic Commission (IOC), a body set up by Unesco in 1960 with a somewhat independent status. WMO became closely involved in IOC's work right from the beginning, participating in its working groups dealing with oceanographic stations, development of codes for transmission of data and telecommunications needs. WMO and IOC later developed a joint programme entitled "Integrated Global Ocean Services System (IGOSS)"* which was instituted in 1977 and has since been twice revised and updated, in 1982 and 1986. IGOSS is a global programme for the acquisition and exchange of ocean data and the dissemination of oceanographic products and services to governmental and commercial (e.g. shipping and fisheries) sectors as well as academic and research circles. Operationally, IGOSS is similar to the WWW programme and comprises (*a*) an observing system; (*b*) a data-processing and services system; and (*c*) telecommunication arrangements to ensure the rapid and reliable collection and exchange of data as well as of processed information, the WWW Global Telecommunication System being used for the exchange of these data and products.

WMO's participation in the meteorological (air-sea interaction) aspects of the IOC-sponsored international investigation of the *El Niño* phenomenon, greatly facilitated by the IGOSS programme, deserves particular mention. The economic implications of this phenomenon are so great that the alarming decline in the catches of anchovy off the Peruvian coast in 1972 led the Peruvian Government to request IOC to organize jointly with FAO and WMO a scientific study of the problem.

Another important field of collaboration between WMO and IOC is that of marine pollution studies, reinforced by the recommendation of the UN Conference on the Human Environment (1972) that the two organizations should promote the monitoring of marine pollution. Substantial support is given by UNEP in this work.

*The initial title "Integrated Global Ocean Station System" was changed to its present title in 1983 to reflect better its service-oriented character.

Other fields of co-operation with Unesco

Among other fields of mutual co-operation between WMO and Unesco, special mention may be made of WMO's collaboration in Unesco's interdisciplinary scientific programme known as the Man and the Biosphere Programme, which commenced in 1971, and of Unesco's collaboration with WMO in the publication of several climatic atlases (for Asia, Europe, North and Central America and South America).

WORLD HEALTH ORGANIZATION (WHO)

Air pollution studies

WHO's concern about the problem of air pollution and its effects on human health, and the important role of meteorology in the study of this problem have led to a close and long-standing collaboration between WMO and WHO in this field. In 1957 WMO was invited to participate in the work of the WHO Expert Committee on Air Pollution and in 1965 in the work of the WHO Scientific Group on Identification and Measurement of Air Pollutants.

The two organizations sponsored jointly, with support from UNEP, a Technical Conference on the Observation and Measurement of Atmospheric Pollution (Helsinki, 1973). Nearly 220 participants from 48 countries, comprising meteorologists, physicists, chemists and public health personnel, were given the opportunity to discuss requirements for the measurement of air pollution on the local, regional and global scales, as well as techniques for the standardization of measurements.

Climate and human health

The study of the effects of weather and climate on human health constitutes one of the challenging areas of meteorology. Collaboration in this field between WMO and WHO, initiated in 1973 through a working group of the Commission for Special Applications of Meteorology and Climatology (later renamed Commission for Climatology), has developed steadily. In 1983 WMO organized a meeting of experts on climate and human health with the participation of WHO. This was followed by a WMO/WHO/UNEP Symposium on Climate and Human Health in Leningrad in 1986.

Other important WMO meetings with human health aspects held with the active collaboration of WHO include the Technical Conference on Urban Climatology and its Applications with Special Regard to Tropical Areas (Mexico City, 1984).

Global water-quality monitoring

The growing demand for water throughout the world on the one hand and a deterioration in the cleanliness and safety of many of the sources of water on the other led to a WHO Global Water-quality Monitoring Project in 1976 within the GEMS of UNEP and in co-operation with Unesco and WMO. The general responsibility of WMO consists in the design of the global network of monitoring stations, choice of parameters to be observed, techniques of measurement and data handling and processing.

INTERNATIONAL CIVIL AVIATION ORGANIZATION (ICAO)

WMO and ICAO maintain close and constant co-operation in keeping under review the requirements of meteorological services for aviation and in the adoption of procedures for the provision of these services, as well as keeping them up to date. The ICAO/WMO Working Arrangements, a comprehensive document approved in 1953, provide the basis for collaboration between the two organizations for dealing with all matters of common interest.

INTERNATIONAL TELECOMMUNICATION UNION (ITU)

Meteorological telecommunications

One of the most essential and permanent tasks of WMO and of its predecessor IMO has been to arrange for an efficient international exchange of weather reports on a round-the-clock basis. Indeed, such is the importance of the rapid collection and exchange of weather reports that meteorological telecommunication has been considered as the life blood of synoptic meteorology. Collaboration with ITU, an organization now over 120 years old, is therefore of long standing, going back to the days of WMO's predecessor, IMO, as stated in Chapter II.

The twelfth Conference of Directors of IMO (Washington, 1947), which drew up and adopted the Convention creating WMO, recommended closer collaboration with ITU, particularly for the collection of weather observations from the oceans, with regard to requirements of frequency for maritime purposes and generally to further meteorological interests in the field of telecommunications. These considerations seem to have proved almost prophetic in the light of the spectacular developments in the systems and techniques for marine observations and data collection. To meet the needs of WMO and IOC, the World Administrative Radio Conferences, convened by ITU, have allocated high-frequency bands for the collection of ocean data from buoys as well as from research ships for both meteorological and oceanographic purposes on a world-wide scale, the utilization of these radio frequencies by the Member States being co-ordinated by WMO and IOC.

The steady development of the Global Telecommunication System of the World Weather Watch, necessitating wider and faster exchange of increasing data, has been due as much to the use of advanced technology as to the fruits of continuous collaboration with ITU. Particular mention may be made of standardization of facsimile apparatus for transmission of weather maps.

World distribution of thunderstorm days

Soon after its creation, WMO received a request from ITU in May 1951 on behalf of the CCIR for world maps of thunderstorm activity, since the CCIR was interested in accurate data concerning the frequency and distribution of thunderstorm activity throughout the world. The WMO Secretariat assembled a large quantity of statistical data from Meteorological Services all over the world, including marine data extracted from several millions of punch-cards, and after analysis of the data, issued a publication entitled "World distribution of thunderstorm days" in two parts. Part 1, issued in 1953, contained tables of mean monthly, seasonal and annual numbers of thunderstorm days for land stations. Part 2, issued in 1955, contained corresponding tables of marine data and a series of 17 world maps showing the monthly, seasonal and annual distribution of thunderstorm days.

INTERNATIONAL MARITIME ORGANIZATION (IMO)*

WMO's long tradition of activities in marine meteorology and IMO's responsibility in matters concerning shipping naturally led

*The Inter-Governmental Maritime Consultative Organization (IMCO) formally changed its title to International Maritime Organization with effect from 22 May 1982.

to close collaboration between them when the latter started its work in 1959. Subsequently, the greatly increased international interest in marine science and its applications as manifested within UN and affecting both organizations has been another factor in the further development of their collaboration.

One of the primary objectives of IMO is to ensure safety at sea. A fruitful result of WMO's co-operation with IMO in the field of safety at sea, carried out through the latter's Maritime Safety Committee and its sub-committees, is the text concerning meteorological support to maritime search and rescue operations for inclusion in a technical annex to the International Convention on Maritime Search and Rescue and in the IMO Search and Rescue Manual. On the WMO side, international procedures for the provision of meteorological information in search and rescue operations to co-ordinating centres have been included in the *Manual on marine meteorological services*, a publication having the status of WMO Technical Regulations.

WMO has participated in planning the Global Maritime Distress and Safety System, now at an advanced stage within IMO, requiring navigational and meteorological warnings to ships by automated direct printing service.

CONCLUDING REMARKS

In the period of some four decades since WMO came into being as a Specialized Agency of the United Nations with a responsibility to collaborate with the latter and its associated organizations, WMO's collaboration has been sought by them in a wide variety of fields and has been extended fully and effectively in all cases. The United Nations and the other organizations concerned, on their part, have also extended similar collaboration to WMO and have assisted in its activities, which has not only contributed greatly to progress in WMO's work, but has also had a profound impact, in some cases, on its programmes and activities.

CHAPTER XVII

ORGANIZATIONAL MACHINERY

Introductory remarks

The experience gained by WMO's predecessor, the non-governmental IMO, provided a valuable basis on which to formulate the structure of the new Organization. Of particular significance was the development by IMO of a system which ensured full and effective international co-operation in the science and practice of meteorology in all its many aspects. It was a system which was operated by the directors of national Meteorological Services who collectively constituted the highest body of IMO—the Conference of Directors. It was, moreover, a very economical system since the directors gave their services and support without cost to the Organization and, for the first 55 years of its existence, without even the assistance of a Secretariat. Another important feature was that the directors of Meteorological Services of territories which were not independent States also participated fully in the work of IMO. Thus a highly effective, economical and wide-ranging system of international co-operation was already in operation when the decision to create WMO was taken.

The transformation of IMO into a governmental body clearly required some changes. The new status of the Organization was such that ultimate authority on all matters had evidently to be vested in the Member States rather than the directors of Meteorological Services. Nevertheless, the lessons learnt and the experience gained from the earlier system were well recognized at the governmental level and were fully taken into account when the Convention and the General Regulations of WMO were drawn up. More detailed information on this subject is given later in this chapter. Suffice it, at this stage, to say that the system developed by IMO was maintained with only essential adjustments. The structure given to WMO was basically similar to that of its predecessor which had stood the test of time. Moreover, the basic scientific and technical character of the highly successful non-governmental IMO was preserved in its successor, a governmental body, through a number of significant provisions in the Convention and the General Regulations of WMO (see the later sections dealing with Congress, the Executive Council and Regional Associations respectively).

As a result, the new Organization was able to function smoothly and efficiently from the outset—and this, in turn, ensured that it was in a position to take full advantage of the unprecedented scientific and technological advances which were soon to follow.

The Convention

The essential elements of the structure of WMO are prescribed in its basic document—the Convention. The text of the Convention is given in Appendix A, together with explanations of the relatively few amendments which have been made to the text since it was first approved. The main amendments refer to the increased role of WMO in the field of international hydrology (Seventh Congress, 1975), the progressive increase (see page 152) in the number of members of the executive body (Third Congress, 1959), Fourth Congress (1963), Fifth Congress (1967), Eighth Congress (1979) and Ninth Congress (1983) and the change in its title from "Executive Committee" to "Executive Council" (Ninth Congress (1983).

As will be seen from Appendix A, the Convention defines the conditions of Membership and specifies the various bodies which comprise the Organization, namely:

(1) The Congress;

(2) The Executive Council;

(3) The Regional Associations;

(4) The Technical Commissions;

(5) The Secretariat.

Each of these items is relevant to the present chapter and further information on each therefore follows, starting with the Membership of the Organization.

Membership

Membership of the Organization is defined in Article 3 of the Convention (Appendix A) from which it will be seen that it is not confined to independent States but includes also territories or groups of territories maintaining their own Meteorological Services. In this respect, WMO follows the system developed by IMO, thereby giving due recognition to the importance, from the meteorological point of view, of co-operation with such territories and to the part which they can play in international meteorology. With the exception of the four subjects specified in Article 11 of the Convention on which only Member States may vote, all Members have equal rights.

When the WMO Convention entered into force, on 23 March 1950, the Organization consisted of 52 Members (32 States and 20 Territories or groups of Territories). The Membership has now (April 1990) reached the impressive number of 161 (156 States and five Territories), showing that it can truly be called a world-wide organization.

The substantial changes in the Membership since the creation of WMO, as regards both the total number of Members and the numbers in each of the two categories of Membership, are presented in the table below, showing the Membership at the time of each Congress. The numbers in brackets indicate the representation at the particular Congress.

Congress		Membership		
		Total	States	Territories
I	(1951)	66 (54)	46 (42)	20 (12)
II	(1955)	88 (83)	62 (62)	26 (21)
III	(1959)	101 (88)	78 (69)	23 (19)
IV	(1963)	120 (102)	105 (92)	15 (10)
V	(1967)	129 (113)	117 (106)	12 (7)
VI	(1971)	135 (123)	122 (114)	13 (9)
VII	(1975)	141 (130)	129 (120)	12 (10)
VIII	(1979)	149 (128)	143 (123)	6 (5)
IX	(1983)	157 (138)	152 (134)	5 (4)
X	(1987)	160 (138)	155 (134)	5 (4)

These figures indicate directly the rapid increase in Membership and reflect indirectly the progress achieved in dependent territories becoming independent States. Due no doubt to their association with WMO in their former status, many of them acceded promptly to the Convention, becoming Members of the Organization in their new capacity. The present list of Members of the Organization is given in Appendix C.

The recognition in 1972 of the People's Republic of China as the legitimate authority of that country was a very important decision. It is of interest to note that following such recognition first by the United Nations and subsequently by Specialized Agencies including WMO, the Secretary-General of WMO was in fact the first executive head of the UN system to be officially invited to China. The formal notification of recognition was sent from WMO to the Government of the People's Republic of China on 25 February 1972. The speed of subsequent developments is shown by the fact that less than one month later, the Secretary-General was leaving China after a visit of one week's duration. This early interest on the part of the People's Republic of China in the work of WMO has since been followed by full and active co-operation and, as will be seen from Appendix D, the head of the State Meteorological Administration of that country was in fact elected to the high office of President of the Organization in 1987.

Congress

Congress is specified in the Convention as "the general assembly of delegates representing Members and as such is the supreme body of the Organization". It is worth noting that the Convention also prescribes that each Member should designate as principal delegate to Congress the Director of its Meteorological or Hydrological Service showing thereby the importance of the scientific and technical aspects of the work of the highest decision-making body of the Organization. The responsibilities of Congress are likewise contained in the Convention: they may be said to cover all major aspects of the work of the Organization, but by far the most important task of each Congress is to adopt the policy, programme and budget for a four-year period beginning with the year immediately following the Congress. A full explanation of how First Congress successfully set WMO on its course is given in Chapter II. The remarkable overall progress made by WMO since then is recorded in subsequent chapters which illustrate how well successive Congresses have fulfilled their respective responsibilities.

Congress is required to meet at "intervals as near as possible to four years", a requirement which has been met from the outset. As regards the venues of such gatherings, with one exception all Congresses have been convened in Geneva, the exception being the First Congress (1951), which was held in Paris. Information on the number of delegations attending each Congress has already been given in the previous section.

The Convention provides for the establishment by Congress of the General, Technical, Financial and Staff Regulations. Some information in this respect is given later in this chapter.

An important responsibility of Congress is to elect the President and Vice-Presidents of the Organization and the Members of the Executive Council other than the Presidents of Regional Associations*, who then hold office until the following Congress. In the intervening period the President and the Vice-Presidents of the Organization act in these same capacities in the Executive Council.

*The Presidents of Regional Associations, who are also members of the Executive Council, are elected in the former capacity by the respective Associations.

Among the provisions in the Convention reflecting the scientific and technical character of the Organization is one which stipulates that elegibility to all the offices mentioned above is confined to Directors of Meteorological or Hydrometeorological Services of Members of the Organization. The General Regulations accordingly provide for arrangements to replace those holding these offices when they cease to be Directors of their national Services, without having to wait for the next session of the constituent body concerned.

The President and the Vice-Presidents of the Organization play important roles in conducting its affairs. The President plays a particularly important role in the period between successive sessions of Congress and of the Executive Council in view of the authority vested in him by the Regulations. The names of the Presidents and the period for which each has served in this capacity are shown in Appendix D. It may be added that such is the respect and esteem in which the Presidents are held that, at the termination of the term of office of each, an oil portrait is mounted in the Secretariat. Their predecessors in IMO have been similarly honoured. These portraits thus serve as a constant reminder of the long history of the international co-operation in meteorology and of the many distinguished scientists who have been chosen to serve as leaders of such co-operation.

Another important responsibility of Congress is to elect the Secretary-General for the coming four-year period. The names of the Secretaries-General so selected are also shown in Appendix D. A brief note on each is also given in this chapter. In their honour, their portraits also hang in the Secretariat.

Executive Council

As the name indicates this is the main executive body of the Organization. Its responsibilities therefore cover the whole wide range of WMO activities. It has always comprised a specified number of directors of national Services, duly elected by Congress, taking into account the principles of regional representation embodied in the Convention, as well as the presidents of all the Regional Associations. All members of the Executive Council are elected in their personal capacity. Presidents of Technical Commissions, regional hydrological advisers (see next section) and others involved in the work of the Organization are also invited to attend individual sessions as appropriate. Thus, as with Congress, the record of achievement of the Organization is an indication of the good work which the executive body has accomplished.

The increasing scale of work undertaken by the Executive Council and the increase in the number of Member countries of the Organization have been reflected in the increase in the number of elected members of this body. The original total membership was 15. This number was increased to 18 in 1959 by Third Congress, then to 21 in 1963 by Fourth Congress, to 24 (including a Third Vice-President) by Fifth Congress (1967), to 29 by Eight Congress (1979) and finally to 36 by Ninth Congress (1973). The principle of regional representation was appropriately maintained at each of these occasions.

The Convention stipulates that the Executive Council "shall normally hold a session at least once a year" and this rule has always been followed. In 1951 two sessions were held—the first immediately after First Congress (Paris, March–April 1951) and the second later in the same year (Lausanne, October 1951). The reason for holding the latter only a few months after First Congress was that many decisions had to be taken and many arrangements made, following the creation of WMO as a new governmental body. The session was held in Lausanne simply because this was where the Secretariat of IMO was based at the time when the First Congress was held and the WMO Secretariat was installed there on a temporary basis; arrangements for implementing the Congress decision that the Secretariat should be located in Geneva had still to be made.

The multifarious functions of the Executive Council are given in the Convention, reproduced as Appendix A. These may be broadly summarized as:

- Implementing the decisions of Congress within the financial resources made available;
- Keeping under review all developments which directly or indirectly have a bearing on the effectiveness of the Organization;
- Submitting proposals and recommendations to the subsequent Congress, including the proposed programme and budget for the coming four-year period, on the basis of the Secretary-General's proposals in this respect.

Among the other duties falling to this body, mention must be made of international awards made by the Organization on an annual basis. The most significant of these is the IMO Prize, considered as the highest award by the Organization. As the name indicates, the award was established in honour of IMO, which initiated international collaboration in meteorology in 1873, the cost of the Prize itself, consisting of a gold medal, a parchment scroll and a cash award, being met from the surplus funds of IMO which were taken over by WMO. The arrangements for this award were established by the Executive Council* at its eighth session (1956) on the basis of a decision of Second Congress (1955). They stipulate that "in the selection of a recipient, both scientific eminence and the record of work done in the field of international meteorological organization should be taken into consideration". The first award was made in 1956 to Dr Th. Hesselberg (Norway), a distinguished meteorologist who had played a highly important role in IMO and in the arrangements for the creation of WMO. It is, therefore, very appropriate that his name should head the full list of recipients given in Appendix G.

Mention must also be made of three other awards for which the selection is made by the Executive Council. The WMO Research Award to encourage young scientists working in the field of meteorology, instituted in accordance with a decision of Fifth Congress (1967), has been made annually since 1970. At its thirty-seventh session (1985) the Executive Council established an annual prize for the best scientific paper on meteorological instruments and methods of observation and the first award was made in 1986. The prize, financed from a trust fund, is known as the Professor Dr Vilho Vaisala Award. At its thirty-ninth session (1987) the Council instituted the annual Norbert Gerbier-Mumm Award, financed from an external source, for "an original scientific paper on the influence of meteorology, in a particular field of the physical, natural or human sciences, or conversely on the influence of one of those sciences on meteorology and to stimulate interest in such research, with a view to supporting WMO programmes". The first award was at the Council's fortieth session (1988).

The Executive Council is also responsible for making the arrangements for the IMO Lecture, delivered at each Congress in accordance with a decision of Fourth Congress (1963). The lecture, given by an acknowledged expert in some branch of meteorology, is a condensed version of a study made by him on a specified topic in that field, the text of the study being published as a monograph by the Organization. The IMO Lecture has now become an important feature of each Congress, the first lecture (delivered at Fifth Congress 1967) having been followed by five other lectures at the subsequent quadrennial sessions of Congress. The monographs in this series are referred to in the section on WMO publications.

Presentation of the WMO Research Award for Young Scientists to Mr Quah Leong-Chuan, of Malaysia, by Professor G. O. P. Obasi, Secretary-General of WMO

As already mentioned, the Executive Council meets at yearly intervals. Experience soon showed however that discussions between the President and Vice-Presidents on the one hand and the Secretary-General on the other were necessary between sessions of the Council. This was because of the need to review the progress being made in implementing decisions of the previous session on urgent questions and to consider any unexpected developments calling for prompt attention. The first such meeting was held in 1956 and since then they have become a regular feature of the work of the Organization. The President and Vice-Presidents became known as the Bureau and their gatherings as Bureau meetings. As the activities of the Organization expanded, particularly with the introduction of the World Weather Watch Programme with its appointed World Centres, it was found necessary to include in the Bureau meetings representatives of the World Centres if they were not already represented by the President or a Vice-President. This procedure was introduced in 1971 and has continued since then, but with additional participants as appropriate.

*See note on p. 8.

It is of interest to note that while Bureau meetings are now accepted as an essential supplement to the full sessions of the Executive Council, no specific provision has been made by Congress at any time for the formal establishment of the Bureau in the Convention or the General Regulations. There was, however, a general consensus at Ninth Congress (1983) that the role and composition of the Bureau should continue to be as at present.

Regional Associations

The system of regional bodies was introduced by IMO in 1935 to provide for co-operation among Directors of Meteorological Services in a given geographical region in dealing with special problems of a regional nature. These bodies were termed Regional Commissions. Their value soon became apparent and this was fully recognized when later the WMO Convention was being drawn up. The Convention thus contained provision for the establishment of such bodies, the term Regional Associations then being used. The broad functions of these Associations were laid down in the Convention, while the more detailed arrangements were stipulated in the General Regulations originally approved by First Congress (1951) and reconfirmed without change by all subsequent Congresses. The General Regulations provide for six Regional Associations—for Africa; Asia; South America; North and Central America; South-West Pacific; and Europe.

The General Regulations specify the geographical limits of each Region, as shown in the map in Chapter XV. The WMO Regions, unlike their predecessors under IMO, include ocean as well as land areas. The president of each Regional Association is a member of the Executive Council, as already explained, and one of his duties is to present the views of his Association to the Council as well as to Congress. The General Regulations stipulate that the presidents of the Regional Associations shall be Directors of Meteorological or Hydrometeorological Services and provision exists in the Regulations for their replacement at any time when they cease to be so. Another provision which was included in these Regulations by Tenth Congress (1987) is that each president of a Regional Association should be assisted by a regional hydrological adviser, who should be a representative of a Service responsible for operational hydrology. The fact that the main decisions of First Congress have remained unchanged demonstrates clearly how well the system has served the purposes of the Organization. The information given in the preceding chapters gives similar confirmation. The names of those who have been elected to serve as presidents of the Regional Associations, together with the dates of their terms of office, are given in Appendix E.

The Regional Associations have, as prescribed in the General Regulations, met regularly and the dates and places of all such sessions are also shown in Appendix E.

The present Membership of the Regional Associations is shown below.

Regional Association	Number of Members
Regional Association I (Africa)	53
Regional Association II (Asia)	30
Regional Association III (South America)	13
Regional Association IV (North and Central America)	24
Regional Association V (South-West Pacific)	16
Regional Association VI (Europe)	36

It should be noted that the southern limit of each of the Regional Associations I, III and V is latitude 60°S and for this reason special arrangements have had to be made for the Antarctic, as discussed on page 138.

Technical Commissions

The concept of international groups of meteorologists each dealing with a specialized aspect of their science or its applications emerged at a very early stage in the life of WMO's predecessor, IMO. Indeed it may be said that the need for such groups was one of the basic factors leading to the decision to establish an international organization for meteorology. The system was applied with increasing benefit throughout the life of IMO and it is therefore not surprising that, when the Convention of WMO was being drafted, specific provision was made for the creation of Technical Commissions. The members of these Commissions, who are experts in the fields covered by their respective terms of reference and are designated by their respective governments, offer their services on a voluntary basis.

First Congress (1951) established the following eight Technical Commissions:

- Commission for Bibliography and Publications (CBP);
- Commission for Instruments and Methods of Observation (CIMO);
- Commission for Aerology (CAe);
- Commission for Climatology (CCl);
- Commission for Agricultural Meteorology (CAgM);
- Commission for Maritime Meteorology (CMM);
- Commission for Synoptic Meteorology (CSM);
- Commission for Aeronautical Meteorology (CAeM).

First Congress also laid down the terms of reference of the Commissions except that in respect of CAeM it was decided that the terms of reference would be finalized taking into account the outcome of the negotiations which were to be undertaken for a working arrangement or agreement between WMO and ICAO. The final text of the terms of reference of CAeM was subsequently approved by Second Congress (1955).

The structure and terms of reference of the Technical Commissions have been reviewed by each successive Congress and adjustments have been approved when necessary. Thus at Third Congress (1959) it was decided to discontinue CBP and to replace it by a panel of experts. It may, however, be mentioned that work on two specific items entrusted to CBP was virtually completed by 1959 when a provisional version of the International Meteorological Vocabulary was distributed and a revised version of the relevant section (551.5) of the Universal Decimal Classification was included in the WMO Technical Regulations. At the same Congress a new Commission for Hydrological Meteorology (CHM) was established as a result of the developments described in Chapters IX and XVI.

Fourth Congress (1963) revised completely the terms of reference of CHM and renamed it as Commission for Hydrometeorology (CHy). The terms of reference of the other technical Commissions were also amended.

Fifth Congress (1967) requested the Executive Council* to arrange for a thorough study of the organization of the scientific and technical work of WMO and laid down extensive guidelines for this study. In the interim Congress expanded the terms of

*See note on p. 8.

reference of CSM to enable it to be more closely associated with WWW activities. The same Congress also renamed the Commission for Aerology (CAe) as the Commission for Atmospheric Sciences (CAS).

Sixth Congress (1971), which made an extensive review of the system of Technical Commissions in the light of the study carried out by the Executive Council* at the request of Fifth Congress, classified the Commissions as "Basic Commissions" and "Applications Commissions". Furthermore, the Commissions for Synoptic Meteorology, Maritime Meteorology, Hydrometeorology and Climatology were renamed respectively as Commission for Basic Systems (CBS), Commission for Marine Meteorology (CMM), Commission for Hydrology (CHy) and Commission for Special Applications of Meteorology and Climatology (CoSAMC). CBS, CIMO and CAS were placed under "Basic Commissions" while CAeM, CAgM, CMM, CHy and CoSAMC were placed under "Applications Commissions". The terms of reference of all the Commissions were also modified.

The eight Technical Commissions and their classification in two groups have been retained by the subsequent sessions of Congress, in particular by Ninth Congress (1983), which considered several alternative structures of the system of Technical Commissions, proposed by an EC panel set up in accordance with a decision of Eighth Congress (1979). While retaining the existing structure, Ninth Congress revised substantially the terms of reference of the Commissions. Other changes during the period from the Seventh to the Tenth Congress are mentioned below.

Tenth Congress (1987) added the Commission for Hydrology (CHy) to those under the "Basic Commissions". Also, the name of the Commission for Special Applications of Meteorology and Climatology (CoSAMC) was changed twice, being renamed as Commission for Climatology and Applications of Meteorology (CCAM) by Eighth Congress (1979) and subsequently as Commission for Climatology (CCl) by Ninth Congress (1983). A small but important change in the terms of reference of CHy was made by Eighth Congress (1979), while those of CAeM and CAgM were further amended by Tenth Congress (1987).

The present structure of the commissions is as follows:

I. *Basic Commissions*

Commission	Area
• Commission for Basic Systems (CBS)	Basic operations and facilities
• Commission for Instruments and Methods of Observation (CIMO)	
• Commission for Hydrology (CHy)	
• Commission for Atmospheric Sciences (CAS)	Research in atmospheric sciences

II. *Applications Commissions*

Commission	Area
• Commission for Aeronautical Meteorology (CAeM).	Applications to economic and social activities
• Commission for Agricultural Meteorology (CAgM)	
• Commission for Marine Meteorology (CMM)	
• Commission for Climatology (CCl)	

To explain the reasons for the various adjustments in the Technical Commissions referred to above would involve a repetition of the information given in the previous chapters on the respective subjects they deal with. Suffice it at this point to stress that the Technical Commissions have played an essential role in the overall activities of the Organization and in achieving the progress which the Organization has made in the period under review. They provide the means whereby the leading experts in any particular field of meteorology and hydrology can meet together to exchange knowledge and ideas with a view to making recommendations for the action required to ensure that the purposes and objectives of the Organization can best be served on a truly global basis. The overall progress which WMO has so far achieved bears witness to the effectiveness of the system—which in turn bears witness to the friendly spirit of co-operation which has developed between the experts from the different countries serving in each Commission. It is an indication also of the confidence which governments place in recommendations and proposals emerging from such bodies.

In such circumstances it will be clear that to be elected as president of a Technical Commission of WMO is a great responsibility and at the same time a great honour. The names of those who have served in this capacity are shown in Appendix F. This list contains many of the world's most distinguished meteorologists and hydrologists. The dates and places of all the sessions of the Technical Commissions during the period under review are also given in Appendix F. In general such sessions have been held in different countries at the kind invitation of the Members concerned, but occasionally also in Geneva (or in the case of CAeM, in the ICAO headquarters in Montreal) under arrangements made by that Organization.

Panels and working groups

Any description of the organizational structure of WMO would be incomplete without a reference to the panels and working groups. The main decisions are of course taken by the constituent bodies (Congress, Executive Council, Regional Associations, Technical Commissions), each acting in its appointed capacity, but the complexities of the many problems involved make it necessary in each case to create small groups of highly specialized experts to study particular problems. The global membership of the Organization, together with the fact that the membership of these groups is not restricted to the members of the constituent body which establishes them, ensures that in creating such specialized groups some of the world's best experts on the subject concerned are appointed to serve on each—and this in turn ensures that the recommendations they put forward are of the highest possible standard.

Examples of working groups or committees set up by Congress to study specific problems during the inter-sessional period are the Working Group on the Convention (Fourth Congress, 1963), the Advisory Committee on Operational Hydrology (Sixth Congress, 1971) and the Financial Advisory Committee (Tenth Congress, 1987).

In the case of the Executive Council, a noteworthy working group which has been in existence since 1964 is that on Antarctic Meteorology, composed of representatives of Members signatory to the Antarctic Treaty, for co-ordination of meteorological activities in the Antarctic, an arrangement necessitated mainly by the fact that the southern limit of the three southernmost WMO Regions is latitude 60°S. The Executive Council also establishes, as and when necessary, panels of experts to keep it advised on important specialized subjects not specifically included within the terms of reference of any Technical Commission. The preceding chapters describing the activities of the Organization under its major programmes contain several examples of such panels and of the

*See note on p. 8.

important work they have carried out. Some of the panels and committees have been intergovernmental in their constitution, i.e. composed of officially designated representatives of certain Member Governments. Examples in this category are the Intergovernmental Panel on the FGGE (also known as the Global Weather Experiment), the Tropical Experiment Board and the Tropical Experiment Council (both for the GATE) and, more recently, the Intergovernmental TOGA Board for the Tropical Ocean and Global Atmosphere programme in support of the WCRP.*

As regards Regional Associations and Technical Commissions, an indication of the part played by their working groups is given by their numbers—at present nearly 90 in the case of Regional Associations and about 150 for Technical Commissions. These numbers include also rapporteurs who sometimes constitute a working group, otherwise considered as "one-man working groups". Many of the valuable WMO publications in the *Technical Note* series have been prepared by working groups or rapporteurs of Technical Commissions.

It would be difficult to pay adequate tribute to the various panels and working groups and to the persons serving on them, who have dedicated, despite their pressing national duties and responsibilities, their time and knowledge to WMO on a voluntary basis. Indeed, it may truly be said that, throughout the period under review, much of the strength of the Organization in the whole range of its activities, scientific, technical and administrative, has rested with these bodies.

Secretaries-General

The first Secretary-General, appointed by First Congress (1951), was Dr Gustav Swoboda, who was a very well-known figure in IMO circles, having been the Chief of the IMO Secretariat from 1938 until the time when WMO came into being as its successor. Dr Swoboda, who was originally of Czechoslovak nationality and later became a Swiss citizen, had been a member of some of the IMO Technical Commissions while serving as Head of the Forecasting Services in the State Meteorological Institute in Prague. He had, therefore, an intimate knowledge of the international meteorological activities carried out under the aegis of IMO and was moreover closely associated with the entire process which led to the creation of WMO.

Dr Swoboda's term of office ended on 14 August 1955. The development of WMO during its early years owes a great deal to his leadership. In particular, he will be remembered for building up a small but highly efficient Secretariat within the very modest limits set by First Congress.

In 1955 Mr D. A. Davies**, Director of the then British East African Meteorological Department in Nairobi, was appointed by Second Congress to succeed Dr Swoboda. He was already a prominent figure within WMO, having been elected in 1951, and subsequently re-elected in 1953, as president of Regional Association I (Africa). In this capacity he was also a member of the Executive Council.***

Dr Davies, who took up his appointment as Secretary-General on 1 August 1955, was thereafter unanimously re-appointed to this office no fewer than five times by successive quadrennial sessions of Congress: Third Congress (1959), Fourth Congress (1963), Fifth Congress (1967), Sixth Congress (1971) and Seventh Congress (1975). At the time of his retirement at the end of 1979, he had thus held this office for over twenty-four years, the longest period of service as Executive Head of any organization within the United Nations system. The present volume bears testimony to the remarkable development in WMO's activities during his tenure of office. In a resolution sponsored by 107 Member countries and adopted by acclamation, Eighth Congress (1979) placed on record "his outstanding services to WMO and hence to the cause of international co-operation and understanding" and gave him the honorary title "Secretary-General Emeritus". It would be appropriate to add that a well-known meteorological journal wrote at that time in the following terms: "During the twenty-four years Dr D. A. Davies served as Secretary-General, responsibilities of WMO grew commensurately with our science and its services. The manifold activities bear the imprint of the man whose name had, to many, become synonymous with that of the Organization."

Among the many distinctions received by Sir Arthur Davies** special mention may be made of the United Nations Peace Medal in 1979 (shortly before his retirement) and in 1985, WMO's highest honour, the IMO Prize.

Professor A. C. Wiin-Nielsen (Denmark), who was appointed as Secretary-General by Eighth Congress (1979), took up his appointment on 1 January 1980, succeeding Dr Davies. Professor Wiin-Nielsen, who is a reputed scientist, was the first Director of the prestigious European Centre for Medium-range Weather Forecasts in Reading (United Kingdom) from 1974 until his appointment as Secretary-General of WMO. He was a member of the joint WMO/ICSU Scientific Committee for the planning of the Global Weather Experiment and the editor of the *Compendium of meteorology for use by Class I and Class II meteorological personnel*, a WMO publication intended for students receiving an advanced-level education in meteorology.

The present Secretary-General, Professor G. O. P. Obasi (Nigeria), who was appointed to this office by Ninth Congress (1983) and subsequently unanimously re-appointed by Tenth Congress (1987) for another four-year term commencing in 1988, succeeded Professor Wiin-Nielsen on 1 January 1984. Professor Obasi, who held the post of Director, Education and Training Department in the WMO Secretariat from 1978 until his appointment as Secretary-General, had earlier been the Dean of the Faculty of Science and Professor of Meteorology at the University of Nairobi (Kenya). He had also taken a prominent part in international scientific activities, including those under the auspices of WMO (notably GARP and the Commission for Atmospheric Sciences).

The Secretariat

The basic functions of the Secretariat are defined in Articles 20–22 of the Convention (Appendix A). In brief, they provide for the appointment by Congress of a Secretary-General, who in turn is responsible for the appointment of the technical and administrative staff of the Secretariat with the approval of the Executive Council. The Secretary-General is responsible to the President of the Organization for the technical and administrative work of the Secretariat.

The General Regulations lay down in more detail the general functions of the Secretariat as follows:

(1) To serve as the administrative, documentary and information centre of the Organization;

(2) To perform day-to-day programme-management functions under the guidance of the Executive Council and in close co-operation with the Technical Commissions in

---

*The TOGA Board reports to the Executive Councils of WMO and IOC.

**During the course of his career, Mr Davies was awarded several honorary doctorates. He received a knighthood from the British Government in 1980 after his retirement from WMO.

***See note on p. 8.

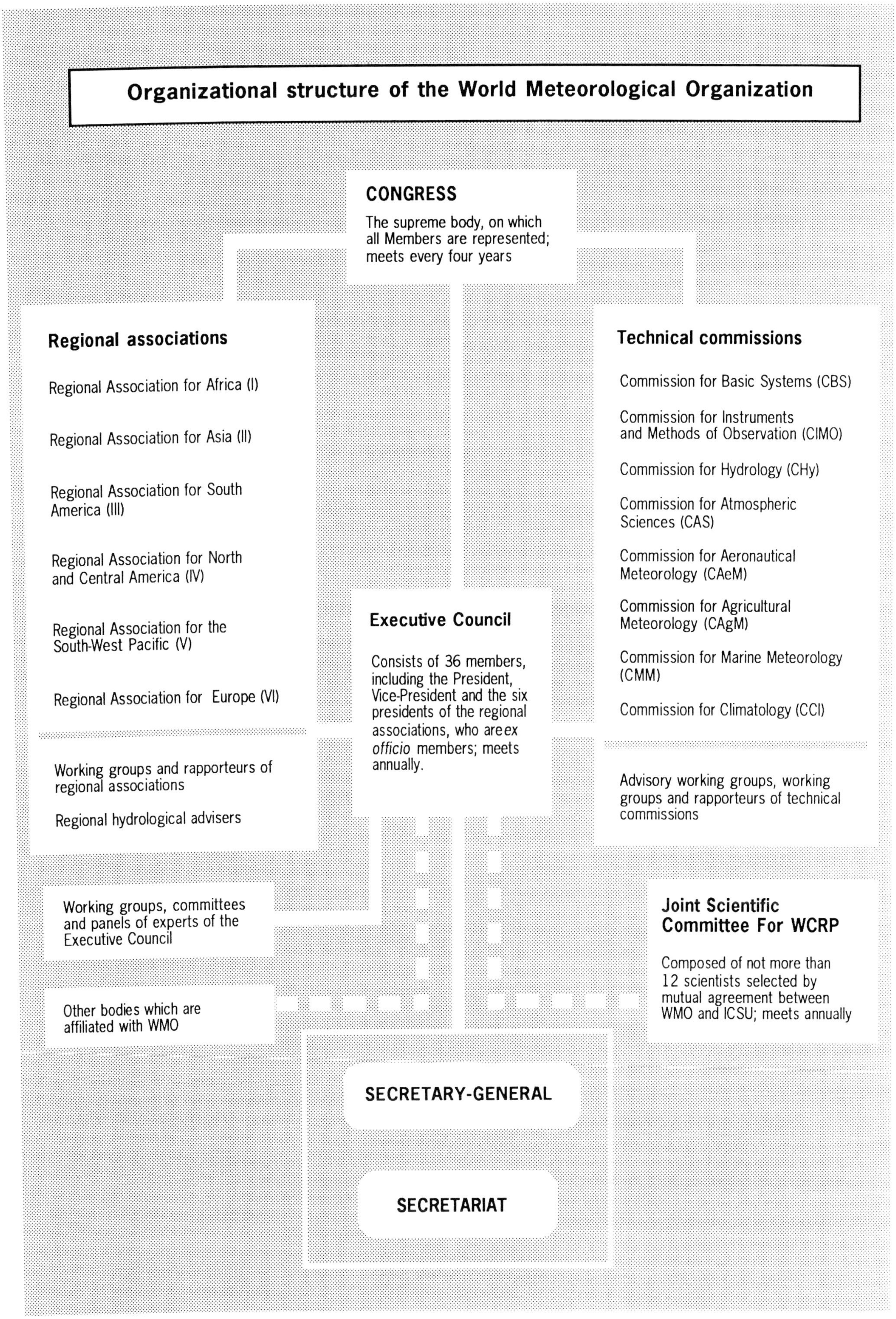
Organizational structure of the World Meteorological Organization
CONGRESS
The supreme body, on which all Members are represented; meets every four years
Regional associations
Regional Association for Africa (I)
Regional Association for Asia (II)
Regional Association for South America (III)
Regional Association for North and Central America (IV)
Regional Association for the South-West Pacific (V)
Regional Association for Europe (VI)
Working groups and rapporteurs of regional associations
Regional hydrological advisers
Executive Council
Consists of 36 members, including the President, Vice-President and the six presidents of the regional associations, who are ex officio members; meets annually.
Technical commissions
Commission for Basic Systems (CBS)
Commission for Instruments and Methods of Observation (CIMO)
Commission for Hydrology (CHy)
Commission for Atmospheric Sciences (CAS)
Commission for Aeronautical Meteorology (CAeM)
Commission for Agricultural Meteorology (CAgM)
Commission for Marine Meteorology (CMM)
Commission for Climatology (CCl)
Advisory working groups, working groups and rapporteurs of technical commissions
Working groups, committees and panels of experts of the Executive Council
Other bodies which are affiliated with WMO
Joint Scientific Committee For WCRP
Composed of not more than 12 scientists selected by mutual agreement between WMO and ICSU; meets annually
SECRETARY-GENERAL
SECRETARIAT

connection with the scientific and technical programmes approved by Congress;

(3) To make technical studies as directed by Congress or the Executive Council;

(4) To organize and perform secretarial duties at sessions of Congress, the Executive Council, the Regional Associations and the Technical Commissions within the limits of the appropriate provisions of these Regulations;

(5) To arrange for the issue with the provisional agenda of an explanatory memorandum summarizing the problems to be discussed in respect of each item on the agenda of each constituent body;

(6) To prepare or edit, arrange for the publication of and distribute the approved publications of the Organization;

(7) To provide an appropriate public relations service for the Organization;

(8) To maintain records of the extent to which each Member implements the decisions of the Organization;

(9) To maintain files of the correspondence of the Secretariat;

(10) To carry out the duties allocated to the Secretariat in the Convention and the regulations of the Organization, and such other work as Congress, the Executive Council and the President of the Organization may decide.

The WMO Secretariat had at its disposal from the outset a small group of staff to ensure continuity in the work taken over from IMO and to serve as a nucleus for undertaking its new responsibilities pending the recruitment of the staff approved by First Congress, which necessarily took time.

While the basic functions of the Secretariat have not changed greatly since the beginning of WMO, the volume and nature of the work which the Secretariat is called upon to perform has, on the other hand, changed significantly as a result of the progressive development of the programmes of the Organization. Such development made it necessary to expand the staff of the Secretariat significantly over the first 30 years of its existence. Thereafter, however, financial restraints precluded further expansion and the staff situation has for several years been static.

The geographical representation on the Secretariat staff changes from time to time with the departure of some staff members and the recruitment of new ones. Since the creation of WMO over 80 nationalities have been represented on the staff.

To avoid possible misunderstanding it should perhaps be stressed that the initial increase in the Secretariat in no way signified a departure from the basic principle whereby the technical strength and the technical control of the affairs of the Organization both rest with the Member countries. The increase was in fact due to a variety of other reasons. For example, WMO as a Specialized Agency of the United Nations became able to participate in the technical co-operation programmes of that body and this in turn involved the Secretariat in a new form of activity which it alone could accept. The creation of WMO's own Voluntary Assistance Programme in 1967 (later known as the Voluntary Co-operation Programme) added further to this Secretariat responsibility. Then again, on the non-technical side, the responsibilities of the Secretariat in translation and interpretation duties increased considerably. At the outset the main requirement was for such facilities in English, French and Spanish whereas at the present time the languages used are Arabic, Chinese, English, French, Russian and Spanish. The increase in Membership of the Organization increased the work of the Secretariat in many other ways. The inevitable contacts with the Secretariats of the United Nations and the other Specialized Agencies have also expanded the responsibilities of the Secretariat.

Turning now to the question of the location and nature of the office accommodation of the Secretariat, prompt and efficient action was called for from the outset. Initially, the Secretariat was located in a few rented rooms in Lausanne in which the small IMO Secretariat had been based during World War II.

As mentioned in Chapter II, after an exceptionally lengthy debate, First Congress made a choice among Geneva, Lausanne and Paris and accepted the invitation of the Swiss authorities to locate the WMO Secretariat in Geneva, where of course the European Office of the United Nations* itself, as well as several of the UN Specialized Agencies, was already based. Congress also decided that the Secretariat should, in the first instance, move from Lausanne to occupy one of the temporary buildings offered free of charge by the Canton of Geneva. A group of temporary huts located in Campagne Rigot, Avenue de la Paix, adjacent to the Palais des Nations where the UN Office was based, was selected for this purpose by the Executive Council** at its second session (Lausanne, 1951) and the Secretary-General signed an agreement with the Canton of Geneva on 6 November 1951 for the temporary occupancy of this accommodation. The WMO Secretariat moved in December 1951.

The establishment of the WMO Secretariat on the soil of the Swiss Confederation was formally marked by an Agreement between the Swiss Federal Council and the Organization, concluded on 10 March 1955, governing its legal status, with a guarantee of certain rights similar to those for other UN Specialized Agencies with their headquarters in Switzerland. The nature as well as the limited size of the accommodation in the huts soon made it necessary for more adequate arrangements to be made, a situation already foreseen by First Congress, which had given necessary directives in this matter. Discussions were started with the Geneva authorities and, having regard to the very limited financial resources of WMO, an arrangement was made whereby the Canton of Geneva would construct a building suitable to house the WMO Secretariat and located in close proximity to the UN Office. When completed the WMO Secretariat would occupy this building and pay an agreed rental to the Canton, with the possibility of purchase at any time. Additional accommodation, on a rental basis, had to be obtained about this time in the vicinity of the Campagne Rigot huts, first in 1956 for the Technical Assistance Unit and the IGY Meteorological Data Centre, and then in 1958 in the building known as the "International Centre" for another part of the Secretariat.

The site selected for the new building for the Secretariat was on Avenue Giuseppe Motta. The foundation stone of the building was laid on 14 May 1958 during a short ceremony in the presence of the members of the Executive Council** who were then attending its tenth session and representatives of the Swiss federal authorities (Federal Councillor Mr G. Leport) and of the Cantonal authorities of Geneva (Mr J. Dutoit, President of the *Conseil d'Etat*). The Secretariat moved into the new building upon its completion during June 1960. The building was formally inaugurated on 12 July 1960 at a ceremony held in the main conference hall. The ceremony, which marked also the tenth

*Now called the United Nations Office in Geneva.

**See note on p. 8.

anniversary of the Organization, was attended by all members of the Executive Council who were in Geneva to attend its twelfth session, Federal Councillor Mr W. Spühler, representing the Swiss federal authorities, representatives of the Cantonal and municipal authorities of Geneva (Mr J. Treina, President of the *Conseil d'Etat*, Mr J. Babel, President of the *Grand Conseil* and Mr F. Cottier, Mayor of Geneva), Mr P. P. Spinelli, Director of the European Office of the UN, representing the Secretary-General of UN, and representatives of many of the diplomatic missions and of other Specialized Agencies. The same building has served as the WMO Headquarters since that day.

The increase of staff and responsibilities referred to in the preceding paragraphs made it necessary to rent additional office accommodation outside the WMO building in the years 1967–1968 for occupation until the construction of an extension to the building. By this time however the financial arrangements for constructing accommodation for international organizations in Geneva had changed and it was necessary for WMO itself to construct the extension, using a loan from a newly created Swiss body, "Foundation for Buildings for International Organizations". Fortunately, the original agreement permitted WMO to purchase the original building and indeed to use the money already paid as rental for this purpose. The idea was approved in principle by Fifth Congress (1967) and detailed proposals were considered by the Executive Council* and finally approved by Members by a postal ballot in January 1968. The financial arrangements included in these proposals involved a loan of Sfr 6.5 million to cover the cost of construction of the extension, to be repaid over 30 years, and annual payments to the Canton of Geneva over the same period, for the purchase of the building. The construction of the extension, which was started in 1969, was completed in 1970. Thus WMO became the owner of the whole building, but with of course a substantial loan still to be repaid.

One of the most interesting features of the WMO Headquarters is the large collection of gifts from Member countries, some of a practical nature and others of great artistic merit.

It should be mentioned that the WMO Secretariat also includes two small Regional Offices located outside Geneva, each with two Professional staff. The Regional Office for Africa has been located in Bujumbura (Burundi) since 1981, and the Regional Office for Latin America in Asunción (Paraguay) since 1978 following their transfer from Geneva. The latter office is now called the Regional Office for the Americas (WMO Regions III and IV).

The Secretary-General is the head of the Secretariat. He has been assisted from the outset by a Deputy Secretary-General and, since 1985, by an Assistant Secretary-General also. The names of those who have served as Secretary-General are listed in Appendix D, while short biographical notes are found earlier in this chapter.

WMO publications

As stated in the previous sections, one of the duties of the Secretariat is to prepare or edit and arrange for the publication and distribution of the approved publications of the Organization.

The publications fall into two broad categories: mandatory publications and programme-supporting publications. The first category comprises *Basic Documents* (including regulatory material of various kinds), operational publications, official records, WMO *Guides*, *Annual Reports* and the WMO *Bulletin*.

Inserting the basic documents of WMO into the foundation stone of the WMO building (May, 1958). Left to right: *Mr. G. Friedrich, Chef du Service des Bâtiments; Mr. D. A. Davies, Secretary-General; Mr. E. Martin, architect*

The second category comprises WMO *Technical Notes* and several other series within the fields covered by the major WMO programmes. The development in these programmes has led to a corresponding growth in the publications and new series have been introduced over the years, such as WWW reports of various kinds, training publications, *Operational Hydrology Reports*, *Special Environmental Reports*, *Reports on Marine Science Affairs*, as well as joint publications with ICSU (on the GARP and more recently, on the WCRP). Titles of some specific publications have been mentioned elsewhere in the present volume. A few other publications are mentioned below in view of their particular importance.

WMO Bulletin

The WMO *Bulletin*, which has been issued quarterly since its first number came out in April 1952 (in accordance with a decision of First Congress), provides a summary of the activities of the Organization and of developments in international meteorology and hydrology of interest to Members and to others concerned with the application of meteorology and hydrology to human activities. The *Bulletin* was at first issued in English and French. Its success as a medium of information on the activities of WMO and on other noteworthy developments of interest to Members led Fifth Congress (1967) to decide that it should also be issued in Russian and Spanish. The *Bulletin* has accordingly been issued in the four languages as from its July 1968 number.

WMO Technical Notes

The WMO *Technical Note* series was started in 1954, mainly to provide a wider distribution of a selection of papers containing authoritative reviews of the state of knowledge in specific branches of meteorology, prepared for sessions of constituent bodies. The scope of this series has since been somewhat enlarged and a total of 193 publications have been issued so far as *Technical Notes*. Many of them have been prepared by working groups of Technical Commissions (see page 156) or by individual authors and contain reviews of theoretical know-

*See note on p. 8.

ledge as well as descriptions of techniques for the application of meteorological information to specific fields. Scientists and operational experts of international reputation have contributed to these publications. Results of important projects undertaken by the Secretariat have also appeared in *Technical Notes*, the titles of some of which have been mentioned elsewhere in this publication.

International Cloud Atlas

The *International Cloud Atlas*, first published in 1956, made a landmark both as a work of reference and as a publication of great practical value, since observations of cloud are reported from weather stations all over the world several times a day and uniformity in the observations is therefore important.

The preparation of the *Atlas*, which represented an enormous task, was undertaken by a small group within the then Commission for Synoptic Meteorology following a request by First Congress (1951) to this body to complete a project intitiated by IMO for the revision of an earlier version of the *Atlas* which had been issued some twenty years before.

The new *Atlas* was published in two volumes: Volume I contained a detailed descriptive and explanatory text and Volume II an impressive collection of 224 plates (101 in colour and 123 in black and white), each accompanied by a short text. There were also companion volumes in the form of abridged atlases for meteorological observers at surface stations and for observers in aircraft.

A revised version of Volume I was published in 1975, but with the sub-title *Manual on the observation of clouds and other meteors*, as certain parts of the text were intended to constitute Annex I to WMO Technical Regulations (see page 161) and Congress had already decided that any annex to the Technical Regulations should be called a *Manual*. Those parts of this new edition having the status of Technical Regulations were distinguished from the rest by an appropriate form of presentation.

A revised version of Volume II of the *Atlas* was published in 1987, containing 196 photographs, 161 of which are in colour, each photograph being accompanied by an explanatory legend.

IMO Lecture monographs

The IMO Lectures, instituted in 1967 (see page 153), have resulted in a series of valuable monographs mentioned below in chronological order:

The nature and theory of the general circulation of the atmosphere, by Edward N. Lorenz; *Radiation processes in the atmosphere*, by K. Ya. Kondratyev; *The atmospheric boundary layer*, by R. W. Stewart; *Climatic changes and their effects on the biosphere*, by B. Bolin; *Monsoons*, by P. K. Das.

The latest monograph, entitled *Dispersion processes in large-scale weather prediction*, by N. A. Phillips, is in press.

General Regulations

The General Regulations, which are adopted by Congress (see page 152), are intended to lay down rules in sufficient detail within the provisions of the Convention to ensure the smooth working of the Organization, covering all its constituent bodies. The General Regulations adopted by First Congress (1951) have been reviewed by each subsequent session of Congress in the light of the experience gained in their application and amendments and additions have been approved by each Congress. The first set of General Regulations adopted in 1951 contained 142 Regulations, while the current version contains 193 Regulations.

Examples of the application of some of the provisions of the General Regulations of particular importance to the functioning of the Organization are given below.

Permanent Representatives of Members

Each Member designates a Permanent Representative, who should be the Director of the Meteorological or Hydrological Service, to act on technical matters for the Member between sessions of Congress, a procedure which has greatly facilitated the scientific and technical work of the Organization.

Use of postal ballots

The principle of Members taking decisions by postal ballot in the period between sessions of constituent bodies (which, with the exception of the Executive Council, are normally held at four-yearly intervals) was adopted by Second Congress (1955) in order to speed up the work of the Organization. The relevant provisions have since been used with advantage on many occasions. A few examples of major decisions taken in this way are given below:

- Decisions on the applications for Membership of the Organization, under Articles 3 (*c*) and 3 (*e*) of the Convention, resulting in the admission of several countries over the years;
- The utilization of surplus from the first financial period (1951–1955) as temporary working capital to finance the IGY Meteorological Data Centre in the Secretariat (1956);
- Acceptance of the terms of the offer by the Canton of Geneva to construct a building for the WMO Secretariat, to be occupied on a rental basis but with the possibility of later purchase (1957);
- Approval of the proposals for the purchase of the WMO building and for the construction of an extension by means of a loan (1968);
- Approval of the plan for the utilization of the New Development Fund of US $1.5 million during the fourth financial period (1964–1967) which had been agreed to, in principle, by Fourth Congress (1963) subject to approval of the plan (1964);
- Approval of supplementary estimates to meet unforeseen programme activities of an urgent character agreed to by the Executive Council;
- Approval of the WMO/IAEA Agreement (1959);
- Adoption of important resolutions and recommendations by Regional Associations and Technical Commissions from time to time.
- The decision to follow the United Nations in recognizing the People's Republic of China (1972).

Co-operation with non-Member countries

One of the aims of WMO being to facilitate world-wide co-operation in meteorology, the importance of co-operation with non-Member countries was realized from the outset. Thus, as part of the general policy adopted by First Congress (1951) it was decided that non-Member countries having Meteorological Services should be invited to send observers to sessions of Congress and appropriate Regional

Associations, while the Directors of these Services should be represented at sessions of Technical Commissions. Seven non-Member countries out of seventeen which had been invited, after consultation with Members, were present at Second Congress (1955). The principle of invitation to non-Member countries with the approval of Members, which was incorporated in the revised version of General Regulations as adopted by the same Congress, has still remained in force except for an amendment introduced by Ninth Congress (1983) whereby approval of Members is not necessary for those non-Member countries which are Members of UN or have been granted observer status with UN. Except for Ninth Congress (1983), non-Member countries have been present at all the other Congress sessions from the Third (1959) to Tenth Congress (1987) inclusive.

Another important aspect of co-operation with non-Member countries in the early days of WMO concerns those countries without a Meteorological Service. Through WMO's technical assistance missions within the framework of the then Expanded Programme of Technical Assistance (EPTA) of UN, assistance was given to many of them in setting up networks of meteorological observing stations, the first step towards the establishment of a Meteorological Service. As a result, some of these countries later became Members of WMO.

Languages

The practice as regards the use of different languages in the work of the Organization has changed progressively since First Congress. The General Regulations adopted at that Congress laid down the *official languages* as English, French, Russian and Spanish, but only two of them were designated, depending on the constituent bodies, as their *working languages*. It was Fourth Congress (1963) which decided that the four *official and working languages* should be English, French, Russian and Spanish, to which Chinese and Arabic were added respectively by Seventh Congress (1975) and Eighth Congress (1979).

The implementation of the decisions mentioned above has been going on on a step-by-step basis as regards interpretation at sessions of constituent bodies, documentation for these sessions, publication of their official records and of the *Basic Documents* of the Organization.

Technical Regulations and Guides

Technical Regulations

The WMO Convention provides for the establishment by Congress of Technical Regulations. Second Congress (1955) adopted these Regulations, prepared on the basis of the principles and a detailed plan which had been laid down for this purpose by First Congress (1951). It was stated by Second Congress that these Regulations were designated to facilitate co-operation in meteorology between Members and to meet international requirements in the various fields of application of meteorology through uniformity and standardization in meteorological practices and procedures. It was also then decided that the Technical Regulations should comprise *standard* practices and procedures (those which it is *necessary* that Members follow or implement) and *recommended* practices and procedures (those which it is *desirable* that Members follow or implement).

The Technical Regulations, which form part of WMO's *Basic Documents*, have since then been kept under constant review by the Technical Commissions within their respective spheres in the light of new developments, and proposals for changes have been considered at each Congress on the basis of recommendations of the Executive Council. Several editions of the Technical Regulations have had to be issued over the years and the original form of presentation of these Regulations has also undergone changes as a result of developments in WMO's programmes. The basic decisions of Second Congress as stated above, however, still remain valid except for the inclusion of hydrology within their scope and these decisions appear in the Technical Regulations. Important changes in these Regulations are described below.

Following the adoption by Sixth Congress (1971) of Technical Regulations in operational hydrology, these were published in a separate volume in the same year; the title of this volume was changed to "Hydrology" in the next edition in accordance with the relevant decisions of Seventh Congress (1975).

The decision of Sixth Congress (1971) to publish *Manuals* with the status of Technical Regulations and forming annexes to them, has led to a series of such *Manuals* in the field of WWW activities.

The Technical Regulations are now arranged in the following way:

Volume I: *General*, comprising six sections: World Weather Watch; Climatology; Meteorological services (marine activities, agriculture, international air navigation*, environmental pollution, meteorological bibliography and publications); Hydrology**; Education and training; Units and procedures used in international meteorological research programmes and during special observational periods;

Volume II: *Meteorological service for international air navigation;*

Volume III: *Hydrology*.

Annexes to Technical Regulations, published separately and having the status of Technical Regulations

I. International Cloud Atlas, Vol. I, Manual on the observation of clouds and other meteors***

II. Manual on Codes, Vol. I

III. Manual on the Global Telecommunication System, Vol. I

IV. Manual on the Global Data-processing System, Vol. I

V. Manual on the Global Observing System, Vol. I

VI. Manual on marine meteorological services

Guides

Second Congress (1955) decided that in addition to Technical Regulations, appropriate *Guides* should be published describing in detail the practices, procedures, methods and techniques in specific fields which would enable Members to comply with the Technical Regulations and in otherwise developing their Meteorological Services. Congress agreed that the WMO "Guide to international meteorological instrument and observing practice" (a former IMO publication) met this need in the particular field. It therefore became a model for other *Guides* which were to follow.

*Issued separately as Volume II. The material is prepared jointly by WMO and ICAO (see p 149).

**Issued separately as Volume III.

***Certain parts of this publication constitue Annex I to Technical Regulations.

A large number of *Guides* are now available. Many of them have had several editions with revised and expanded material and with a change in the title in one or two cases. For example, the *Guide* mentioned in the paragraph above has seen five editions and its title has been amended twice.

All the *Guides* so far published are listed below. The Technical Commissions concerned and their groups of experts deserve high tribute for the preparation of the *Guides*.

- *Guide to meteorological instruments and methods of observation*
- *Guide to climatological practices*
- *Guide to agricultural meteorological practices*
- *Guide to hydrological practices* (in two volumes)
- *Guide on the Global Data-processing System*
- *Guide on the automation of meteorological telecommunication centres*
- *Guide to marine meteorological services*
- *Guide on the Global Observing System*
- *Guide on the automation of data-processing centres*
- *International operations handbook for measurement of background atmospheric pollution*

Financial Regulations

The Financial Regulations, adopted by First Congress (1951) for the administration of the finances of the Organization, have been regularly reviewed at each subsequent session of Congress. The most important amendments have been those adopted by Third Congress (1959) and Tenth Congress (1987). A Publications Fund was established by Third Congress to meet the cost of production of all publications, the income derived from their sale accruing to this Fund.* The publications programme of the Organization has greatly benefited from this arrangement. Tenth Congress adopted the Swiss franc as the unit of currency for budgetary purposes, replacing the United States dollar, in view of the difficulties caused by fluctuations in currency exchange rates since the early 1970s. The same Congress also adopted a budget cycle comprising two bienniums within the four-year financial period, authorizing the Executive Council to approve appropriations for each of the two bienniums comprising the financial period within the maximum expenditures noted by Congress. (Earlier, the Council was authorized to approve the appropriations for each financial year only).

Staff Regulations

The Staff Regulations, adopted by Congress, lay down the broad principles for the employment of the staff of the Secretariat. They were adopted by First Congress (1951) and have been reviewed by each successive Congress; no major amendments have been made.

Quadrennial programme and budget and long-term planning

WMO operates on a four-yearly budgetary system, the budgetary period beginning with the calendar year immediately following a session of Congress constituting a financial period. The programme of activities of the Organization during a financial period and the maximum expenditures for that period are approved by Congress. Appropriations within the maximum expenditures were approved by the Executive Council on an annual basis until the decision of Tenth Congress (1987) that the annual budget cycle should be replaced by a biennial cycle commencing with the tenth financial period.

In addition to the four-yearly programme which may be considered as a medium-term plan, a system of long-term planning for the scientific and technical programmes has been undertaken by the Executive Council following a decision of Ninth Congress (1983), based on the following principle. Each Long-term Plan should cover a period of ten years, but should be replaced at four-yearly intervals by a new plan adopted by each Congress for the following decade. The second Long-term Plan (1988–1997) was approved by Tenth Congress (1987).

Turning now to the budget of the Organization, the maximum expenditures approved by successive Congresses are given below, it being understood that these figures do not represent the actual expenditures, nor do they take into account certain "flexibility clauses" which Congress has, by tradition, allowed when determining the maximum expenditures. Also, the figures below should not be taken, prima facie, as a true representation of the growth of the budget over certain financial periods. It is not proposed to enter here into the complexities of the subject, which involves many factors—one of which has been the effect of the continuing fluctuations of the exchange rate of United States dollar since the early 1970s (well over four-fifths of WMO's expenditures are in Swiss francs).

WMO's activities under the Technical Co-operation Programme are financed mainly from allocations received from the UNDP (for UNDP projects) and from the Voluntary Co-operation Fund (for VCP projects).

Maximum expenditures approved by each Congress

Session of Congress	Financial period to which the approved maximum expenditures were applicable	Authorized maximum expenditures for the financial period
First (1951)	First financial period (4 April 1951–31 December 1955)	US $1 273 000
Second (1955)	Second financial period (1956–1959)	US $1 700 000
Third (1959)	Third financial period (1960–1963)	US $2 694 484
Fourth (1963)	Fourth financial period (1964–1967)	US $5 373 581
Fifth (1967)	Fifth financial period (1968–1971)	US $11 817 000
Sixth (1968)	Sixth financial period (1972–1975)	US $17 300 000
Seventh (1971)	Seventh financial period (1976–1979)	US $40 542 000
Eighth (1979)	Eighth financial period (1980–1983)	US $74 400 000
Ninth (1983)	Ninth financial period (1984–1987)	US $77 516 400
Tenth (1987)	Tenth.financial period (1988-1991)	SF 170 000 000*

*The budget was adopted in Swiss francs by amending the Financial Regulations (see above).

*The specific Financial Regulation concerning the Publications Fund was subsequently considered unnecessary by Eighth Congress (1979) as the Secretary-General is authorized to establish special accounts subject to reporting thereon to the Executive Council.

Concluding remarks

The structure of WMO, an intergovernmental organization, is basically similar to that of its non-governmental predecessor, IMO, even though its tasks and responsibilities have enlarged considerably and continue to do so. Its Membership has grown to a point making it a truly world-wide organization. Several features stand out prominently in the method of operation of WMO. The Directors of Meteorological Services of its Member countries play a key role in the executive and technical bodies of the Organization (including the highest decision-making body, the Congress). The programmes of the Organization are implemented by the Member countries, which provide the necessary facilities for this purpose. Moreover, co-operation among Members on a wide basis is a *sine qua non* for the execution of the programmes. The Technical Commissions, composed of the experts nominated by Members and working on a voluntary basis, provide the main source of scientific and technical advice to the Organization. The Organization is therefore in a position to operate in an economic manner.

WMO's predecessor, IMO, was one of the oldest institutions in the field of international co-operation. WMO is successfully maintaining this tradition.

CHAPTER XVIII

FURTHER OUTLOOK

Introductory remarks

The origins of international collaboration in meteorology, going back over a century and leading to the establishment of IMO and, over seventy years later, of its successor, WMO, have been described in Chapter I. The growth of WMO and the major aspects of its activities over nearly four decades have been dealt with in the subsequent chapters. They present a story of success and achievement, a story which explains the increasingly important role which WMO now plays on the international scene. The question naturally arises as to what the future holds and in this concluding chapter an attempt is therefore made to assess the future role of WMO.

In the decade of the 1990s international co-operation in meteorology and problems of the global atmosphere will be of unprecedented importance. WMO's programmes and plans for the 1990s show that the Members foresee greatly extended responsibilities for WMO in the changing world scene and that they have taken steps to ensure that the Organization will be fully capable of meeting future needs.

An indication of what lies ahead beyond the 1990s may be drawn from present trends. In the first place, the ever present influence of weather and climate on human affairs will doubtless require the continuation—and indeed the expansion—of existing Meteorological and Hydrological Services throughout the world. Changing needs may well require new types of activity within these Services. Fortunately, scientific and technological advances augur well for the necessary adjustments. The provision of some new services will require new global initiatives and special co-operative arrangements between groups of countries. The spirit of co-operation between the Members of WMO which has prevailed at all times suggests that such arrangements can be readily achieved. It is moreover heartening to note that initiatives in this direction are already being taken by all Member countries, both developed and developing.

As WMO is an intergovernmental body, the future situation in world politics is a relevant factor in considering its future. Past experience suggests, however, that in spite of difficulties in other fields, the spirit of friendly international co-operation will enable solutions to be found in meteorology and hydrology. Closely related to the world political situation is the world economy. Solutions to atmospheric environmental problems must take into account economic disparities and expanded technical co-operation in meteorology, including climatology and hydrology. The WMO relationship with the United Nations has assisted greatly in this field and will doubtless continue to result in important benefits in the future.

The most important development has been the recent recognition by all countries of the need for major steps to be taken for the protection of the atmospheric and hydrological environment on a global scale—and hence for adequate attention to be given to meteorology and hydrology and indeed several other geophysical sciences. WMO will therefore need to play an essential role in such activities in the years to come.

Our thoughts about the future of WMO may be better structured by considering separately the next decade and what follows beyond the year 2000.

The next decade

As far as the next decade is concerned a number of relevant studies are available. Some guidance is also contained in the Second WMO Long-term Plan, which was approved by Tenth Congress in 1987 and covers the period from 1988 to 1997. Furthermore, the first steps have already been taken to prepare WMO's Third Long-term Plan, which will overlap with the current one and will cover the period from 1992 to 2001. In fact many very important recent developments were considered only quite recently at the forty-first session (1989) of the Executive Council.

Considering the role of WMO in the 1990s, the Executive Council concluded that:

- The basic purposes (Article 2 of the Convention) and overall objectives of WMO will remain valid;
- There will be greater emphasis on the interaction between meteorology and the other geophysical disciplines;
- The major environmental issues of the decade related to the global atmosphere will fall increasingly within the purview of WMO.

In the light of this assessment the Executive Council agreed that the following major objectives may be proposed to Eleventh Congress in 1991 for the period to be covered by the Third Long-term Plan (1992–2001):

- World Weather Watch: To consolidate and stengthen the World Weather Watch as the fundamental international system for free and unrestricted exchange of the basic meteorological and related geophysical and environmental data and products needed to support applications and services at the national, regional, and global levels;
- Climate change: To ensure that WMO effectively fulfils its role as the leading international agency responsible for climate monitoring, research and applications, and provides an authoritative international scientific voice on matters related to climate change;
- Environmental quality: To contribute, through scientific monitoring and research, to understanding, arresting and reversing the degradation of the atmospheric (including the stratospheric ozone layer), marine and hydrological environment, and, using WMO capabilities, to provide information and warnings on impending environmental disasters;
- Natural disaster reduction: To contribute, within the framework of the International Decade for Natural Disaster Reduction (IDNDR), and through the implementation of detection, prediction and warning systems, to safety of life and to reduction of the social and economic impact of natural disasters;
- Transportation safety: To enhance the safety of the increasingly mobile global population through improved meteorological and related services in support of air, sea and land transportation systems;

- Food and water: To contribute, through effective application of meteorological and hydrological data and services at national and international levels, to the accurate assessment of water resources, to increased production of food, and to the efficient use and distribution of both food and water for the growing world population;
- Energy use: To contribute, through effective application of meteorological and hydrological data and services, to the planning, development and use of existing and new energy sources which would reduce the rate of increase of greenhouse gases in the atmosphere;
- Bridging the gap: To support the accelerated development of national Meteorological and Hydrological Services of developing countries through a co-ordinated strategic approach to technical co-operation and development;
- Community benefit: To ensure that the general community in all countries better understands the value of, and is better assisted to benefit from, the basic public information, forecast and warning services provides by national Meteorological and Hydrological Services;
- Sustainable development: To contribute, through the meteorological, hydrological and oceanographic monitoring, research and prediction facilities and programmes of national Services, to environmentally sustainable development in all countries.

The forty-first session of the Executive Council also considered the major advances expected during the last decade of the millenium. It was agreed that WMO's responsibility is to provide the authoritative scientific information and advice on the condition and behaviour of the global atmosphere and climate and the factors that affect them.

Important advances foreseen by the Council, which will require particular attention, will result from the following activities:

(*a*) Increased efforts will be made to assist Members in promoting a greater awareness and understanding of the potential economic and social benefits of meteorology and hydrology with a view to improving safety and economic efficiency in all countries.

(*b*) World leaders and the public are becoming deeply concerned at the deterioration of the atmospheric environment and some of its alarming long-term consequences for the health-protecting ozone layer, climate change and long-range transport of air pollutants. WMO is directly involved in the scientific and operational aspects of the problem. Its work in this field has produced the scientific basis for international agreements and action. WMO's role in the atmospheric and hydrological environment, through monitoring programmes and scientific studies, and the application of these scientific results to the development of international agreements must be enhanced.

(*c*) Expansion of marine meteorological activities and of oceanographic observations and services will be a priority area in view of the growing importance of the oceans in climate-related studies requiring oceanographic data of various kinds and in economic and human activities in general.

(*d*) Development of improved systems for warnings against weather-related disasters such as tropical cyclones, floods and droughts in order to minimize their adverse effects will receive high priority.

(*e*) WMO's activities will be strengthened in support of measurement and assessment of water resources and water quality as a major contribution to sustainable development and to the extension of the United Nations Drinking Water Decade.

(*f*) Collaborative development of cost-effective modern technology; its introduction into the operation of national Meteorological and Hydrological Services will be fostered in order to enhance the efficiency of their operation, especially in the collection and processing of data and their dissemination.

(*g*) Members will be encouraged to pool resources through co-operative arrangements for the establishment and operation of major international facilities and services, including special observing systems (drifting buoys, automated observations on board aircraft and ships) and key data-processing facilities to meet special requirements within a Region or Regional Specialized Meteorological Centres to provide services for major sectors of the economy.

(*h*) Pooling of resources will also be initiated to ensure the operational reliability and continuity of meteorological satellites, particular attention being given to increasing the number of direct participants in the global meteorological satellite network, and bringing national Meteorological and Hydrological Services and their countries' space-related organizations into a partnership role.

(*i*) Special efforts will be made to overcome the difficulties experienced by developing countries in the operation of their Meteorological and Hydrological Services so as to achieve greater progress in bridging the gap between the developed and developing countries.

Beyond the year 2000

The global context

The kind of world in which WMO and the national Meteorological and Hydrological Services will find themselves at the dawn of the twenty-first century will no doubt be a key factor on which the ability of WMO to meet its tasks effectively will depend. Some of the features of that world will concern population, agriculture and food production, energy production and consumption, industrial and technological development, the economic situation and, for an organization such as WMO, certain environmental issues of great importance to mankind which have been its preoccupation, in common with other organizations, for more than a decade.

The role of WMO

It is not difficult to foresee that the basic role of WMO since its inception—namely, a planning and co-ordinating role in a field of activity in which international co-operation on a wide basis is an indispensable condition for success—must remain. Another important role, arising mainly from the increasing need for co-operative arrangements between groups of Members to provide essential observational data from vast areas outside national jurisdiction and other facilities which will enable adequate meteorological and hydrological services to be provided in all regions of the world, can also be envisaged. The jointly financed North Atlantic Ocean Stations (NAOS) under the auspices of WMO, an example in this category, has been followed by new observing systems using modern technology (ASDAR, ASAP, instrumented drifting buoys, etc.) and the concept of Regional Meteorological Centres with specialized functions to assist Members within their respective areas, all of which call for consortium arrangements or joint support because of high cost and other considerations. Extended use of such arrangements resulting from their steady introduction in the future is likely to need a more active role on the part of WMO. The experience already being gained in initiating activities of this nature will facilitate the assumption by WMO of any such future role.

Other developments that have already occurred point towards the potential of WMO in providing assistance in the application of international conventions in areas beyond those of traditional applications of meteorology and hydrology to which we have been accustomed. For the implementation of the IAEA Convention (1986) relating to the early notification of a nuclear accident, WMO has accepted the task, involving both scientific and operational aspects, of prompt dissemination of meteorological and hydrological data and forecasts of trans-boundary flows of radioactive material following a nuclear accident. WMO has also been taking an active part in some aspects of the work resulting from the Convention on Long-range Trans-boundary Air Pollution under the auspices of ECE.

An undisputed strength of WMO lies in the international meteorological system based on the principle of free and unrestricted exchange of information between nations, the most striking example of which is provided by the WWW system, in which all Members participate. The world-wide exchange of real-time meteorological information on a round-the-clock basis may indeed be considered as the symbol of the long tradition of international co-operation in meteorology. An agreement has been made for the transmission of relevant information on the WWW Global Telecommunication System to assist IAEA in the case of a nuclear accident and of certain seismic data to support the work of the UN Conference on Disarmament. The possibility of serving similarly, when appropriate, the needs of other disciplines or activities requiring the world-wide transmission of warnings and messages might be contemplated for the welfare of mankind. Such an open policy might also bring the strongest support to strengthening WMO programmes throughout the world.

The decisions taken by the Executive Council at its forty-first session in 1989 were an important step forward in defining WMO's role in connection with global climate issues. Noting Resolution 43/53 of the United Nations General Assembly in which the Secretary-General of WMO was requested to develop with UNEP—namely through the Intergovernmental Panel on Climate Change (IPCC)—the elements of a possible future *convention on climate change*, the Council requested the IPCC to undertake the tasks outlined in the above-mentioned UNGA resolution. The results of this work should be ready for first presentation at the Second World Climate Conference in November 1990. It was also decided that WMO should establish a "Special Fund for Climate and Atmospheric Environment Studies" and that the participation of the developing countries in these studies should be promoted. In brief, the above decisions underline WMO's commitment in the broad area of the scientific and technical aspects of global climate issues.

Bridging the gap

The gap between the meteorological and hydrological services available in industrially developed countries and those in developing countries has been widening, as the cost of necessary equipment and facilities based on high technology is generally beyond the resources of the latter. Scientific advances and new technologies have permitted the development and delivery of new and improved products such as medium-range weather forecasts, quantitative precipitation forecasts, river flow predictions, specially tailored agrometeorological, marine and aviation bulletins and warnings in several regions, but not all are widely available in the developing world since it would be far beyond the resources of some countries to provide the equipment and technology required to produce the full range of valuable products. As an innovative solution to the problem, the concept of Regional Specialized Meteorological Centres within the WWW system has been approved. These centres, to be established through co-operative arrangements between Members and equipped with satellite communications facilities and computers, will be capable of delivering advanced and specialized meteorological products as mentioned above to national Services, which will utilize them after necessary adaptation by specially trained staff to meet their own requirements. In this way it will be possible to reduce the gap between the services provided in different countries. This would also be consistent with the principle, adopted in 1988, that meteorological and hydrological services in every country shall not be below an established minimum level.

Another type of Regional Specialized Meteorological Centre of a somewhat different scope and in some ways of a more advanced character is planned to be set up by pooling resources available within the Region concerned and with support from other sources. The African Centre of Meteorological Applications for Development (ACMAD), to be set up with the active support of ECA, will provide services to the major sectors of the economy, monitor drought situations and undertake studies on African weather and climate. Advanced centres to be created under new arrangements in Latin America would prepare processed data and forecasts for participating Members and assist them in providing advanced climatological services. A Regional Specialized Meteorological Centre, expected to be set up with the support of the Association of the South-East Asian Nations (ASEAN), would provide improved weather forecasts and special products tailored to the requirements of users in the ASEAN group of countries. Another Regional Specialized Meteorological Centre for the benefit of countries in the South Pacific would be mainly concerned with tropical cyclones and their warnings.

Technological advances

Meteorological satellites

A permanent system of meteorological satellites has become indispensable. In addition to providing the essential space-based component of the Global Observing System, these satellites serve other purposes such as the collection and relay of data from other observing platforms.

New generations of satellites with improved instrumentation will have further impacts on relevant WMO activities. In particular, satellites devoted to oceanic studies and research will play an invaluable role in international research programmes such as the World Climate Research Programme, for which an adequate base of global atmospheric and oceanographic observations is of paramount importance. Nevertheless, the development of a relationship between the meteorological and hydrological interests, the users, and those responsible for the planning, design and construction of satellites is considered very important, in order to enable the latter to appreciate fully the particular requirements of meteorology and hydrology. Planning is already under way for the preparation of the future generation of satellites which will provide global meteorological and environmental observations in the first decades of the twenty-first century. The development of these space platforms, which will naturally call for very substantial expenditure of resources, will be largely dominated by industrial policy factors. WMO's aim would therefore be to secure the recognition of meteorological and hydrological objectives in such technological development programmes by bringing the national Meteorological and Hydrological Services of Member countries having space-related organizations into a partnership role so that they may have the opportunity of participating directly in and contributing to a meteorological satellite network. WMO will also promote co-operative arrangements between Members, financially or otherwise (e.g. provision of instrumentation for platforms).

Other technological developments

Technological advances in automated observations and remote sensing which hold out the prospects of global coverage (including large ocean areas, rapid collection, processing and distribution of data and products taking maximum advantage of new technology in interrelated computer telecommunications systems and supercomputers capable of using large quantities of data and producing complex global weather prediction models) are expected to lead to further significant benefits from WWW and to all-round progress in WMO programmes.

The environmental challenge

Environmental problems of a complex and global nature and having wide implications, some of which have influenced WMO's activities for more than a decade now, will continue to need close attention, since the efforts being made to deal with them at the international level, and co-ordinated with UNEP, are necessarily of a long-term nature. Close collaboration with UNEP will therefore remain a prominent feature of WMO's activities with emphasis on monitoring of greenhouse gases and the study of their long-term impact on climate.

Climate change

A major problem facing the world stems from the climatic implications of the steady increase of carbon dioxide and of other greenhouse gases in the atmosphere, leading to a change in radiation balance. One of the most serious consequences of this would be a rise in sea-level and inundation of many coastal areas.

For over 40 years WMO has been engaged in the monitoring of global concentration of atmospheric carbon dioxide. WMO, UNEP and ICSU have jointly undertaken periodic assessments of the role of carbon dioxide and other greenhouse gases in climatic changes and associated impacts. At the second assessment meeting of the three organizations, with the participation of scientists from 29 developed and developing countries (Villach, Austria, October 1985), there was a consensus that, if present trends continue, the combined concentrations of atmospheric carbon dioxide and other greenhouse gases would be radiatively equivalent to a doubling of carbon dioxide from pre-industrial levels, possibly as early as the 2030s, leading to a global warming of between 1.5K and 4.5K, but values outside this range cannot be excluded because of some limitations in the assessment method. It is estimated that a global warming of this extent would lead to a sea-level rise of 20–140 centimetres. In the upper portion of this range, the rise would have direct major effects on coastal areas and estuaries. This estimate is based on the expected thermal expansion of the oceans. A significant melting of the west Antarctic ice sheet, leading to a much larger rise in sea-level, although possible at some future date, is not expected during the next century.

There is little doubt that a future change in climate such as is indicated by climate models for a doubling of the concentration of greenhouse gases could have profound effects on global ecosystems, agriculture, water resources and sea ice.

The statement issued by the assessment meeting contains a number of specific recommendations to governments and funding institutions. The role of the WMO/ICSU World Climate Research Programme in further studies of the problem, as well as the follow-up action to be taken jointly by UNEP, WMO and ICSU, has also been specified.

The study and monitoring of the entire climate system falls within the scope of WMO's World Climate Programme, which should therefore merit high priority.

In response to the world-wide concern manifested in the deliberations of both the Tenth World Meteorological Congress (May, 1987) and the fourteenth session of the Governing Council of UNEP (June, 1987), WMO and UNEP have jointly established an Intergovernmental Panel on Climate Change (IPCC). The responsibilities given to the IPCC embrace a wide range of scientific and public policy issues associated with climate change and the assessment of its impacts.

At its fortieth session (June 1988), the WMO Executive Council proposed that the initial responsibilities of the IPCC should include the development of an overall plan for its activities, including:

(*a*) Identification of uncertainties and gaps in our present knowledge with regard to climate change and its potential impacts, and preparation of a plan of action over the short term in filling these gaps;

(*b*) Identification of information needed to evaluate policy implications of climate change and response strategies;

(*c*) Review of current and planned national/international policies related to the greenhouse-gas issue;

(*d*) Scientific and environmental assessments of all aspects of the greenhouse-gas issue and the transfer of these assessments and other relevant information to governments and intergovernmental organizations to be taken into account in their policies on social and economic development and environmental programmes.

Since then a number of events of great significance, some of which are mentioned below, have taken place, showing the continuing concern of governments over the threat of climate change with its likely consequences and the need for urgent action on both international and national levels. These events have, moreover, shown that the initiative which had been taken by WMO in establishing the IPCC jointly with UNEP was a timely one.

The UN General Assembly at its forty-third session (1988) adopted Resolution 43/53 entitled "Protection of global climate for present and future generations of mankind", when more than 70 countries expressed their concern at the prospect of a global warming and rise in sea-level due to greenhouse gases. It is gratifying to note that the resolution gave recognition to WMO's work on the problem of climate change and endorsed the establishment of the IPCC. As was already mentioned, the resolution requested the Secretary-General of WMO and the Executive Director of UNEP to initiate immediate action leading to a comprehensive review and recommendations with respect to scientific and socio-economic aspects of climate change and supported the development of an appropriate framework convention on global climate change.

It would be appropriate to mention that the first session of the IPCC (Geneva, 9–11 November 1988), at which its work plan was established, in fact took place at a time when the UN General Assembly was discussing the draft of its Resolution 43/53. The IPCC, which thus set about its task with the utmost expedition, is now actively engaged in the study of the questions assigned to it as a first step towards formulating its recommendations for international, national and regional consideration. The General Assembly will, of course, consider the outcome of the IPCC's studies. The first report on these studies will be submitted to WMO and UNEP by September 1990 and will be discussed during the second World Climate Conference (see below) in November before being taken up by the UN General Assembly.

A Ministerial Conference on Protection of the Ozone Layer (London, 5–7 March 1989) was followed successively by a Summit Meeting on Protection of the Global Atmosphere (The Hague, 10–11 March 1989) and a Meeting of the Signatories to the Vienna Convention on Protection of the Ozone Layer and the Montreal Protocol on CFCs (Helsinki, 6 April–5 May 1989). The Hague meeting issued a Declaration signed by 24 Heads of States or their senior representatives for strengthening existing institutions within the UN system or developing new ones. The Helsinki meeting recognized the role of WMO in ozone studies (see also page 88).

The WMO Executive Council, at its forty-first session (1989) reviewed the role of the Organization in relation to global climate issues in the light of major developments at the highest governmental levels and in the UN system. The Council agreed that WMO's role is to work with its 160 Members to provide authoritative scientific information and advice on the condition and behaviour of the global atmosphere and climate and the factors that affect them. To enable WMO to fulfil its role the Council also approved a number of measures aimed at strengthening the relevant activities of the Organization. These measures include a new expanded and strengthened system to be known as WMO Global Atmospheric Watch for ensuring an early warning of changes in concentrations of various atmospheric species (especially the greenhouse gases) and also atmospheric ozone. It was also decided to prepare a climate-change detection project in collaboration with other agencies. The Council agreed that the preparation of a framework of an eventual Convention on Climate should begin in anticipation of the completion of the IPCC's work on the scientific assessment of the climate-change issue and the formulation of response strategies. Finally, the Council emphasized the importance of the full participation of developing countries in activities related to the problem of climate change.

Following the decision of the Tenth World Meteorological Congress in 1987, the WMO Executive Council also made arrangements for holding the Second World Climate Conference in November 1990. The Conference will review all aspects relating to global climate issues and will be a major event of international political and scientific importance.

Finally, the Economic Summit Meeting of the Group of the Seven Major Industrialized Nations (Paris, 15–16 July 1989) devoted considerable attention to environmental issues. Its Declaration included a call for urgent international action for the protection of the environment and specifically supported WMO's role in measurement of greenhouse gases and in the climate change detection project.

It seems reasonable to conclude that as a result of the decisions already taken by WMO's policy-making and executive bodies, the Organization is well prepared to fulfil its mandate on the climate-change issue, working in close co-operation with UNEP, ICSU and other organizations in the best traditions of WMO.

Natural disasters

The United Nations General Assembly has designated the decade 1990–1999 as the International Decade for Natural Disaster Reduction to promote concerted international action to reduce the devastation caused by such disasters. WMO will of course participate fully in the programme of the Decade and benefit from its results. The most important WMO activity in this field relates to the Tropical Cyclone Programme, which itself owes its origin to a resolution of the General Assembly adopted in 1970. There is much to say about the sustained efforts of WMO in carrying out this programme, which concerns more than 50 countries, and about the results so far obtained. The major improvements in predicting formation and tracks of tropical cyclones in all regions of the world affected by them, made possible through the WWW, and the establishment of facilities needed for warning systems such as weather radars through UNDP projects, are of course the main achievements. Assistance to cyclone-prone countries in the application of flood-forecasting techniques and storm-surge prediction is another. The protection of life and property from the destructive effects of tropical cyclones must, however, depend on measures for disaster prevention and preparedness. Here again meteorology and hydrology have a contribution to make by providing to national authorities the necessary data for risk evaluation as a basis for long-term protective measures and of course by an efficient system of cyclone warnings during the emergency caused by an actual cyclone. The Tropical Cyclone Programme therefore also includes a prevention and preparedness element for fostering closer contacts between Meteorological and Hydrological Services on the one hand and the various agencies responsible for disaster prevention and preparedness on the other. Close co-operation is also maintained by WMO with the Office of the United Nations Disaster Relief Co-ordinator (UNDRO) and the League of Red Cross and Red Crescent Societies (LRCS), which have special expertise in disaster mitigation. The contribution of meteorology and hydrology in this area is demonstrated by UNDRO's Disaster Prevention Award in 1988 to the Typhoon Committee of WMO and the Economic and Social Commission for Asia and the Pacific (ESCAP), which is responsible for planning and co-ordination of measures for minimizing typhoon damage in the ESCAP region.

Nevertheless, the fact remains that each year tropical cyclones take a heavy toll of life and cause extensive material damage, and in the final analysis it is the prevention and preparedness measures that are crucial for saving lives and property. A greater involvement of WMO in this area, in collaboration with appropriate international organizations and national agencies, might therefore be usefully contemplated.

Hydrology and water resources

While hydrological forecasting possesses a well-developed methodology, that for forecasting hydrology and its future progress in time is rudimentary. The same applies to the water resources applications of the science to assessment, development and management of this prime resource. Even the recent examples of such forecasts, for instance those appearing in the International Association of Hydrological Sciences *Hydrology 2000* report, or in the proceedings of the joint IAHS/IAHR symposium "Water for the Future", add little to the techniques available for understanding the shaping of the future science. What they do foresee are scenarios which are highly correlated with today's activities in hydrology and water resources. However, more difficult to foretell are those unexpected global changes in technology, economics and society which will have an impact on this area.

The continuing rise in the world's population and the resultant increasing demand for water are relatively easy to foresee. What is not so certain is the limit to the world's available water and how this will be altered regionally by changes in climate and the constraints imposed by environmental considerations and the aims of sustainable development. The rise in population will mean that greater numbers are exposed to risk from the natural hazards, such as floods and droughts, while the increasing pollution of water will add another dimension of difficulty. These and other problems are likely to be faced by the world's Hydrological Services during the 21st century—other problems which are much more mundane, such as lack of finance and of trained staff, and government policies which seek to reduce their public sector responsibilities in favour of the private sector. Of particular concern will be the rise in the cost of collecting ground-based data in relation to costs of data processing and modelling.

These pointers indicate a future Hydrology and Water Resources Programme where instrument network design will increasingly focus on integration of a threshold level of ground-based stations needed for calibration purposes using remote sensing methods. Most data collection is expected to be in real time from automated stations, except in the more difficult areas of sediment and water quality. Various forms of data transmission will be employed and a range of processing will be conducted at different levels as computers become cheaper and more powerful. There will be combinations of geographic information systems, including terrain models and distributed hydrological models with improved meteorological forecasts to produce upgraded hydrological forecasts. Hydrological information will become as readily available as that of any other discipline as the information revolution gathers pace.

The Hydrology and Water Resources Programme will promote such advances and continue to seek development of instruments and techniques for hydrological variables where measurements are currently not available. It will continue to place considerable emphasis on transfer of technology through conventional means such as *Guides*, *Manuals* and other publications and via the technology transfer system currently known as HOMS. The programme will endeavour to seek out the best practice for adoption by the hydrological community at large.

There are however wider external influences which will continue to focus different pressures on the Hydrology and Water Resources Programme. Climate change and its many manifestations, particularly sea-level rise, is one example where the programme will have to continue to accommodate interests which contrast with those deemed previously as central by many. Other environmental problems, accidental pollution for example, will become increasingly important. Groundwater studies, the problems of wetlands and the need to monitor abstractions satisfactorily will become increasingly important. The development of information centres, such as the Global Runoff Data Centre at Coblenz, will take place in other areas of the science. These centres and others such as the HOMS National Reference Centres will become networked as the information revolution advances. Hitherto unforeseen changes will probably have an impact on the programme which are greater than those already forecast. The only certain thing in hydrology and water resources is the uncertainty surrounding this endeavour.

Widening the scientific background

The ICSU International Geosphere–Biosphere Programme, with its wide scope and scientific and practical significance, provides a good example of a major inter-disciplinary Earth science programme to which WMO continues to make its contribution.

In the future, it will be even more clearly understood than today that many global environmental issues need a complex, multi-disciplinary approach with the involvement of a wide range of environmental (or Earth) sciences (meteorology, hydrology, oceanography, seismology, geology, etc.). This will raise some problems. Currently, not all of those sciences mentioned above are adequately covered in the existing system of intergovernmental organizations (UN or otherwise). One may therefore speculate on a possible future scenario whereby a new World Environmental Sciences Organization or World Geophysical Organization would be proposed. Such an organization may, eventually, evolve from the existing ones among which WMO is probably the most widely recognized. WMO should be prepared to respond adequately to any such developments. Apart from this, closest collaboration with ICSU and other scientific organizations will continue to be essential for the success of many of WMO's programmes.

Commercial use of meteorological and hydrological information

The trend in some countries to leave more activity to the private sector and recover more costs in the public sector may spread to other parts of the world, leading to increased commercial exploitation of meteorological and hydrological information. Tenth Congress (1987) reaffirmed the principle of free and unrestricted international exchange of information among the Meteorological and Hydrological Services of Member countries and has emphasized that WMO should continue to promote the adherence to this principle. The Organization will need to remain vigilant in this matter so as to ensure that this basic principle is not threatened.

Resources

WMO's method of operation has by tradition been a remarkably economical one, the programmes being implemented by Member countries through facilities provided by them. The fulfilment of WMO's long-term objectives, for which the use of modern technology will be essential, will call for increased resources at both national and international levels. Members' own resources to implement their national programmes related to the internationally agreed WMO programmes would, in many cases, be insufficient. Technical co-operation funds which can be used to finance certain types of activity might also be either inadequate or inappropriate. New lines of approach are envisaged for the solution to the problem, such as sharing of costs through consortium arrangements for certain activities contributing to global benefit and pooling of resources for the establishment of advanced facilities for the benefit of Members within specific regions. It would still be necessary to explore all possible avenues for mobilizing resources to overcome financial impediments. Meteorological and Hydrological Services are now recognized as essential for economic development, their benefit/cost ratios normally being in the range of 5:1 to 20:1. Their role in the concerted efforts to meet some of the challenges facing the world today has also been demonstrated. The possibility of obtaining support from international financial institutions and intergovernmental regional bodies outside the United Nations engaged in strengthening co-operation for social and economic progress will need to be kept in view.

Concluding remarks

After nearly four decades of existence with an impressive record of achievement to its credit, WMO now faces new challenges and opportunities. The challenges are posed by the growing needs for specialized meteorological and hydrological services closely related to social and economic development and by the environmental problems which, because of their alarming long-term consequences, have led to concerted international action in which meteorology and hydrology and other geophysical sciences have an important role. The opportunities, which at the same time constitute an impetus, are provided by such technological advances as improved techniques of satellite and automated observations offering a realistic approach to global data coverage, means for rapid collection of data and their processing, and the new generation of computers capable of running global weather-prediction models. All these factors will have an impact on one or other of WMO's manifold activities; they may also lead in due course to consideration of a possible adjustment to the Convention and even to the title of the Organization.

WMO's role will be wider than hitherto if the challenges and opportunities indicated above are be taken into account. The major advances that will result will prepare the Organization for the more distant future extending into the twenty-first century.

The fulfilment of such long-term objectives will depend particularly on a number of factors. The first and foremost is perhaps the ability of all Members to participate fully in the basic programme, the World Weather Watch, because of the very high degree of interdependence of Members in this programme. It will therefore be essential to reduce the disparity among Members in order to build an effective global and regional WWW system. New lines of approach are envisaged to achieve this aim, in particular to ensure a minimum level of meteorological and hydrological services in each country.

Close international co-operation, which among other things forms the basis of future plans, will be essential. Fortunately, the Organization is well placed in this respect. An international spirit has prevailed within the Organization throughout its history, the high degree of international collaboration being reflected in its activities. Collaboration with ICSU and other scientific bodies, which has characterized WMO's policy all along, will also be essential, particularly because of the growing importance of inter-disciplinary activities.

Adequate resources and their effective use will be another important requirement. While technological developments have given new tools essential for the provision of services, the problem of meeting the costs also needs to be taken into account. Here again new lines of approach are being considered, such as sharing of costs and development of co-operative arrangements among Members. Other avenues will still need to be explored. The method of operation of WMO is a financially economical one, as will be clear from the preceding chapters, the regular budget of the Organization being meant only to provide an international framework for the overall activities, which are of an inherently international character.

The responsiveness of the Organization to the changing requirements of time, which cannot be foreseen, as well as the adaptability to any new arrangements necessary to meet those requirements, will be an important condition for success in the future. Past evidence in this respect has been positive.

Finally, the need for continued full support of Members is self-evident. The confidence shown by them in the Organization in the past leaves no doubt that their support will be forthcoming in full measure.

The past has given WMO a solid basis on which to build the future and the Organization has every reason to contemplate with optimism and hope an even more important role in the work of the UN system.

APPENDIX A

WMO CONVENTION

The text of the WMO Convention as originally approved is reproduced below. At subsequent sessions of Congress, various Articles were amended. Each of these is marked by a superior reference number or by an asterisk referring to an explanatory note at the end of the text.

CONVENTION OF THE WORLD METEOROLOGICAL ORGANIZATION

Authentic text[†]

With a view to co-ordinating, standardizing, and improving world meteorological activities and to encouraging an efficient exchange of meteorological information between countries in the aid of human activities the contracting States agree to the present Convention, as follows:[1]

PART I

Establishment

ARTICLE 1

The World Meteorological Organization (hereinafter called the Organization) is hereby established.

PART II

ARTICLE 2 [1,2,3]

Purposes

The purposes of the Organization shall be:

(*a*) To facilitate world-wide co-operation in the establishment of networks of stations for the making of meteorological observations or other geophysical observations related to meteorology and to promote the establishment and maintenance of meteorological centres charged with the provision of meteorological services;

(*b*) To promote the establishment and maintenance of systems for the rapid exchange of weather information;

(*c*) To promote standardization of meteorological observations and to ensure the uniform publication of observations and statistics;

(*d*) To further the application of meteorology to aviation, shipping, agriculture, and other human activities; and

(*e*) To encourage research and training in meteorology and to assist in co-ordinating the international aspects of such research and training.

PART III

Membership

ARTICLE 3

Members

The following may become Members of the Organization by the procedure set forth in the present Convention:

(*a*) Any State represented at the Conference of Directors of the International Meteorological Organization convened at Washington, DC, on September 22, 1947, as listed in Annex I attached hereto, and which signs the present Convention and ratifies it in accordance with Article 32, or which accedes thereto, in accordance with Article 33;

(*b*) Any Member of the United Nations having a Meteorological Service by acceding to the present Convention in accordance with Article 33;

(*c*) Any State, fully responsible for the conduct of its international relations and having a Meteorological Service, not listed in Annex I of the present Convention and not a Member of the United Nations, after the submission of a request for membership to the Secretariat of the Organization and after its approval by two-thirds of the Members of the Organization as specified in paragraphs (*a*), (*b*) and (*c*) of this Article by acceding to the present Convention in accordance with Article 33;

(*d*) Any territory or group of territories maintaining its own Meteorological Service and listed in Annex II attached hereto, upon application of the present Convention on its behalf, in accordance with paragraph (*a*) of Article 34, by the State or States responsible for its international relations and represented at the Conference of Directors of the International Meteorological Organization convened at Washington, DC, on September 22, 1947, as listed in Annex I of the present Convention;

(*e*) Any territory or group of territories, not listed in Annex II of the present Convention, maintaining its own Meteorological Service but not responsible for the conduct of its international relations, on behalf of which the present Convention is applied in accordance with paragraph (*b*) of Article 34, provided that the request for membership is presented by the Member responsible for its international relations, and secures approval by two-thirds of the Members of the Organization as specified in paragraphs (*a*), (*b*) and (*c*) of this Article;

(*f*) Any trust territory or group of trust territories maintaining its own Meteorological Service and administered by the United Nations to which the United Nations applies the present Convention in accordance with Article 34.

Any request for membership in the Organization shall state in accordance with which paragraph of this Article membership is sought.

PART IV

Organization

ARTICLE 4 [1,*]

(*a*) The Organization shall comprise:

(1) The World Meteorological Congress (hereinafter called the Congress);

(2) The Executive Committee;

(3) Regional Meteorological Associations (hereinafter called the Regional Associations);

[†] For comparative purposes, orthography and the use of capitals have been standardized throughout this text to conform to present usage within the Organization.

(4) Technical Commissions;

(5) The Secretariat.

(*b*) There shall be a President and two Vice-Presidents of the Organization who shall also be President and Vice-Presidents of the Congress and of the Executive Committee.

PART V

Eligibility [1]

ARTICLE 5 [1,2,*]

(*a*) Eligibility for election to the offices of President and Vice-President of the Organization, of president and vice-president of the Regional Associations, and for membership, subject to the provisions of Article 13 (*c*) of the present Convention, on the Executive Committee should be confined to the Directors of Meteorological Services of Members of the Organization.

(*b*) In the performance of their duties, the officers of the Organization and the members of the Executive Committee should regard themselves as representatives of the Organization rather than as representatives of particular Members thereof.

PART VI

The World Meteorological Congress

ARTICLE 6 [1,2]

Composition

(*a*) The Congress is the supreme body of the Organization and shall be composed of delegates representing Members. Each Member shall designate one of its delegates, who should be the Director of its Meteorological Service, as its principal delegate.

(*b*) With a view to securing the widest possible technical representation, any Director of a Meteorological Service or any other individual may be invited by the President to be present at and participate in the discussions of the Congress.

ARTICLE 7 [1,2,*]

Functions

The functions of the Congress shall be:

(*a*) To determine general regulations, subject to the provisions of the present Convention, prescribing the constitution and the functions of the various bodies of the Organization;

(*b*) To determine its own rules of procedure;

(*c*) To elect the President and Vice-Presidents of the Organization, and other members of the Executive Committee, in accordance with the provisions of Article 10 (*a*) (4) of the present Convention. Presidents and vice-presidents of Regional Associations and Technical Commissions shall be elected in accordance with the provisions of Articles 18 (*e*) and 19 (*c*), respectively, of the present Convention;

(*d*) To adopt technical regulations covering meteorological practices and procedures;

(*e*) To determine general policies for the fulfilment of the purposes of the Organization as set forth in Article 2 of the present Convention;

(*f*) To make recommendations to Members on matters within the purposes of the Organization;

(*g*) To refer to any other body of the Organization any matter within the provisions of the present Convention upon which such body is empowered to act;

(*h*) To consider the reports and activities of the Executive Committee and to take such action in regard thereto as the Congress may determine;

(*i*) To establish Regional Associations in accordance with the provisions of Article 18; to determine their geographical limits, co-ordinate their activities, and consider their recommendations;

(*j*) To establish Technical Commissions in accordance with the provisions of Article 19; to define their terms of reference, co-ordinate their activities, and consider their recommendations;

(*k*) To determine the location of the Secretariat of the Organization;

(*l*) To take any other appropriate action to further the purposes of the Organization.

ARTICLE 8

Execution of Congress decisions

(*a*) All Members shall do their utmost to implement the decisions of the Congress.

(*b*) If, however, any Member finds it impracticable to give effect to some requirement in a technical resolution adopted by Congress, such Member shall inform the Secretary-General of the Organization whether its inability to give effect to it is provisional or final, and state its reasons therefor.

ARTICLE 9 [1,*]

Meetings

Meetings of the Congress shall be convened by decision of the Congress or of the Executive Committee at intervals not exceeding four years.

ARTICLE 10 [1,2,3,*]

Voting

(*a*) Each Member shall have one vote in decisions of the Congress, except that only Members of the Organization which are States, as specified in paragraphs (*a*), (*b*) and (*c*) of Article 3 of the present Convention (hereinafter referred to as "Members which are States"), shall be entitled to vote on any of the following subjects:

(1) Amendment or interpretation of the present Convention or proposals for a new Convention;

(2) Membership of the Organization;

(3) Relations with the United Nations and other intergovernmental organizations;

(4) Election of the President and Vice-Presidents of the Organization and of the members of the Executive Committee other than the presidents and vice-presidents of the Regional Associations.

(*b*) Decisions of the Congress shall be by two-thirds majority of the votes cast for and against, except that elections of individuals to serve in any capacity in the Organization shall be by simple majority of the votes cast. The provisions of this paragraph, however, shall not apply to decisions taken in accordance with Articles 3, 25, 26, and 28 of the present Convention.

ARTICLE 11 [1]

Quorum

A majority of the Members shall be required to constitute a quorum for meetings of the Congress. For those meetings of the Congress at which decisions are taken on the subjects enumerated in paragraph (*a*) of Article 10, a majority of the Members which are States shall be required to constitute a quorum.

ARTICLE 12 [1]

First meeting of the Congress

The first meeting of the Congress shall be convened by the President of the International Meteorological Committee of the International Meteorological Organization as soon as practicable after the coming into force of the present Convention.

PART VII

The Executive Committee

ARTICLE 13 [1,2,3,4,5,6,*]

Composition

The Executive Committee shall consist of:

(*a*) The President and Vice-Presidents of the Organization;

(*b*) The presidents of Regional Associations, or in the event that presidents cannot attend, alternates as provided for in the General Regulations;

(*c*) Directors of Meteorological Services of Members of the Organization or their alternates, equal in number to the number of Regions, provided that not more than one-third of the members of the Executive Committee, including the President and Vice-Presidents of the Organization, shall come from one Region.

ARTICLE 14 [1,2,3,4,*]

Functions

The Executive Committee is the executive body of the Congress and its functions shall be:

(*a*) To supervise the execution of the resolutions of the Congress;

(*b*) To adopt resolutions arising out of recommendations of the Technical Commissions on matters of urgency affecting the Technical Regulations, provided that all Regional Associations concerned are given an opportunity to express their approval or disapproval before adoption by the Executive Committee;

(*c*) To provide technical information, counsel, and assistance in the field of meteorology;

(*d*) To study and make recommendations on any matter affecting international meteorology and the operation of meteorological services;

(*e*) To prepare the agenda for the Congress and to give guidance to the Regional Associations and Technical Commissions in the preparation of their agenda;

(*f*) To report on its activities to each session of the Congress;

(*g*) To administer the finances of the Organization in accordance with the provisions of Part XI of the present Convention;

(*h*) To perform such other functions as may be conferred on it by the Congress or by the present Convention.

ARTICLE 15 [1,*]

Meetings

The Executive Committee shall meet at least once a year. The time and place of the meeting shall be determined by the President of the Organization, taking account of the views of the other members of the Committee.

ARTICLE 16 [1,*]

Voting

Decisions of the Executive Committee shall be by two-thirds majority of the votes cast for and against. Each member of the Executive Committee shall have only one vote, notwithstanding that he may be a member in more than one capacity.

ARTICLE 17 [1,*]

Quorum

The quorum shall consist of a majority of the members of the Executive Committee.

PART VIII

Regional Associations

ARTICLE 18 [1,*]

(*a*) Regional Associations shall be composed of the Members of the Organization, the networks of which lie in or extend into the Region.

(*b*) Members of the Organization shall be entitled to attend the meetings of Regional Associations to which they do not belong, take part in the discussions, present their views upon questions affecting their own Meteorological Service, but shall not have the right to vote.

(*c*) Regional Associations shall meet as often as necessary. The time and place of the meeting shall be determined by the presidents of the Regional Associations in agreement with the President of the Organization.

(*d*) The functions of the Regional Associations shall be:

(i) To promote the execution of the resolutions of Congress and the Executive Committee in their respective Regions;

(ii) To consider matters brought to their attention by the Executive Committee;

(iii) To discuss matters of general meteorological interest and to co-ordinate meteorological and associated activities in their respective Regions;

(iv) To make recommendations to Congress and the Executive Committee on matters within the purposes of the Organization;

(v) To perform such other functions as may be conferred on them by the Congress.

(*e*) Each Regional Association shall elect its president and vice-president.

PART IX

Technical Commissions

ARTICLE 19*

(*a*) Commissions consisting of technical experts may be established by the Congress to study and make recommendations to the Congress and the Executive Committee on any subject within the purposes of the Organization.

(*b*) Members of the Organization have the right to be represented on the Technical Commissions.

(*c*) Each Technical Commission shall elect its president and vice-president.

(*d*) Presidents of Technical Commissions may participate without vote in the meetings of the Congress and of the Executive Committee.

PART X

The Secretariat

ARTICLE 20

The permanent Secretariat of the Organization shall be composed of a Secretary-General and such technical and clerical staff as may be required for the work of the Organization.

ARTICLE 21*

(*a*) The Secretary-General shall be appointed by the Congress on such terms as the Congress may approve.

(*b*) The staff of the Secretariat shall be appointed by the Secretary-General with the approval of the Executive Committee in accordance with regulations established by the Congress.

ARTICLE 22

(*a*) The Secretary-General is responsible to the President of the Organization for the technical and administrative work of the Secretariat.

(*b*) In the performance of their duties, the Secretary-General and the staff shall not seek or receive instructions from any authority external to the Organization. They shall refrain from any action which might reflect on their position as international officers. Each Member of the Organization on its part shall respect the exclusively international character of the responsibilities of the Secretary-General and the staff and not seek to influence them in the discharge of their responsibilities to the Organization.

PART XI

Finances

ARTICLE 23 [1,*]

(*a*) The Congress shall determine the maximum expenditures which may be incurred by the Organization on the basis of estimates submitted by the Secretary-General and recommended by the Executive Committee.

(*b*) The Congress shall delegate to the Executive Committee such authority as may be required to approve the annual expenditures of the Organization within the limitations determined by the Congress.

ARTICLE 24

The expenditures of the Organization shall be apportioned among the Members of the Organization in the proportions determined by the Congress.

PART XII

Relations with the United Nations

ARTICLE 25 [1]

The Organization shall be brought into relationship with the United Nations pursuant to Article 57 of the Charter of the United Nations, subject to the approval of the terms of the agreement by two-thirds of the Members which are States.

PART XIII

Relations with other organizations

ARTICLE 26 [1,*]

(*a*) The Organization shall establish effective relations and co-operate closely with such other intergovernmental organizations as may be desirable. Any formal agreement entered into with such organizations shall be made by the Executive Committee, subject to approval by two-thirds of the Members which are States.

(*b*) The Organization may on matters within its purposes make suitable arrangements for consultation and co-operation with non-governmental international organizations and, with the consent of the government concerned, with national organizations, governmental or non-governmental.

(*c*) Subject to approval by two-thirds of the Members which are States, the Organization may take over from any other international organization or agency, the purpose and activities of which lie within the purposes of the Organization, such functions, resources, and obligations as may be transferred to the Organization by international agreement or by mutually acceptable arrangements entered into between competent authorities of the respective organizations.

PART XIV

Legal status, privileges and immunities

ARTICLE 27 [1]

(*a*) The Organization shall enjoy in the territory of each Member such legal capacity as may be necessary for the fulfilment of its purposes and for the exercise of its functions.

(*b*) (i) The Organization shall enjoy in the territory of each Member to which the present Convention applies such privileges and immunities as may be necessary for the fulfilment of its purposes and for the exercise of its functions.

(*b*) (ii) Representatives of Members and officials of the Organization shall similarly enjoy such privileges and immunities as are necessary for the independent exercise of their functions in connection with the Organization.

(*c*) Such legal capacity, privileges, and immunities shall be defined in a separate agreement to be prepared by the Organization in consultation with the Secretary-General of the United Nations and concluded between the Members which are States.

PART XV

Amendments

ARTICLE 28

(*a*) The text of any proposed amendment to the present Convention shall be communicated by the Secretary-General to Members of the Organization at least six months in advance of its consideration by the Congress.

(*b*) Amendments to the present Convention involving new obligations for Members shall require approval by the Congress, in accordance with the provisions of Article 10 of the present Convention, by a two-thirds majority vote, and shall come into force on acceptance by two-thirds of the Members which are States for each such Member accepting the amendment and thereafter for each remaining such Member on acceptance by it. Such amendments shall come into force for any Member not responsible for its own international relations upon the acceptance on behalf of such a Member by the Member responsible for the conduct of its international relations.

(*c*) Other amendments shall come into force upon approval by two-thirds of the Members which are States.

PART XVI

Interpretation and disputes

ARTICLE 29

Any question or dispute concerning the interpretation or application of the present Convention which is not settled by negotiation or by the Congress shall be referred to an independent arbitrator appointed by the President of the International Court of Justice, unless the parties concerned agree on another mode of settlement.

PART XVII

Withdrawal

ARTICLE 30

(*a*) Any Member may withdraw from the Organization on twelve months' notice in writing given by it to the Secretary-General of the Organization, who shall at once inform all the Members of the Organization of such notice of withdrawal.

(*b*) Any Member of the Organization not responsible for its own international relations may be withdrawn from the Organization on twelve months' notice in writing given by the Member or other authority responsible for its international relations to the Secretary-General of the Organization, who shall at once inform all the Members of the Organization of such notice of withdrawal.

PART XVIII

Suspension

ARTICLE 31

If any Member fails to meet its financial obligations to the Organization or otherwise fails in its obligations under the present Convention, the Congress may by resolution suspend it from exercising its rights and enjoying privileges as a Member of the Organization until it has met such financial or other obligations.

PART XIX

Ratification and accession

ARTICLE 32

The present Convention shall be ratified by the signatory States and the instruments of ratification shall be deposited with the Government of the United States of America, which will notify each signatory and acceding State of the date of deposit thereof.

ARTICLE 33 [1]

Subject to the provisions of Article 3 of the present Convention, accession shall be effected by the deposit with the Government of the United States of America of an instrument of accession, which shall take effect on the date of its receipt by the Government of the United States of America, which will notify each signatory and acceding State thereof.

ARTICLE 34

Subject to the provisions of Article 3 of the present Convention,

(*a*) Any contracting State may declare that its ratification of, or accession to, the present Convention includes any territory or group of territories for the international relations of which it is responsible.

(*b*) The present Convention may at any time thereafter be applied to any such territory or group of territories upon a notification in writing to the Government of the United States of America and the present Convention shall apply to the territory or group of territories on the date of the receipt of the notification by the Government of the United States of America, which will notify each signatory and acceding State thereof.

(*c*) The United Nations may apply the present Convention to any trust territory or group of trust territories for which it is the administering authority. The Government of the United States of America will notify all signatory and acceding States of any such application.

PART XX

Entry into force

ARTICLE 35

The present Convention shall come into force on the thirtieth day after the date of the deposit of the thirtieth instrument of ratification or accession. The present Convention shall come into force for each State ratifying or acceding after that date on the thirtieth day after the deposit of its instrument of ratification or accession.

The present Convention shall bear the date on which it is opened for signature and shall remain open for signature for a period of 120 days thereafter.

IN WITNESS WHEREOF the undersigned, having been duly authorized by their respective governments, have signed the present Convention.

DONE at Washington this eleventh day of October 1947, in the English and French languages, each equally authentic, the original of which shall be deposited in the archives of the Government of the United States of America. The Government of the United States of America shall transmit certified copies thereof to all the signatory and acceding States.

The signatures of the delegates of the countries mentioned below follow here.

SIGNATORY COUNTRIES

The Convention, which was opened for signature on October 11, 1947, at Washington and remained open for signature for a period of 120 days thereafter, has been signed on behalf of the following countries:

ARGENTINA
AUSTRALIA
BELGIUM
(including the Belgian Congo)
BRAZIL
BURMA
CANADA
CHILE
CHINA
COLOMBIA
CUBA
CZECHOSLOVAKIA
DENMARK
DOMINICAN REPUBLIC
ECUADOR
EGYPT
FINLAND
FRANCE
GREECE
GUATEMALA
HUNGARY
ICELAND
INDIA
IRELAND
ITALY
MEXICO
KINGDOM OF THE NETHERLANDS
NEW ZEALAND
NORWAY
PAKISTAN
PARAGUAY
POLAND
PORTUGAL
REPUBLIC OF THE PHILIPPINES
SIAM
SWEDEN
SWITZERLAND
TURKEY
UNION OF SOUTH AFRICA
UNITED KINGDOM OF GREAT BRITAIN AND NORTHERN IRELAND
UNITED STATES OF AMERICA
URUGUAY
YUGOSLAVIA

ANNEX I

States represented at the Conference of Directors of the International Meteorological Organization convened at Washington DC, on September 22, 1947

Argentina
Australia
Belgium
Brazil
Burma
Canada
Chile
China
Colombia
Cuba
Czechoslovakia
Denmark
Dominican Republic
Ecuador
Egypt
Finland
France
Greece
Guatemala
Hungary
Iceland
India
Ireland
Italy
Mexico
Netherlands
New Zealand
Norway
Pakistan
Paraguay
Philippines
Poland
Portugal
Rumania
Siam
Sweden
Switzerland
Turkey
Union of South Africa
Union of Soviet Socalist Republics
United Kingdom of Great Britain and Northern Ireland
United States of America
Uruguay
Venezuela
Yugoslavia

ANNEX II

Territories or groups of territories which maintain their own Meteorological Services and of which the States responsible for their international relations are represented at the Conference of Directors of the International Meteorological Organization convened at Washington, DC, September 22, 1947

Anglo-Egyptian Sudan
Belgian Congo
Bermuda
British East Africa
British Guiana
British West Africa
Cameroons
Cape Verde Islands
Ceylon
Curacao
French Equatorial Africa
French Oceanic Colonies
French Somaliland
French Togoland
French West Africa
Hong Kong
Indo-China
Jamaica
Madagascar
Malaya
Mauritius
Morocco (not including the Spanish Zone)
Netherlands Indies
New Caledonia
Palestine
Portuguese East Africa
Portuguese West Africa
Rhodesia
Surinam
Tunisia

EXPLANATORY NOTES

Some of the amendments to the Convention necessitated the renumbering of certain Articles. Article numbers in the text in italics are those valid at the time of the amendment; where a change in numbering has taken place, the original number is also given (bracketed) for ease of reference. Superior figures indicate the order of amendment. An asterisk refers to the general amendment approved by Ninth Congress.

Amendments approved by Third Congress (1959)

Article 10 [1], provision (a) (2), was amended to read:

(2) Requests for membership in the Organization;

Article 13 [1] was amended to read:

ARTICLE 13

Composition

The Executive Committee shall consist of:

(*a*) The President and the Vice-Presidents of the Organization;

(*b*) The presidents of Regional Associations, who can be replaced at sessions by their alternates, as provided for in the General Regulations;

(*c*) Nine Directors of Meteorological Services of Members of the Organization, who can be replaced at sessions by alternates, provided:

(i) That these alternates shall be as provided for in the General Regulations;

(ii) That no more than one-third of the members of the Executive Committee, comprising the President and the Vice-Presidents of the Organization, the presidents of Regional Associations and the nine elected Directors, shall come from one Region, this Region being determined in the case of each Member in accordance with the General Regulations.

Amendments approved by Fourth Congress (1963)

Article 2 [1], provision (b), was amended to read:

(*b*) To promote the establishment and maintenance of systems for the rapid exchange of meteorological information;

Part V - Eligibility [1], Article 5 [1], was amended to read:

PART V

Officers of the Organization and members of the Executive Committee

ARTICLE 5

(*a*) Eligibility for election to the offices of President and Vice-Presidents of the Organization, of president and vice-president of the Regional Associations, and for membership, subject to the provisions of Article 12 (*c*) (ii) of the Convention, of the Executive Committee shall be confined to persons who are designated as the Directors of their Meteorological Service by the Members of the Organization for the purpose of this Convention.

(*b*) In the performance of their duties, all officers of the Organization and members of the Executive Committee shall act as representatives of the Organization and not as representatives of particular Members thereof.

Article 6 [1] was amended to read:

ARTICLE 6

Composition

(*a*) The Congress is the general assembly of delegates representing Members and as such is the supreme body of the Organization.

(*b*) Each Member shall designate one of its delegates, who should be the Director of its Meteorological Service, as its principal delegate at Congress.

(*c*) With a view to securing the widest possible technical representation, any Director of a Meteorological Service or any other individual may be invited by the President to be present and to participate in the discussions of the Congress in accordance with the provisions of the General Regulations (hereinafter referred to as "Regulations").

Article 7 [1] was amended to read:

ARTICLE 7

Functions

In addition to functions set out in other Articles of the Convention, the primary duties of the Congress shall be:

(*a*) To determine general policies for the fulfilment of the purposes of the Organization as set forth in Article 2;

(*b*) To make recommendations to Members on matters within the purposes of the Organization;

(*c*) To refer to any body of the Organization any matter within the provisions of the Convention upon which such a body is empowered to act;

(*d*) To determine regulations prescribing the procedures of the various bodies of the Organization, and in particular, the General, Technical, Financial and Staff Regulations;

(*e*) To consider the reports and activities of the Executive Committee and to take appropriate action in regard thereto;

(*f*) To establish Regional Associations in accordance with the provisions of Article 17; to determine their geographical limits, co-ordinate their activities, and consider their recommendations;

(*g*) To establish Technical Commissions in accordance with the provisions of Article 18; to define their terms of reference, co-ordinate their activities, and consider their recommendations;

(*h*) To determine the location of the Secretariat of the Organization;

(*i*) To elect the President and Vice-Presidents of the Organization, and members of the Executive Committee other than the presidents of the Regional Associations.

Congress may also take any other appropriate action on matters affecting the Organization.

Article 10 [2] was amended to read:

ARTICLE 10

Voting

(*a*) In a vote to Congress each Member shall have one vote. However, only Members of the Organization which are States (hereinafter referred to as "Members which are States") shall be entitled to vote or to take a decision on the following subjects:

(1) Amendment or interpretation of the Convention or proposals for a new Convention;

(2) Requests for Membership of the Organization;

(3) Relations with the United Nations and other intergovernmental organizations;

(4) Election of the President and Vice-Presidents of the Organization and of the members of the Executive Committee other than the presidents of the Regional Associations.

(*b*) Decisions shall be by a two-thirds majority of the votes cast for and against, except that elections of individuals to serve in any capacity in the Organization shall be by simple majority of the votes cast. The provisions of this paragraph, however, shall not apply to decisions taken in accordance with Articles 3, 24, 25 and 27 of the Convention.

Article 11 [1] was amended to read:

ARTICLE 11

Quorum

The presence of delegates of a majority of the Members shall be required to constitute a quorum for meetings of the Congress. For those meetings of the Congress at which decisions are taken on the subjects enumerated in paragraph (*a*) of Article 10, the presence of delegates of a majority of the Members which are States shall be required to constitute a quorum.

Article 12 [1] was deleted.

Article 13 [2] was renumbered to read Article 12 - Composition, and provision (c) was amended to read:

(*c*) Twelve Directors of Meteorological Services of Members of the Organization, who can be replaced at sessions by alternates, provided:

(i) That these alternates shall be as provided for in the General Regulations;

(ii) That not more than seven and not less than two members of the Executive Committee, comprising the President and Vice-Presidents of the Organization, the presidents of Regional Associations and the twelve elected Directors, shall come from one Region, this Region being determined in the case of each member in accordance with the General Regulations.

Article 14 [1] was amended to read:

ARTICLE 13

Functions

The Executive Committee is the executive body of the Organization.

In addition to functions set out in other Articles of the Convention, the primary functions of the Executive Committee shall be:

(*a*) To implement the decisions taken by the Members of the Organization either in Congress or by means of correspondence and to conduct the activities of the Organization in accordance with the intention of such decisions;

(*b*) To consider and, where necessary, take action on behalf of the Organization on resolutions and recommendations of Regional Associations and Technical Commissions in accordance with the procedures laid down in the Regulations;

(*c*) To provide technical information, counsel, and assistance in the field of meteorology;

(*d*) To study and make recommendations on any matter affecting international meteorology and the operation of Meteorological Services;

(*e*) To prepare the agenda for the Congress and to give guidance to the Regional Associations and Technical Commissions in the preparation of their agenda;

(*f*) To report on its activities to each session of Congress;

(*g*) To administer the finances of the Organization in accordance with the provisions of Part XI of the Convention.

The Executive Committee may also perform such other functions as may be conferred on it by the Congress or by Members collectively.

Article 15 [1] was amended to read:

ARTICLE 14

Sessions

(*a*) The Executive Committee shall normally hold a session at least once a year, at a place and on a date to be determined by the President of the Organization, after consultation with other members of the Committee.

(*b*) An extraordinary session of the Executive Committee shall be convened according to the procedures contained in the Regulations, after receipt by the Secretary-General of requests from a majority of the members of the Executive Committee. Such a session may also be convened by agreement between the President and the two Vice-Presidents of the Organization.

Article 16 was renumbered Article 15.

Article 17 [1] was amended to read:

ARTICLE 16

Quorum

The presence of two-thirds of the members shall be required to constitute the quorum for meetings of the Executive Committee.

(Articles 18 to 22 were renumbered 17 to 21 respectively.)

PART XI - Finances, Article 23 [1], was amended to read:

PART XI

Finances

ARTICLE 22

(*a*) The Congress shall determine the maximum expenditure which may be incurred by the Organization on the basis of the estimates submitted by the Secretary-General, after prior examination by, and with the recommendations of, the Executive Committee.

(*b*) The Congress shall delegate to the Executive Committee such authority as may be required to approve the annual expenditures of the Organization within the limitations determined by the Congress.

PART XII - Relations with the United Nations, Article 25 [1], was amended to read:

PART XII

Relations with the United Nations

ARTICLE 24

The Organization shall be in relationship with the United Nations pursuant to Article 57 of the Charter of the United Nations. Any agreement concerning such relationship shall require approval by two-thirds of the Members which are States.

Article 26 [1] was renumbered Article 25 and provision (a) was amended to read:

(*a*) The Organization shall establish effective relations and co-operate closely with such other intergovernmental organizations as may be desirable. Any formal agreement entered into with such organizations shall be made by the Executive Committee, subject to approval by two-thirds of the Members which are States, either in Congress or by correspondence.

Article 27 [1] was amended to read:

PART XIV

Legal status, privileges and immunities

ARTICLE 26

(*a*) The Organization shall enjoy in the territory of each Member such legal capacity as may be necessary for the fulfilment of its purposes and for the exercise of its functions.

(*b*) (i) The Organization shall enjoy in the territory of each Member to which the present Convention applies such privileges and immunities as may be necessary for the fulfilment of its purposes and for the exercise of its functions.

(ii) Representatives of Members, officers and officials of the Organization as well as members of the Executive Committee shall similarly enjoy such privileges and immunities as are necessary for the independent exercise of their functions in connection with the Organization.

(*c*) In the territory of any Member which is a State and which has acceded to the Convention on the Privileges and Immunities of the Specialized Agencies adopted by the General Assembly of the United Nations on 21 November 1947 such legal capacity, privileges and immunities shall be those defined in the said Convention.

(Articles 28 to 35 were renumbered 27 to 34 respectively.)

Amendments approved by Fifth Congress (1967)

Article 2 [2], provision (d), was amended to read:

(*d*) To further the application of meteorology to aviation, shipping, water problems, agriculture, and other human activities; and

Article 4 [1], provision (b), was amended to read:

(*b*) There shall be a President and three Vice-Presidents of the Organization who shall also be a President and Vice-Presidents of the Congress and of the Executive Committee.

After Article 4, a new Article was inserted:

ARTICLE 5

The activities of the Organization and the conduct of its affairs shall be decided by the Members of the Organization.

(*a*) Such decisions shall normally be taken by Congress in session;

(*b*) However, except on matters reserved in the Convention for decisions by Congress, decisions may also be taken by Members by correspondence, when urgent action is required between sessions of Congress. Such a vote shall be taken upon receipt by the Secretary-General of the request of a majority of the Members of the Organization, or when so decided by the Executive Committee. Such votes shall be conducted in accordance with Articles 11 and 12 of the Convention and with the General Regulations (hereinafter referred to as the "Regulations").

(All following Articles were renumbered accordingly.)

Article 10 (9) [1] was amended to read:

ARTICLE 10

Sessions

(*a*) Congress shall normally be convened at intervals as near as possible to four years, at a place and on a date to be decided by the Executive Committee;

(*b*) An extraordinary Congress may be convened by decision of the Executive Committee;

(*c*) On receipt of requests for an extraordinary Congress from one-third of the Members of the Organization, the Secretary-General shall conduct a vote by correspondence and if a simple majority of the Members are in favour, an extraordinary Congress shall be convened.

Article 11 (10) [3]: Article 10 (c) was added to the list of Articles in the last sentence.

Article 133, provision (c), was amended to read:

(*c*) Fourteen Directors of Meteorological Services of Members of the Organization, who can be replaced at sessions by alternates, provided:

Article 14 [2] was amended to read:

ARTICLE 14

Functions

The Executive Committee is the executive body of the Organization and is responsible to Congress for the co-ordination of the programmes of the Organization and for the utilization of its budgetary resources in accordance with the decisions of Congress.

In addition to functions set out in other Articles of the Convention, the primary functions of the Executive Committee shall be:

(*a*) To implement the decisions taken by the Members of the Organization either in Congress or by means of correspondence and to conduct the activities of the Organization in accordance with such decisions;

(*b*) To examine the programme and budget estimates for the following financial period prepared by the Secretary-General and to present its observations and its recommendations thereon to Congress;

(The former provisions (b) to (g) were accordingly renumbered (c) to (h).)

Article 16 [1]: The existing text was numbered (a) and the following added:

(*b*) Between sessions, the Executive Committee may vote by correspondence. Such votes shall be conducted in accordance with Articles 16 (*a*) and 17 of the Convention.

Article 33 [1] was amended to read:

ARTICLE 33

Subject to the provisions of Article 3 of the present Convention, accession shall be effected by the deposit of an instrument of accession with the Government of the United States of America, which shall notify each Member of the Organization thereof.

Amendments approved by Seventh Congress (1975)

The preamble [1] was amended to read:

With a view to co-ordinating, standardizing and improving world meteorological and related activities, and to encouraging an efficient exchange of meteorological and related information between countries in the aid of human activities, the contracting States agree to the present Convention, as follows:

Article 2 [3] was amended to read:

ARTICLE 2

Purposes

The purposes of the Organization shall be:

(*a*) To facilitate world-wide co-operation in the establishment of networks of stations for the making of meteorological observations as well as hydrological and other geophysical observations related to meteorology, and to promote the establishment and maintenance of centres charged with the provision of meteorological and related services;

(*b*) To promote the establishment and maintenance of systems for the rapid exchange of meteorological and related information;

(*c*) To promote standardization of meteorological and related observations and to ensure the uniform publication of observations and statistics;

(*d*) To further the application of meteorology to aviation, shipping, water problems, agriculture and other human activities;

(*e*) To promote activities in operational hydrology and to further close co-operation between Meteorological and Hydrological Services; and

(*f*) To encourage research and training in meteorology and, as appropriate, in related fields and to assist in co-ordinating the international aspects of such research and training.

Part V, Article 6 (5) [2]*, was amended to read:*

PART V

Officers of the Organization and members of the Executive Committee

ARTICLE 6

(*a*) Eligibility for election to the offices of President and Vice-Presidents of the Organization, of president and vice-president of the Regional Associations, and for membership, subject to the provisions of Article 13 (*c*) (ii) of the Convention, of the Executive Committee, shall be confined to persons who are designated as Directors of their Meteorological or Hydrometeorological Services by the Members of the Organization for the purpose of this Convention, as provided for in the Regulations.

(*b*) In the performance of their duties, all officers of the Organization and members of the Executive Committee shall act as representatives of the Organization and not as representatives of particular Members thereof.

Article 7 (6) [2] *was amended to read:*

ARTICLE 7

Composition

(*a*) The Congress is the general assembly of delegates representing Members and as such is the supreme body of the Organization.

(*b*) Each Member shall designate one of its delegates, who should be the Director of its Meteorological or Hydrometeorological Service, as its principal delegate at Congress.

(*c*) With a view to securing the widest possible technical representation, any Director of a Meteorological or Hydrometeorological Service or any other individual may be invited by the President to be present at and to participate in the discussions of Congress in accordance with the provisions of the Regulations.

Article 8 (7) [2] *was amended to read:*

ARTICLE 8

Functions

In addition to the functions set out in other Articles of the Convention, the primary duties of Congress shall be:

(*a*) To determine general policies for the fulfilment of the purposes of the Organization as set forth in Article 2;

(*b*) To make recommendations to Members on matters within the purposes of the Organization;

(*c*) To refer to any body of the Organization any matter within the provisions of the Convention upon which such a body is empowered to act;

(*d*) To determine regulations prescribing the procedures of the various bodies of the Organization, in particular the General, Technical, Financial and Staff Regulations;

(*e*) To consider the reports and activities of the Executive Committee and to take appropriate action in regard thereto;

(*f*) To establish Regional Associations in accordance with the provisions of Article 18; to determine their geographical limits, co-ordinate their activities, and consider their recommendations;

(*g*) To establish Technical Commissions in accordance with the provisions of Article 19; to define their terms of reference, co-ordinate their activities, and consider their recommendations;

(*h*) To establish any additional bodies it may deem necessary;

(*i*) To determine the location of the Secretariat of the Organization;

(*j*) To elect the President and Vice-Presidents of the Organization and members of the Executive Committee other than the presidents of the Regional Associations.

Congress may also take any other appropriate action on matters affecting the Organization.

Article 13 [4]*, provision (c), was amended to read:*

(*c*) Fourteen Directors of Meteorological or Hydrometeorological Services of Members of the Organization, who can be replaced at sessions by alternates, provided:

(i) That these alternates shall be as provided for in the Regulations;

(ii) That not more than seven and not less than two members of the Executive Committee, comprising the President and Vice-Presidents of the Organization, the presidents of Regional Associations and the fourteen elected Directors, shall come from one Region, this Region being determined in the case of each member in accordance with the Regulations.

Article 14 [3] *was amended to read:*

ARTICLE 14

Functions

The Executive Committee is the executive body of the Organization and responsible to Congress for the co-ordination of the programme of the Organization and for the utilization of its budgetary resources in accordance with the decision of Congress.

In addition to functions set out in other Articles of the Convention, the primary functions of the Executive Committee shall be:

(*a*) To implement the decisions taken by the Members of the Organization either in Congress or by means of correspondence and to conduct the activities of the Organization in accordance with the intention of such decisions;

(*b*) To examine the programme and budget estimates for the following financial period prepared by the Secretary-General and to present its observations and its recommendations thereon to Congress;

(*c*) To consider and, where necessary, take action on behalf of the Organization on resolutions and recommendations of Regional Associations and Technical Commissions in accordance with the procedures laid down in the Regulations;

(*d*) To provide technical information, counsel and assistance in the fields of activity of the Organization;

(*e*) To study and make recommendations on any matter affecting international meteorology and related activities of the Organization;

(*f*) To prepare the agenda for Congress and to give guidance to the Regional Associations and Technical Commissions in the preparation of their agenda;

(*g*) To report on its activities to each session of Congress;

(*h*) To administer the finances of the Organization in accordance with the provisions of Part XI of the Convention.

The Executive Committee may also perform such other functions as may be conferred on it by Congress or by Members collectively.

Article 18 [1] was amended to read:

ARTICLE 18

(*a*) Regional Associations shall be composed of the Members of the Organization, the networks of which lie in or extend into the Region.

(*b*) Members of the Organization shall be entitled to attend the meetings of Regional Associations to which they do not belong, to take part in the discussions and to present their views upon questions affecting their own Meteorological or Hydrometeorological Service but shall not have the right to vote.

(*c*) Regional Associations shall meet as often as necessary. The time and place of the meeting shall be determined by the Presidents of the Regional Associations in agreement with the President of the Organization.

(*d*) The function of the Regional Associations shall be:

(i) To promote the execution of the resolutions of Congress and the Executive Committee in their respective Regions;

(ii) To consider matters brought to their attention by the Executive Committee;

(iii) To consider matters of general interest and to co-ordinate meteorological and related activities in their respective Regions;

(iv) To make recommendations to Congress and the Executive Committee on matters within the purposes of the Organization;

(v) To perform such other functions as may be conferred on them by Congress.

(*e*) Each Regional Association shall elect its president and vice-president.

Amendment approved by Eighth Congress (1979)

Article 13 [5], provision (c), was amended to read:

(*c*) Nineteen Directors of Meteorological or Hydrometeorological Services of Members of the Organization, who can be replaced at sessions by alternates, provided:

(i) That these alternates shall be as provided for in the Regulations;

(ii) That not more than seven and not less than two members of the Executive Committee, comprising the President and Vice-Presidents of the Organization, the presidents of Regional Associations and the nineteen elected Directors, shall come from one Region, this Region being determined in the case of each member in accordance with the Regulations.

Amendments approved by Ninth Congress (1983)

Article 13 [6], provision (c), was amended to read:

(*c*) Twenty-six Directors of Meteorological or Hydrological Services of Members of the Organization, who can be replaced at sessions by alternates, provided:

(i) That these alternates shall be as provided for in the Regulations;

(ii) That not more than nine and not less than three members of the Executive Council, comprising the President and Vice-Presidents of the Organization, the presidents of Regional Associations and the twenty-six elected Directors, shall come from one Region, this Region being determined in the case of each member in accordance with the Regulations.

Article 14 [4], provision (f), was amended to read:

(*f*) To prepare the agenda for Congress and to give guidance to the Regional Associations and Technical Commissions in the preparation of their work programme.

General amendment:* "Executive Committee" was replaced by "Executive Council" throughout the Convention.

APPENDIX B

AGREEMENT BETWEEN UN AND WMO

AGREEMENT BETWEEN THE UNITED NATIONS AND THE WORLD METEOROLOGICAL ORGANIZATION

Preamble

In consideration of the provisions of Article 57 of the Charter of the United Nations and of Article 25 of the Convention of the World Meteorological Organization, the United Nations and the World Meteorological Organization agree as follows:

Article I

The United Nations recognizes the World Meteorological Organization (hereinafter called "the Organization") as the Specialized Agency responsible for taking such action as may be appropriate under its basic instrument for the accomplishment of the purposes set forth therein.

Article II

Reciprocal representation

(1) The United Nations shall be invited to send representatives to participate, without vote, in the deliberations of all Congresses and meetings of the Executive Committee and Regional Associations. It shall also, after appropriate consultation, be invited to send representatives to attend meetings of the Technical Commissions or any other meetings convened by the Organization with the right to participate without vote in the discussion of items of interest to the United Nations.

(2) The Organization shall be invited to send representatives to attend meetings of the Economic and Social Council of the United Nations (hereinafter called "the Council"), of its commissions and committees and to participate, without vote, in the deliberations thereof with respect to items on the agenda in which the Organization may be concerned.

(3) The Organization shall be invited to send representatives to attend the meetings of the General Assembly during which questions within the competence of the Organization are under discussion for purposes of consultation, and to participate, without vote, in the deliberations of the main committees of the General Assembly with respect to items concerning the Organization.

(4) The Organization shall be invited to send representatives and to attend meetings of the Trusteeship Council of the United Nations and to participate, without vote, in the deliberations thereof, with respect to items on its agenda relating to meteorological matters.

(5) Written statements presented by the Organization shall be distributed by the Secretariat of the United Nations to members of the General Assembly, the Council and its commissions, and the Trusteeship Council as appropriate. Similarly, written statements presented by the United Nations shall be distributed by the Organization to its Members.

Article III

Proposal of agenda items

Subject to such preliminary consultation as may be necessary, the Organization shall include on the agenda of its Congresses and meetings of the Executive Committee, Regional Associations and Technical Commissions or, as the case may be, shall submit to its Members, items proposed to it by the United Nations. Similarly, the Council, its commissions and committees and the Trusteeship Council shall include on their agenda items proposed by the Organization.

Article IV

Recommendations of the United Nations

(1) The Organization, having regard to the obligations of the United Nations to promote the objectives set forth in Article 55 of the Charter, and the functions and power of the Economic and Social Council under Article 62 of the Charter to make or initiate studies and reports with respect to international economic, social, cultural, educational, health and related matters and to make recommendations concerning these matters to the specialized agencies concerned, and having regard also to the responsibility of the United Nations, under Articles 58 and 63 of the Charter, to make recommendations for the co-ordination of the policies and activities of such specialized agencies, agrees to arrange for the submission as soon as possible to its appropriate organ or to its Members for such action as may seem proper, of all formal recommendations which the United Nations may make to it.

(2) The Organization agrees to enter into consultation with the United Nations upon request with respect to such recommendations and in due course to report to the United Nations on the action taken by the Organization or by its Members to give effect to such recommendations, or on the other results of their consideration.

(3) The Organization agrees to co-operate in whatever further measures may be necessary to make co-ordination of the activities of specialized agencies and those of the United Nations fully effective. In particular, it agrees to co-operate with any body or bodies which the Council may establish for the purpose of facilitating such co-ordination and furnish such information as may be required for the carrying out of this purpose.

Article V

Exchange of information and documents

(1) Subject to such arrangements as may be necessary for the safeguarding of confidential material, the fullest and promptest exchange of information and documents shall be made between the United Nations and the Organization to meet the requirements of each.

(2) Without prejudice to the generality of the provisions of the preceding paragraph:

(*a*) The Organization shall submit to the United Nations an annual report on its activities;

(*b*) The Organization shall comply to the fullest extent practicable with any request which the United Nations may make for the furnishing of special reports, studies or information, subject to the conditions set forth in Article XIII;

(*c*) The Secretary-General of the United Nations shall, upon request, consult with the Secretary-General of the Organization regarding the provision to the Organization of such information as may be of special interest to it.

Article VI

Assistance to the United Nations

The Organization agrees to co-operate with and to render all possible assistance to the United Nations, its principal and subsidiary organs, in accordance with the United Nations Charter and the World Meteorological Convention, taking fully into account the particular position of the individual Members of the Organization which are not members of the United Nations.

Article VII

Relations with the International Court of Justice

(1) The Organization agrees to furnish any information which may be requested by the International Court of Justice in pursuance of Article 34 of the Statute of the Court.

(2) The General Assembly authorizes the Organization to request advisory opinions of the International Court of Justice on legal questions arising within the scope of its competence other than questions concerning the mutual relationships of the Organization with the United Nations or with other specialized agencies.

(3) Such requests may be addressed to the Court by the Congress or the Executive Committee acting in pursuance of an authorization by the Congress.

(4) When requesting the International Court of Justice to give an advisory opinion, the Organization shall inform the Council of the request.

Article VIII

Headquarters and regional offices

(1) The Organization agrees to consult with the United Nations before making any decision concerning the location of its permanent headquarters.

(2) Having due regard to the special needs of world meteorology, any regional or branch office which the Organization may establish shall so far as is practicable be closely associated with such regional or branch offices as the United Nations or other specialized agencies may establish.

Article IX

Personnel arrangements

(1) The United Nations and the Organization agree to develop as far as practicable common personnel standards, methods and arrangements designed to avoid serious discrepancies in terms and conditions of employment, to avoid competition in recruitment of personnel, and to facilitate any mutually desirable interchange of personnel in order to obtain the maximum benefit from their services.

(2) The United Nations and the Organization agree to co-operate to the fullest extent possible in achieving these ends and to consult in regard to the participation of the Organization in the work of the International Civil Service Advisory Board and the United Nations Joint Staff Pension Fund.

(3) The United Nations and the Organization agree further to consult as to the desirability of concluding a special agreement extending the competence of the United Nations Administrative Tribunal to the Organization.

Article X

Statistical services

(1) The United Nations and the Organization agree to strive for maximum co-operation, the elimination of all undesirable duplication between them, and the most efficient use of their technical personnel in their respective collection, analysis, publication, standardization, improvement and dissemination of statistical information. They agree to combine their efforts to secure the greatest possible usefulness and utilization of statistical information and to minimize the burdens placed upon national governments and other organizations from which such information may be collected.

(2) The Organization recognizes the United Nations as the central agency for the collection, analysis, publication, standardization, improvement and dissemination of statistics serving the general purposes of international organizations.

(3) In view of the fact that meteorological statistics of universal application to scientific research, aviation, shipping, agriculture, health and other human activities can best be derived from data collected and compiled by or through the Organization, the United Nations recognizes the Organization as the specialized agency responsible in conformity with Article 2 of its Convention for the collection, analysis, publication, standardization, improvement and dissemination of statistics in the field of meteorology and its applications, and for the supply of such statistics to other specialized agencies without prejudice to the right of the United Nations to concern itself with such statistics so far as it may be essential for its own purposes or for the improvement of statistics throughout the world. All decisions as to the form in which its service documentation shall be compiled rest with the Organization.

(4) The United Nations shall, in consultation with the Organization and with the other specialized agencies where appropriate, develop administrative instruments and procedures through which effective statistical co-operation may be secured between the United Nations and the specialized agencies and among the specialized agencies themselves.

(5) It is recognized as important that the collection of meteorological statistical information shall not be duplicated by the United Nations or any of its other specialized agencies whenever it is practicable for any of them to utilize information or material which the Organization has or can make available.

(6) In order to build up a central collection of statistical information for general use, it is agreed that data supplied to the Organization for incorporation in its basic statistical series or special reports should, so far as practicable, be made available to the United Nations upon request.

(7) It is agreed that data supplied to the United Nations by other sources than that of the Organization for incorporation in its basic statistical series or special reports or for other purposes should, so far as practicable and appropriate, be made available to the latter upon request.

Article XI

Administrative and technical services

(1) The United Nations and the Organization recognize the desirability, in the interest of the most efficient use of personnel and resources, of avoiding, wherever possible, the establishment of competitive or overlapping services, and agree when necessary to consult thereon to achieve these ends.

(2) Arrangements shall be made between the United Nations and the Organization in regard to the registration and deposit of official documents.

(3) Officials of the Organization shall have the right to use the "laissez-passer" of the United Nations in accordance with special arrangements to be negotiated between the Secretary-General of the United Nations and the competent authorities of the Organization.

ARTICLE XII

Budgetary and financial arrangements

(1) The Organization recognizes the desirability of establishing close budgetary and financial relationships with the United Nations in order that the administrative operations of the United Nations and of the specialized agencies shall be carried out in the most efficient and economical manner possible and that the maximum measure of co-ordination and uniformity with respect to these operations shall be secured.

(2) The United Nations and the Organization agree to co-operate to the fullest extent possible in achieving these ends and, in particular, shall if it appears expedient to both organizations consult together concerning the desirability of making appropriate arrangements for the inclusion of the budget of the Organization within a general budget of the United Nations. Any such arrangements shall be defined in a supplementary agreement between the two organizations.

(3) Pending the conclusion of any such agreement, the following arrangement shall govern budgetary and financial relationships between the United Nations and the Organization:

(*a*) In the preparation of the budget of the Organization, the Secretariat of the Organization shall consult with the Secretary-General of the United Nations with a view to achieving, in so far as practicable, uniformity in presentation of the budgets of the United Nations and of the specialized agencies for the purpose of providing a basis for comparison of the several budgets;

(*b*) The Organization agrees to transmit its budget or budgetary estimates to the United Nations by 1 July of the preceding year or such other date as may be agreed upon by the United Nations and the Organization. The General Assembly shall examine the budget or budgetary estimates of the Organization and may make such recommendations as it may consider necessary;

(*c*) Representatives of the Organization shall be entitled to participate, without vote, in the deliberations of the General Assembly or any committee thereof or established by it, at all times when the budget of the Organization or general administrative or financial questions affecting the Organization are under consideration;

(*d*) The United Nations may undertake the collection of contributions from those Members of the Organization which are also members of the United Nations in accordance with such arrangements as may be defined by a later agreement between the United Nations and the Organization;

(*e*) The United Nations shall, upon its own initiative or upon the request of the Organization, arrange for studies to be undertaken concerning other financial and fiscal questions of interest to the Organization and to other specialized agencies with a view to the provision of common services and the securing of uniformity in such matters;

(*f*) The Organization agrees to conform, as far as may be practicable, to standard practices and forms recommended by the United Nations.

ARTICLE XIII

Financing of special services

(1) In the event of the Organization being faced with the necessity for incurring substantial extra expenses as a result of any request which the United Nations may make for special reports, studies or assistance in accordance with Article VI or with any other provisions of this agreement, the Organization shall consult with the United Nations prior to incurring such expense with a view to determining the most equitable manner in which such expense shall be borne.

(2) Consultation between the United Nations and the Organization shall similarly take place with a view to making such arrangements as may be found equitable for covering the costs of central administrative, technical or fiscal services or facilities or other special assistance requested by the Organization and provided by the United Nations.

ARTICLE XIV

Inter-agency agreements

(1) The Organization agrees to inform the Council of the nature and scope of any formal agreement contemplated between the Organization and any other specialized agency or other inter-governmental organization or international non-governmental organization, and further to inform the Council of the details of any such agreement when concluded.

(2) The United Nations agrees to inform the Organization of the nature and scope of any formal agreement contemplated by any other specialized agencies on matters which might be of concern to the Organization and further will inform the Organization of the details of any such agreement, when concluded.

ARTICLE XV

Liaison

(1) The United Nations and the Organization agree to the foregoing provisions in the belief that they will contribute to the maintenance of effective liaison between the two organizations. They affirm their intention of taking whatever further measures may be necessary to this end.

(2) The liaison arrangements provided for in this agreement shall apply, as far as appropriate, to the relations between such branch and regional offices as may be established by the two organizations as well as between their central headquarters.

ARTICLE XVI

Implementation of the agreement

The Secretary-General of the United Nations and the appropriate authority of the Organization may enter into such supplementary arrangements for the implementation of this agreement as may be found desirable.

ARTICLE XVII

Revision

On six months' notice given on either part, this agreement shall be subject to revision by agreement between the United Nations and the Organization.

ARTICLE XVIII

Entry into force

This agreement shall come into force on its approval by the General Assembly of the United Nations and by the Organization in accordance with Article 25 of the World Meteorological Convention.

APPENDIX C

MEMBERS OF WMO*

States (156)

Afghanistan
Albania
Algeria
Angola
Antigua and Barbuda
Argentina
Australia
Austria
Bahamas
Bahrain
Bangladesh
Barbados
Belgium
Belize
Benin
Bolivia
Botswana
Brazil
Brunei Darussalam
Bulgaria
Burkina Faso
Burundi
Byelorussian SSR
Cambodia
Cameroon
Canada
Cape Verde
Central African Republic
Chad
Chile
China
Colombia
Comoros
Congo
Costa Rica
Côte d'Ivoire
Cuba
Cyprus
Czechoslovakia
Democratic People's Rep. of Korea
Democratic Yemen
Denmark
Djibouti
Dominica
Dominican Republic
Ecuador
Egypt
El Salvador
Ethiopia
Fiji
Finland
France
Gabon
Gambia
German Democratic Rep.
Germany, Fed. Rep. of
Ghana
Greece
Guatemala
Guinea
Guinea-Bissau
Guyana
Haiti
Honduras
Hungary
Iceland
India
Indonesia
Iran, Islamic Rep. of
Iraq
Ireland
Israel
Italy
Jamaica
Japan
Jordan
Kenya
Kuwait
Lao People's Dem. Rep.
Lebanon
Lesotho
Liberia
Libyan Arab Jamahiriya
Luxembourg
Madagascar
Malawi
Malaysia
Maldives
Mali
Malta
Mauritania
Mauritius
Mexico
Mongolia
Morocco
Mozambique
Myanmar
Nepal
Netherlands
New Zealand
Nicaragua
Niger
Nigeria
Norway
Oman
Pakistan
Panama
Papua New Guinea
Paraguay
Peru
Philippines
Poland
Portugal
Qatar
Republic of Korea
Romania
Rwanda
Saint Lucia
Sao Tome and Principe
Saudi Arabia
Senegal
Seychelles
Sierra Leone
Singapore
Solomon Islands
Somalia
South Africa **
Spain
Sri Lanka
Sudan
Suriname
Swaziland
Sweden
Switzerland
Syrian Arab Republic
Thailand
Togo
Trinidad and Tobago
Tunisia
Turkey
Uganda
Ukrainian SSR
Union of Soviet Socialist Republics
United Arab Emirates
United Kingdom of Great Britain and Northern Ireland
United Rep. of Tanzania
United States of America
Uruguay
Vanuatu
Venezuela
Viet Nam
Yemen
Yugoslavia
Zaire
Zambia
Zimbabwe

Territories (5)

British Caribbean Territories
French Polynesia
Hong Kong
Netherlands Antilles
New Caledonia

---

* (as at 30 April 1990)
** Suspended by Resolution 38 (Cg-VII) from exercising its rights and enjoying its privileges as a Member of WMO

APPENDIX D

PRESIDENTS AND SECRETARIES-GENERAL OF WMO

President	Elected at Congress:
Dr F. W. Reichelderfer (USA)	I (1951)
Mr A. Viaut (France)	II (1955)
Mr A. Viaut (France)	III (1959)
Dr A. Nyberg (Sweden)	IV (1963)
Dr A. Nyberg (Sweden	V (1967)
Mr M. F. Taha (Egypt)	VI (1971)
Mr M. F. Taha (Egypt)	VII (1975)
Dr R. L. Kintanar (Philippines)	VIII (1979)
Dr R. L. Kintanar (Philippines	IX (1983)
Mr Zou Jingmeng (China)	X (1987)

Secretary-General	Elected at Congress:
Dr G. Swoboda (Switzerland)	I (1951)
Dr D. A. Davies* (UK)	II 1955)
Dr D. A. Davies (UK)	III (1959)
Dr D. A. Davies (UK)	IV (1963)
Dr D. A. Davies (UK)	V (1967)
Dr D. A. Davies (UK)	VI (1971)
Dr D. A. Davies (UK)	VII (1975)
Professor A. C. Wiin-Nielsen (Denmark)	VIII (1979)
Professor G. O. P. Obasi (Nigeria)	IX (1983)
Professor G. O. P. Obasi (Nigeria)	X (1987)

*Now Sir Arthur Davies; by decision of Eighth Congress (1979), Secretary-General Emeritus

APPENDIX E

PRESIDENTS AND SESSIONS OF REGIONAL ASSOCIATIONS

Regional Association I (Africa)

President	*Country*	*Date*
D. A. Davies	British East Africa	1951-55
J. Ravet	Madagascar	1955-62
Ramanisarivo	Madagascar	1962-65
N. A. Akingbehin	Nigeria	1965-67
M. Seck	Senegal	1967-73
C. A. Abayomi	Nigeria	1973-79
S. Mbele-Mbong	United Republic of Cameroon	1979-81
S. B. Mpata	Malawi	1981-82
W. Degefu	Ethiopia	1982

Session	*Place*	*Date*
First	Tananarive (Madagascar)	1953
Second	Las Palmas (Canary Islands)	1957
Third	Addis Ababa (Ethiopia)	1962
Fourth	Lagos (Nigeria)	1965
Fifth	WMO Headquarters, Geneva	1969
Sixth	WMO Headquarters, Geneva	1973
Seventh	Nairobi (Kenya)	1978
Eighth	Cairo (Egypt)	1982
Ninth	Harare (Zimbabwe)	1986

Regional Association II (Asia)

President	*Country*	*Date*
V. V. Sohoni	India	1951-53
A. A. Solotoukhine	USSR	1953-54
S. Basu	India	1954-59
Po E	Burma	1959-62
S. N. Naqvi	Pakistan	1962-65
M. H. Ganji	Iran	1965-70
A. P. Navai	Iran	1970-75
D. Tubdendorj	Mongolia	1975
C. Charoen-Rajapark	Thailand	1975-76
A. G. J. Al-Sultan	Iraq	1976-82
I. D. T. de Mel	Sri Lanka	1982
U Thu Ta	Burma	1982-84
S. A. A. Kazmi	Pakistan	1984-86
I. Hussain Al-Majed	Qatar	1986

Session	*Place*	*Date*
First	New Delhi (India)	1955
Second	Rangoon (Burma)	1959
Third	Bangkok (Thailand)	1962
Fourth	Tehran (Iran)	1965
Fifth	Tokyo (Japan)	1970
Sixth	Colombo (Sri Lanka)	1975
Seventh	WMO Headquarters, Geneva	1980
Eighth	WMO Headquarters, Geneva	1984
Ninth	Beijing (China)	1988

Regional Association III (South America)

President	*Country*	*Date*
F. X. R. de Souza	Brazil	1951-56
R. V. Garcia	Argentina	1957
J. L. Vieira Maldonado	Brazil	1957-61
T. Farina Sanchez	Paraguay	1961-63
F. L. Fernandez	Argentina	1963-66
A. Garcia S.	Ecuador	1966-70
S. Bravo Flores	Chile	1970-73
G. Echeverri Ossa	Colombia	1973
O. Piccone Ocampo	Peru	1973-74
R. Venerando Pereira	Brazil	1974-79
V. L. Gomez	Ecuador	1979-80
F. Roll Fuenzalida	Chile	1980-82
C. A. Grezzi	Uruguay	1982-86
T. R. Prado Fernandez	Venezuela	1986-88
E. A. Rousseau	Chile	1989

Session	*Place*	*Date*
First	Rio de Janeiro (Brazil)	1953
Second	Caracas (Venezuela)	1957
Third	Palais des Nations, Geneva	1963
Fourth	Quito (Ecuador)	1966
Fifth	Bogotá (Colombia)	1970
Sixth	Buenos Aires (Argentina)	1974
Seventh	Brasilia (Brazil)	1978
Eighth	Montevideo (Uruguay)	1982
Ninth	Lima (Peru)	1986

Regional Association IV (North and Central America)

President	*Country*	*Date*
A. Thomson	Canada	1951-58
F. W. Reichelderfer	USA	1958-63
P. D. McTaggart-Cowan	Canada	1963-64
Elliott Coen	Costa Rica	1964-66
J. R. H. Noble	Canada	1966-73
C. Urrutia Evans	Guatemala	1973-77
D. O. Vickers	Jamaica	1977-78
S. Aguilar Anguiano	Mexico	1978-85
G. Lizano Vindas	Costa Rica	1985
C. E. Berridge	British Caribbean Territories	1985

Session	*Place*	*Date*
First	Toronto (Canada)	1953
Second	Washington, DC (USA)	1958
Third	Palais des Nations, Geneva	1963
Fourth	Asheville (USA)	1966
Fifth	Palais des Nations, Geneva	1971
Sixth	Guatemala City (Guatemala)	1973
Seventh	Mexico City (Mexico)	1977
Eighth	Havana (Cuba)	1981
Ninth	San Jose (Costa Rica)	1985
Tenth	Tegucigalpa (Honduras)	1989

Regional Association V (South West-Pacific)

President	*Country*	*Date*
M. A. F. Barnett	New Zealand	1951-55
C. Del Rosario	Philippines	1955-58
L. J. Dwyer	Australia	1958-62
J. L. Giovannelli	French Polynesia	1962-67
K. Rajendram	Singapore	1967-74
R. L. Kintanar	Philippines	1974-78
Ho Tong Yuen	Malaysia	1978-86
J. Hickman	New Zealand	1986-88
H. P. A. Jaafar	Brunei Darussalam	1988

Session	*Place*	*Date*
First	Melbourne (Australia)	1954
Second	Manila (Philippines)	1958
Third	Noumea (New Caledonia)	1962
Fourth	Wellington (New Zealand)	1966
Fifth	Kuala Lumpur (Malaysia)	1970
Sixth	Manila (Philippines)	1974
Seventh	Jakarta (Indonesia)	1978
Eighth	Melbourne (Australia)	1982
Ninth	Wellington (New Zealand)	1986

Regional Association VI (Europe)

President	*Country*	*Date*
J. Lugeon	Switzerland	1951-56
A. Nyberg	Sweden	1956-63
M. Perovic	Yugoslavia	1963-69
R. Schneider	Switzerland	1969-75
R. Czelnai	Hungary	1975-81
A. W. Kabakibo	Syrian Arab Republic	1981-84
L. A. Mendes Victor	Portugal	1984-87
E. J. Jatila	Finland	1987

Session	*Place*	*Date*
First	Zurich (Switzerland)	1952
Second	Dubrovnik (Yugoslavia)	1956
Third	Madrid (Spain)	1960
Fourth	Paris (France)	1965
Fifth	Varna (Bulgaria)	1969
Extraordinary	Lucerne (Switzerland)	1972
Sixth	Bucharest (Romania)	1974
Extraordinary	Budapest (Hungary)	1976
Seventh	Prague (Czechoslovakia)	1978
Eighth	Rome (Italy)	1982
Ninth	Potsdam (German Democratic Republic)	1986

APPENDIX F

PRESIDENTS AND SESSIONS OF TECHNICAL COMMISSIONS

Commission for Bibliography and Publications (CBP)

President	*Country*	*Date*
M. Mezin	France	1951-57
A. Vandenplas	Belgium	1957-59

Session	*Place*	*Date*
First	Paris (France)	1953
Second	Paris (France)	1957

Commission for Instruments and Methods of Observation (CIMO)

President	*Country*	*Date*
J. Patterson	Canada	1951-53
A. Perlat	France	1953-57
L. S. Mathur	India	1962-67
V. D. Rockney	USA	1967-73
H. Treussart	France	1973-81
S. P. Huovila	Finland	1981

Session	*Place*	*Date*
First	Toronto (Canada)	1953
Second	Paris (France)	1957
Third	New Delhi (India)	1962
Fourth	Tokyo (Japan)	1965
Fifth	Versailles (France)	1969
Sixth	Helsinki (Finland)	1973
Seventh	Hamburg (Federal Republic of Germany)	1977
Eighth	Mexico City (Mexico)	1981
Ninth	Ottawa (Canada)	1985
Tenth	Brussels (Belgium)	1989

Commission for Atmospheric Sciences (CAS)*

President	*Country*	*Date*
J. Van Mieghem	Belgium	1951-57
R. C. Sutcliffe	United Kingdom	1957-61
G. P. Cressman	USA	1961-65
J. F. Gabites	New Zealand	1965-68
J. S. Sawyer	United Kingdom	1968-73
W. L. Godson	Canada	1973-77
A. Villevieille	France	1977-82
F. Mesinger	Yugoslavia	1982

Session	*Place*	*Date*
First	Toronto (Canada)	1953
Second	Paris (France)	1957
Third	Rome (Italy)	1961
Fourth	Brussels (Belgium)	1965
Fifth	Washington, DC (USA)	1970
Sixth	Versailles (France)	1973
Seventh	Manila (Philippines)	1978
Eighth	Melbourne (Australia)	1982
Ninth	Sofia (Bulgaria)	1986

* The Commission for Aerology (CAe) was renamed the Commission for Atmospheric Sciences (CAS) by Fifth Congress (1967).

Commission for Climatology (CCl)*

President	Country	Date
C. W. Thornthwaite	USA	1951-57
R. G. Veryard	United Kingdom	1957-60
C. C. Boughner	Canada	1960-69
H. E. Landsberg	USA	1969-78
M. K. Thomas	Canada	1978-82
J. L. Rasmussen	USA	1982-89
W. J. Maunder	New Zealand	1989

Session	Place	Date
First	Washington, DC (USA)	1953
Second	Washington, DC (USA)	1957
Third	London (United Kingdom)	1960
Fourth	Stockholm (Sweden)	1965
Fifth	WMO Headquarters, Geneva	1969
Sixth	Bad Homburg (Federal Republic of Germany)	1973
Seventh	WMO Headquarters, Geneva	1978
Eighth	Washington, DC (USA)	1982
Ninth	International Conference Centre, Geneva	1985
Tenth	Cascais, Lisbon (Portugal)	1989

Commission for Agricultural Meteorology (CAgM)

President	Country	Date
J. J. Burgos	Argentina	1951-58
P. M. Austin Bourke	Ireland	1958-62
L. P. Smith	United Kingdom	1962-71
W. Baier	Canada	1971-79
N. Gerbier	France	1979-85
A. Kassar	Tunisia	1985

Session	Place	Date
First	Paris (France)	1953
Second	Warsaw (Poland)	1958
Third	Toronto (Canada)	1962
Fourth	Quezon City, Manila (Philippines)	1967
Fifth	WMO Headquarters, Geneva	1971
Sixth	Washington, DC (USA)	1974
Seventh	Sofia (Bulgaria)	1979
Eighth	WMO Headquarters, Geneva	1983
Ninth	Madrid (Spain)	1986

Commission for Marine Meteorology (CMM)**

President	Country	Date
C. E. N. Frankcom	United Kingdom	1951-56
H. Thomsen	Denmark	1956-60
J. A. van Duijnen Montijn	Netherlands	1960-64
K. T. McLeod	Canada	1964-68
S. L. Tierney	Ireland	1968-72
J. M. Dury	Belgium	1972-76
K. P. Vasiliev	USSR	1976-84
F. Gérard	France	1984

(continued)

* The name of this Commission was changed to Commission for Special Applications of Meteorology and Climatology (CoSAMC) by Sixth Congress (1971). Eighth Congress (1979) changed the name to Commission for Climatology and Applications of Meteorology (CCAM). Ninth Congress (1983) renamed it Commission for Climatology (CCl).

** The Commission for Maritime Meteorology (CMM) was renamed the Commission for Marine Meteorology (CMM) by Sixth Congress (1971).

CMM *(continued)*

Session	*Place*	*Date*
First	London (United Kingdom)	1952
Second	Hamburg (Federal Republic of Germany)	1956
Third	Utrecht (Netherlands)	1960
Fourth	WMO Headquarters, Geneva	1964
Fifth	Kingston, Rhode Island (USA)	1968
Sixth	Tokyo (Japan)	1972
Seventh	WMO Headquarters, Geneva	1976
Eighth	Hamburg (Federal Republic of Germany)	1981
Ninth	International Conference Centre, Geneva	1984
Tenth	Paris (France)	1989

Commission for Basic Systems (CBS)*

President	*Country*	*Date*
W. Bleeker	Netherlands	1951-58
P. H. Kutschenreuter	USA	1958-62
S. N. Sen	India	1962-68
N. G. Leonov	USSR	1970-72
O. Lonnqvist	Sweden	1972-78
J. R. Neillon	USA	1978-88
A. A. Vasiliev	USSR	1988

Session	*Place*	*Date*
First	Washington, DC (USA)	1953
Second	New Delhi (India)	1958
Third	Washington, DC (USA)	1962
Fourth	Wiesbaden (Federal Republic of Germany)	1966
Fifth	Palais des Expositions, Geneva	1970
Sixth	Belgrade (Yugoslavia)	1974
Extraordinary	Partly in WMO Headquarters Geneva, and partly in ITU Headquarters, Geneva	1976
Seventh	Washington, DC (USA)	1978
Extraordinary	International Conference Centre, Geneva	1980
Eighth	International Conference Centre, Geneva	1983
Extraordinary	Hamburg (Federal Republic of Germany)	1985
Ninth	International Conference Centre, Geneva	1988

Commission for Aeronautical Meteorology (CAeM)

President	*Country*	*Date*
A. H. Nagle	USA	1951-59
A. Silva de Sousa	Portugal	1959-64
W. A. Dwyer	Australia	1964-67
N. A. Lieurance	USA	1967-71
P. Duvergé	France	1971-76
R. R. Dodds	Canada	1976-82
J. Kastelein	Netherlands	1982

Session	*Place*	*Date*
First	ICAO Headquarters, Montreal	1954
Second	ICAO Headquarters, Montreal	1959
Third	Paris (France)	1964
Fourth	ICAO Headquarters, Montreal	1967
Fifth	WMO Headquarters, Geneva	1971
Sixth	ICAO Headquarters, Montreal	1976
Seventh	ICAO Headquarters, Montreal	1982
Eighth	International Conference Centre, Geneva	1986

* The Commission for Synoptic Meteorology (CSM) was renamed the Commission for Basic Systems (CBS) by Sixth Congress (1971).

Commission for Hydrology (CHy)*

President	*Country*	*Date*
M. A. Kohler	USA	1960-68
E. G. Popov	USSR	1968-76
R. H. Clark	Canada	1976-84
O. Starosolszky	Hungary	1984

Session	*Place*	*Date*
First	Washington, DC (USA)	1961
Second	Warsaw (Poland)	1964
Third	WMO Headquarters, Geneva	1968
Fourth	Buenos Aires (Argentina)	1972
Fifth	Ottawa (Canada)	1976
Sixth	Madrid (Spain)	1980
Seventh	International Conference Centre, Geneva	1984
Eighth	International Conference Centre, Geneva	1988

* The Commission for Hydrological Meteorology (CHM), established by Third Congress (1959), was renamed the Commission for Hydrometeorology (CHM) by Fourth Congress (1963). The name was subsequently changed to Commission for Hydrology (CHy) by Sixth Congress (1971).

APPENDIX G

WINNERS OF THE IMO PRIZE

1956	Dr. Th. Hesselberg (Norway)
1957	Professor C. G. Rossby (Sweden and USA)
1958	Mr. E. Gold (UK)
1959	Professor J. Bjerknes (Norway and USA)
1960	Professor J. Van Mieghem (Belgium)
1961	Professor K. R. Ramanathan (India)
1962	Dr. A. Ångström (Sweden)
1963	Dr. R. C. Sutcliffe (UK)
1964	Dr. F. W. Reichelderfer (USA)
1965	Professor S. Petterssen (Norway and USA)
1966	Professor T. Bergeron (Sweden)
1967	Professor K. Ya. Kondratyev (USSR)
1968	Sir Graham Sutton (UK)
1969	Professor E. H. Palmén (Finland)
1970	Dr. R. Th. A. Scherhag (Federal Republic of Germany)
1971	Professor J. G. Charney (USA)
1972	Academician V. A. Bugaev (USSR)
1973	Dr. C. H. B. Priestley (Australia) and Mr. J. S. Sawyer (UK)
1974	Professor J. Smagorinsky (USA)
1975	Dr. W. L. Godson (Canada)
1976	Academician E. K. Fedorov (USSR)
1977	Dr. G. P. Cressman (USA)
1978	Dr. A. E. G. E. Nyberg (Sweden)
1979	Professor H. E. Landsberg (USA)
1980	Dr. R. M. White (USA)
1981	Professor B. Bolin (Sweden)
1982	Dr. W. J. Gibbs (Australia)
1983	Mr. M. F. Taha (Egypt) and Professor J. J. Burgos (Argentina)
1984	Dr. T. F. Malone (USA)
1985	Sir Arthur Davies (UK)
1986	Professor H. Flohn (Federal Republic of Germany)
1987	Professor M. I. Budyko (USSR)
1988	Professor F. Kenneth Hare (Canada)
1989	Professor P. R. Pisharoty (India)

LIST OF ABBREVIATIONS

ACC – Administration Committee on Co-ordination (UN)
ACMAD – African Centre of Meteorological Applications for Development
AFS – Area Forecast System
AGRHYMET – Regional training centre for agrometeorology and operational hydrology and their applications (programme for the Sahelian countries)
ALPEX – Alpine Experiment
AMTEX – Air Mass Transformation Experiment
APT – Automatic picture transmission
ARGOS – Data relay and platform location system
ASAP – Automated Shipboard Aerological Programme
ASDAR – Aircraft-to-satellite data relay
ASEAN – Association of the South-East Asian Nations

BAPMoN – Background Air Pollution Monitoring Network

CAC – Co-ordinating and Advisory Committee (AGRHYMET)
CAe – Commission for Aerology
CAeM – Commission for Aeronautical Meteorology
CAgM – Commission for Agricultural Meteorology
CAMA – Civil Aviation and Meteorological Authority (Yemen Arab Republic)
CARS – Climate Applications Referral System
CAS – Commission for Atmospheric Sciences
CBP – Commission for Bibliography and Publications
CBS – Commission for Basic Systems
CCDP – Climate Change Detection Project
CCl – Commission for Climatology
CFC – Chlorofluorocarbons
CGIAR – Consultative Group for International Agricultural Research
CHM – Commission for Hydrological Meteorology
CHy – Commission for Hydrology
CILSS – Permanent Inter-State Committee for Drought Control in the Sahel
CIMAe – Comité international de météorologie aéronautique
CIMO – Commission for Instruments and Methods of Observation
CLICOM – Climate computing
CMM – Commission for Marine Meteorology
CoSAMC – Commission for Special Applications of Meteorology and Climatology
COSPAR – Committee for Space Research
CSM – Climate system monitoring
CSM – Commission for Synoptic Meteorology
CSWI – Commission for Synoptic Weather Information

EC – Executive Committee (renamed Executive Council in 1983)
ECA – Economic Commission for Africa
ECAFE – Economic Commission for Asia and the Far East (now ESCAP)
ECE – Economic Commission for Europe
ECLA – Economic Commission for Latin America (now ECLAC)
ECLAC – Economic Commission for Latin America and the Caribbean
ECMWF – European Centre for Medium-range Weather Forecasting
ECOSOC – Economic and Social Council
ECWA – Economic Commission for Western Asia (now ESCWA)
EPTA – Expanded Programme of Technical Assistance (UN)
ESCAP – Economic and Social Commission for Asia and the Pacific
ESCWA – Economic and Social Commission for Western Asia

FAO – Food and Agriculture Organization of the United Nations
FGGE – First GARP Global Experiment (also known as the Global Weather Experiment)
FINNIDA – Finnish International Development Agency
FIT – Funds in trust

GARP – Global Atmospheric Research Programme
GATE – GARP Atlantic Tropical Experiment
GAW – Global Atmospheric Watch
GDM – General Department of Meteorology (Yemen Arab Republic)
GDPS – Global Data-processing System
GEMS – Global Environmental Monitoring System (UNEP)
GESAMP – Group of Experts on the Scientific Aspects of Marine Pollution
GOES – Geostationary Operational Environmental Satellite
GO_3OS – Global Ozone Observing System
GOS – Global Observing System
GTS – Global Telecommunication System

HNRC – HOMS National Reference Centre
HOMS – Hydrological Operational Multipurpose Subprogramme
HRM – HOMS Reference Manual

IAEA – International Atomic Energy Agency
IAHR – International Association for Hydraulic Research
IAHS – International Association of Hydrological Sciences
IAMAP – International Association of Meteorology and Atmospheric Physics
IASH – International Association of Scientific Hydrology (now IAHS)
IATA – International Air Transport Association
ICAO – International Civil Aviation Association
ICES – International Council for the Exploration of the Sea
ICRISAT – International Crop Research Institute for the Semi-Arid Tropics
ICS – International Chamber of Shipping
ICSPRO – Inter-Secretariat Committee for Scientific Problems Relating to Oceanography
ICSU – International Council of Scientific Unions
IDNDR – International Decade for Natural Disaster Reduction
IFAD – International Fund for Agricultural Development
IFALPA – International Federation of Airline Pilots Associations
IGC – International Geophysical Co-operation
IGOSS – Integrated Global Ocean Station System (now Integrated Global Ocean Services System)
IGY – International Geophysical Year
IHD – International Hydrological Decade
IHP – International Hydrological Programme
IIOE – International Indian Ocean Expedition
IMC – International Meteorological Committee
IMCO – Inter-governmental Maritime Consultative Organization (now International Maritime Organization)
IMO – International Maritime Organization
IMO – International Meteorological Organization (predecessor of WMO)
IMOP – Instruments and Methods of Observation Programme
INFOCLIMA – World Climate Data Information Referral Service
INMARSAT – International Maritime Satellite System
INSIVUMEH – National Institute for Seismology, Vulcanology, Meteorology and Hydrology (Guatemala)
IOC – Intergovernmental Oceanographic Commission (Unesco)
IPCC – Intergovernmental Panel on Climate Change
IQSY – International Years of the Quiet Sun
ISLSCP – International Satellite Land-Surface Climatology Project
ISS – Integrated WWW System Study
ITU – International Telecommunication Union
IUGG – International Union of Geodesy and Geophysics

LEPOR – Long-term and Expanded Programme of Oceanic Exploration and Research
LRCS – League of Red Cross and Red Crescent Societies

MED-POL – Mediterranean Action Plan, Pollution Monitoring and Research Programme
MONEX – Monsoon Experiment

NAOS – North Atlantic Ocean Stations
NASA – National Aeronautics and Space Administration (USA)
NDF – New Development Fund
NMC – National Meteorological Centre
NWP – Numerical Weather Prediction

OHP	– Operational Hydrology Programme
OPE	– Office for Projects Execution (UNDP)
OPEX	– Operational and executive
OWSE	– Operational WWW Systems Evaluation
PEP	– Precipitation Enhancement Project
PICAO	– Provisional International Civil Aviation Organization
POLEX	– Polar Experiment
RMC	– Regional Meteorological Centre (now RSMC)
RMTC	– Regional Meteorological Training Centre
RSMC	– Regional/Specialized Meteorological Centre
RTH	– Regional Telecommunication Hub
SADCC	– Southern African Development Co-ordination Committee
SATCC	– Southern African Transport and Communications Commission
SCOPE	– Scientific Committee on Problems of the Environment
SCOR	– Scientific Committee on Oceanic Research
SST	– Supersonic transport
SWCC	– Second World Climate Conference
TCDC	– Technical co-operation among developing countries
TCP	– Tropical Cyclone Programme
TOGA	– Tropical Ocean and Global Atmospheric Programme
TOPEX	– Typhoon Operational Experiment
TRUCE	– Tropical Urban Climate Experiment
UN	– United Nations
UNDP	– United Nations Development Programme
UNDRO	– Office of the United Nations Disaster Relief Co-ordinator
UNDTCD	– United Nations Department of Technical Co-operation for Development
UNEP	– United Nations Environment Programme
Unesco	– United Nations Educational, Scientific and Cultural Organization
UNSCEAR	– United Nations Scientific Committee on the Effects of Atomic Radiation
UNSO	– United Nations Sahelian Office
VAP	– Voluntary Assistance Programme (now VCP)
VCP	– Voluntary Co-operation Programme
VCP(ES)	– Voluntary Co-operation Programme (Equipment and Services)
VCP(F)	– Voluntary Co-operation Programme (Fund)
WAFS	– World Area Forecast System
WAMEX	– West African Monsoon Experiment
WCAP	– World Climate Applications Programme
WCDP	– World Climate Data Programme
WCIP	– World Climate Impact Studies Programme
WCP	– World Climate Programme
WCRP	– World Climate Research Programme
WHO	– World Health Organization
WMC	– World Meteorological Centre
WMO	– World Meteorological Organization
WOCE	– World Ocean Circulation Experiment
WRR	– World Radiometer Reference
WWW	– World Weather Watch

Illustrations by courtesy: p. 21 – Météorologie Nationale, France; p. 22 (*left*) – National Space Development Agency of Japan; (*right*) – Meteorological Satellite Centre of Japan; p. 23 – National Meteorological Service of Argentina; p. 24 – NOAA; p. 25 – NOAA; p. 30 (*left*) – NASA; (*right*) – S. Klemm; p. 31 – UN/Carolyn Redanius; p. 33 – Korean Meteorological Service; p. 34 (*left*) – NOAA; (*right*) – WMO/Vickers; p. 35 – LRCS; p. 40 – FAO; p. 41 – Météorologie Nationale, France; p. 42 – T. Gaskell; p. 43 – NCAR; p. 46 – Hong Kong Government; p. 47 – ASECNA; p. 51 – E.A. Mellaart; p. 52 – UN; p. 53 – WHO; p. 54 – UN; p. 55 – Unesco; p. 70 – FAO; p. 72 – J. E. Gherbesi; p. 79 – WMO/Bianco; p. 80 – US FGGE Project Office, NOAA; p. 81 – NOAA; p. 87 – WHO/P. Almasy; p. 88 – FAO/C. Boscardi; p. 94 – WHO/Takahara; p. 98 – Guilde du Livre; p. 104 (*left*) – UN; (*right*) – CNRS, France; p. 107 – Stayffer and Weftek; p. 112 – WMO; p. 118 – WMO; p. 120 – Korean Meteorological Service; p. 122 – WMO; p. 129 – UN; p. 134 (*left*) – National Meteorological Service of Argentina; (*right*) – WMO/Bianco; p. 136 – WMO/G. Rudder; p. 153 – WMO/Bianco.

Printed in Switzerland